THE PHYSIC
VIBRATION

The Physics of
Vibrations and Waves

3rd Edition

H. J. Pain, *Formerly of Department of Physics,*
Imperial College, London

JOHN WILEY & SONS LIMITED

Chichester · New York · Brisbane · Toronto · Singapore

Reprinted April 1987
Reprinted December 1988
Reprinted December 1989
Reprinted March 1992

Library of Congress Cataloging in Publication Data:
Pain, H. J. (Herbert John)
 The physics of vibrations and waves.
 Includes index.
 1. Vibration. 2. Waves. I. Title.
 QC231.P3 1983   531'.1133   83-5880
 ISBN 0 471 90181 4 (cloth)
 ISBN 0 471 90182 2 (paper)

British Library Cataloguing in Publication Data:
Pain, H. J.
 The physics of vibrations and waves.—3rd ed.
 1. Waves   2. Vibration
 I. Title
 531'.1133   QC157
 ISBN 0 471 90181 4 (Cloth)
 ISBN 0 471 90182 2 (Paper)

Set on Linotron and printed and bound in Great Britain
by Courier International Limited, East Kilbride

Introduction to 1st Edition

The opening session of the physics degree course at Imperial College includes an introduction to vibrations and waves where the stress is laid on the underlying unity of concepts which are studied separately and in more detail at later stages. The origin of this short textbook lies in that lecture course which the author has given for a number of years. Sections on Fourier transforms and non-linear oscillations have been added to extend the range of interest and application.

At the beginning no more than school-leaving mathematics is assumed and more advanced techniques are outlined as they arise. This involves explaining the use of exponential series, the notation of complex numbers and partial differentiation and putting trial solutions into differential equations. Only plane waves are considered and, with two exceptions, Cartesian coordinates are used throughout. Vector methods are avoided except for the scalar product and, on one occasion, the vector product.

Opinion canvassed amongst many undergraduates has argued for a 'working' as much as for a 'reading' book; the result is a concise text amplified by many problems over a wide range of content and sophistication. Hints for solution are freely given on the principle that an undergraduate gains more from being guided to a result of physical significance than from carrying out a limited arithmetical exercise.

The main theme of the book is that a medium through which energy is transmitted via wave propagation behaves essentially as a continuum of coupled oscillators. A simple oscillator is characterized by three parameters, two of which are capable of storing and exchanging energy, whilst the third is energy dissipating. This is equally true of any medium.

The product of the energy storing parameters determines the velocity of wave propagation through the medium and, in the absence of the third parameter, their ratio governs the impedance which the medium presents to the waves. The energy dissipating parameter introduces a loss term into the impedance; energy is absorbed from the wave system and it attenuates.

This viewpoint allows a discussion of simple harmonic, damped, forced and coupled oscillators which leads naturally to the behaviour of transverse waves on a string, longitudinal waves in a gas and a solid, voltage and current waves on a transmission line and electromagnetic waves in a dielectric and a conductor. All are amenable to this common treatment, and it is the wide validity of relatively few physical principles which this book seeks to demonstrate.

May 1968   H. J. PAIN

Preface to 2nd Edition

The main theme of the book remains unchanged but an extra chapter on Wave Mechanics illustrates the application of classical principles to modern physics.

Any revision has been towards a simpler approach especially in the early chapters and additional problems. Reference to a problem in the course of a chapter indicates its relevance to the preceding text. Each chapter ends with a summary of its important results.

Constructive criticism of the first edition has come from many quarters, not least from successive generations of physics and engineering students who have used the book; a second edition which incorporates so much of this advice is the best acknowledgement of its value.

June 1976   H. J. PAIN

Preface to 3rd Edition

Since this book was first published the physics of optical systems has been a major area of growth and this development is reflected in the present edition. Chapter 10 has been rewritten to form the basis of an introductory course in optics and there are further applications in chapters 7 and 8.

The level of this book remains unchanged.

January 1983

H. J. PAIN

Contents

Chapter Synopsis

Chapter 1 Simple and Damped Simple Harmonic Motion

Simple harmonic motion of mechanical and electrical oscillators—Vector representation of simple harmonic motion—Superposition of two SHMs by vector addition—Superposition of two perpendicular SHMs—polarization, Lissajous figures—Superposition of many SHMs—Complex number notation and use of exponential series—Damped motion of mechanical and electrical oscillators—Heavy damping – critical damping—Damped simple harmonic oscillations—Amplitude decay – logarithmic decrement – relaxation time—Energy decay – Q-value—Rate of energy decay equal to work rate of damping force—Summary of important results.

Chapter 2 The Forced Oscillator

The vector operator i—Electrical and mechanical impedance—Transient and steady state behaviour of a forced oscillator—Variation of displacement and velocity with frequency of driving force—Frequency dependence of phase angle between force and (a) displacement, (b) velocity—Vibration Insulation—Power supplied to oscillator—Q value as a measure of power absorption bandwidth—Q value as amplification factor of low frequency response—Effect of transient term—Summary of important results.

Chapter 3 Coupled Oscillations

Spring coupled pendulums—Normal coordinates and normal modes of vibration—Inductance coupling of electrical oscillators—Coupling of many oscillators on a loaded string—Wave motion as the limit of coupled oscillations—Summary of important results.

Chapter 4 Transverse Wave Motion

Notation of partial differentiation—Particle and phase velocities—The wave equation—Transverse waves on a string—The string as a forced oscillator—Characteristic impedance of a string—Reflexion and transmission of transverse waves at a boundary—Impedance matching – insertion of quarter wave element—Standing waves on a string of fixed length—Normal modes and eigenfrequencies—Energy in a normal mode of oscillation—Wave groups – group velocity – dispersion—Wave group of many components – Bandwidth Theorem—Transverse waves in a periodic structure (crystal)—Doppler Effect—Summary of important results.

Chapter 5   Longitudinal Waves

Wave equation—Sound waves in gases—Energy distribution in sound waves – Intensity—Specific acoustic impedance.—Longitudinal waves in a solid – Young's Modulus – Poisson's ratio—Longitudinal waves in a periodic structure—Reflexion and transmission of sound waves at a boundary—Summary of important results.

Chapter 6   Waves on Transmission Lines

Ideal Transmission Line—Wave equation – velocity of voltage and current waves—Characteristic Impedance—Reflexion at end of terminated line—Standing waves in short circuited line—Real transmission line with energy losses—Propagation constant – attenuation coefficient—Diffusion equation – diffusion coefficients—Attenuation – wave equation plus diffusion effects—Summary of important results.

Chapter 7   Electromagnetic Waves

Permeability and permittivity of a medium—Maxwell's equations – displacement current—Wave equations for electric and magnetic field vectors in a dielectric—Poynting vector—Impedance of a dielectric to e.m. waves—Energy density of e.m. waves—Electromagnetic waves in a conductor—Effect of conductivity adds diffusion equation to wave equation—Propagation and attenuation of e.m. waves in a conductor – skin depth—Ratio of displacement current to conduction current as a criterion for dielectric or conducting behaviour—Relaxation time of a conductor—Impedance of a conductor to e.m. waves—Reflexion and transmission of e.m. waves at a boundary—Normal incidence—Oblique incidence and Fresnel's equations—Reflexion from a conductor—Connection between impedance and refractive index—Summary of important results.

Chapter 8   Waves in More than One Dimension

Plane wave representation in 2 and 3 dimensions—Wave equation in 2 dimensions—Wave guide – reflection of a 2 dimensional wave at rigid boundaries—Normal modes and method of separation of variables for 1, 2 and 3 dimensions—Normal modes in 2 dimensions on a rectangular membrane – degeneracy—Normal modes in 3 dimensions—Number of normal modes per unit frequency interval per unit volume – application to Planck's Radiation Law and Debye's Theory of Specific Heats—Reflexion and Transmission of an é.m. wave in 3 dimensions – Snell's Law—Total internal reflexion and evanescent waves—Summary of important results.

Chapter 9   Fourier Methods
Fourier series for a periodic function—Fourier series for any interval—Application to a plucked string – energy in normal modes—Application to rectangular velocity pulse on a string – Bandwidth Theorem—Fourier Integral of a single pulse—Fourier Transforms – application to optical diffraction—Summary of important results

Chapter 10   Waves in Optical Systems
Fermat's Principle—Laws of reflexion and refraction—Wavefront propagation through a thin lens and a prism—Optical systems – power of an optical surface—Magnification—Power of a thin lens—Optical Helmholtz Equation—Interference—Division of wavefront—Two equal sources—Young's slit experiment—Spatial coherence—Dipole radiation—Linear array of N equal sources—Division of amplitude—Fringes of constant inclination and thickness—Newton's Rings—Michelson's spectral interferometer – structure of spectral lines—Fabry–Perot interferometer – Finesse—Resolving power—Free spectral range—Fraunhofer diffraction—Slit—N slits—Missing orders—Transmission diffraction grating—Resolving power—Bandwidth Theorem—Rectangular aperture – circular aperture—Fresnel diffraction – straight edge—Cornu spiral—Slit—Circular aperture – zone plate—Holography—Summary of important results.

Chapter 11   Non-linear Oscillations
Anharmonic oscillations—Free vibrations of finite amplitude pendulum—Non linear restoring force – forced vibrations—Thermal expansion of a crystal—Electrical 'relaxation' oscillator—Non linear acoustic effects – shock waves in a gas—No summary.

Chapter 12   Wave Mechanics
Historical review—De Broglie matter waves and wavelength—Heisenberg's Uncertainty Principle—Schrödinger's time independent wave equation—The wave function—Infinite potential well in 1 dimension—Quantization of energy – zero point energy—Probability density – normalization—Infinite potential well in 3 dimensions—Density of energy states – Fermi energy level—The potential step—The finite square potential well—The harmonic oscillator—Summary of important results.

Table of Constants

Charge on electron	$1 \cdot 602 \times 10^{-19}$	coulombs
Rest mass of electron	$9 \cdot 1 \times 10^{-31}$	kilograms
Atomic mass unit	$1 \cdot 66 \times 10^{-27}$	kilograms
1 newton	10^5	dynes
1 electron volt	$1 \cdot 6 \times 10^{-19}$	joules
Planck's constant h	$6 \cdot 62 \times 10^{-34}$	joule sec
Boltzmann's constant k	$1 \cdot 38 \times 10^{-23}$	joules/degree
	$8 \cdot 61 \times 10^{-5}$	electron volt/degree
Avogadro's number	$6 \cdot 022 \times 10^{23}$	per mole
Velocity of light c	3×10^8	metres/sec
Permeability of free space μ_0	$4\pi \times 10^{-7}$	henries/metre
Permittivity of free space ϵ_0	$(36\pi \times 10^9)^{-1}$	farads/metre

Table showing how energy storing processes in a medium govern the wave velocity and the impedance. Potential energy is stored in medium via parameter C and kinetic or inductive energy is stored by ρ or L.

Type of Wave	(Velocity)^2	Impedance	Symbols	
transverse on string	T/ρ	ρc	T tension ρ linear density c wave velocity	
longitudinal in gas	$\gamma P/\rho = B/\rho = (\rho C)^{-1}$	$\rho c = \sqrt{\rho/C}$	γ specific heat ratio P gas pressure B bulk modulus C compressibility c wave velocity	
voltage and current on transmission line	$(L_0 C_0)^{-1}$	$\sqrt{L_0/C_0}$	L_0 inductance C_0 capacitance	per unit length
electromagnetic waves in a dielectric	$(\mu\epsilon)^{-1}$	$\sqrt{\mu/\epsilon}$	μ permeability (henries/metre) ϵ permittivity (farads/metre)	

Chapter 1

Simple and Damped Simple Harmonic Motion

At first sight the eight physical systems in fig. 1.1 appear to have little in common.

1.1(a) is a simple pendulum, a mass m swinging at the end of a light rigid rod of length l.

1.1(b) is a flat disc supported by a rigid wire through its centre and oscillating through small angles in the plane of its circumference.

1.1(c) is a mass fixed to a wall via a spring of stiffness s sliding to and fro in the x direction on a frictionless plane.

1.1(d) is a mass m at the centre of a light string of length $2l$ fixed at both ends under a constant tension T. The mass vibrates in the plane of the paper.

1.1(e) is a frictionless U-tube of constant cross-sectional area containing a length l of liquid, density ρ, oscillating about its equilibrium position of equal levels in each limb.

1.1(f) is an open flask of volume V and a neck of length l and constant cross-sectional area A in which the air of density ρ vibrates as sound passes across the neck.

1.1(g) is a hydrometer, a body of mass m floating in a liquid of density ρ with a neck of constant cross-sectional area cutting the liquid surface. When depressed slightly from its equilibrium position it performs small vertical oscillations.

1.1(h) is an electrical circuit, an inductance L connected across a capacitance C carrying a charge q.

All of these systems are simple harmonic oscillators which, when slightly disturbed from their equilibrium or rest position, will oscillate with simple harmonic motion. This is the most fundamental vibration of a single particle or one-dimensional system. A small displacement x from its equilibrium position sets up a restoring force which is proportional to x acting in a direction towards the equilibrium position.

1

(a)

$$m\ddot{x} + mg\frac{x}{l} = 0$$
$$ml\ddot{\theta} + mg\theta = 0$$
$$\omega^2 = g/l$$

(b)

$$I\ddot{\theta} + c\theta = 0$$
$$\omega^2 = \frac{c}{I}$$

$$mg\sin\theta \approx mg\theta$$
$$\approx mg\frac{x}{l}$$

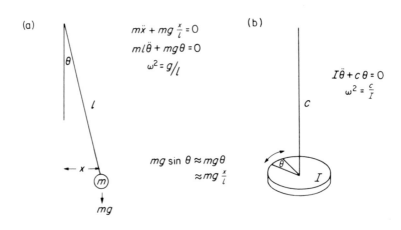

(c)

$$m\ddot{x} + sx = 0$$
$$\omega^2 = s/m$$

(d)

$$m\ddot{x} + 2T\frac{x}{l} = 0$$
$$\omega^2 = \frac{2T}{lm}$$

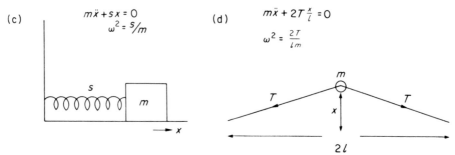

(e)

$$\rho l\ddot{x} + 2\rho gx = 0$$
$$\omega^2 = 2g/l$$

(f)

$$\rho A l\ddot{x} + \frac{\gamma \rho x A^2}{V} = 0$$
$$\omega^2 = \frac{\gamma \rho A}{l\rho V}$$

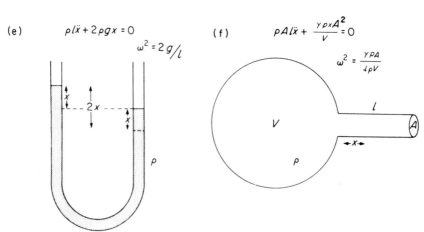

(g)

(h)

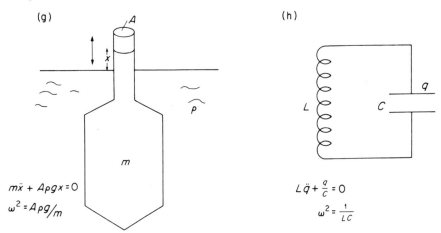

$m\ddot{x} + A\rho g x = 0$

$\omega^2 = A\rho g/m$

$L\ddot{q} + \dfrac{q}{C} = 0$

$\omega^2 = \dfrac{1}{LC}$

Fig. 1.1. Simple harmonic oscillators with their equations of motion and angular frequencies ω of oscillation. (a) A simple pendulum. (b) A torsional pendulum. (c) A mass on a frictionless plane connected by a spring to a wall. (d) A mass at the centre of a string under constant tension T. (e) A fixed length of non-viscous liquid in a U-tube of constant cross-section. (f) An acoustic Helmholtz resonator. (g) A hydrometer mass m in a liquid of density ρ. (h) An electrical LC resonant circuit

Thus, this restoring force F may be written

$$F = -sx$$

where s, the constant of proportionality, is called the stiffness and the negative sign shows that the force is acting against the direction of increasing displacement and back towards the equilibrium position. A constant value of the stiffness restricts the displacement x to small values (this is Hooke's Law of Elasticity). The stiffness s is obviously the restoring force per unit distance (or displacement) and has the dimensions

$$\frac{\text{force}}{\text{distance}} \equiv \frac{MLT^{-2}}{L}$$

The equation of motion of such a disturbed system is given by the dynamic balance between the forces acting on the system, which by Newton's Law is

$$\text{mass times acceleration} = \text{restoring force}$$

or

$$m\ddot{x} = -sx$$

where the acceleration

$$\ddot{x} = \frac{d^2 x}{dt^2}$$

This gives

$$m\ddot{x} + sx = 0$$

or

$$\ddot{x} + \frac{s}{m}x = 0$$

where the dimensions of

$$\frac{s}{m} \quad \text{are} \quad \frac{MLT^{-2}}{ML} = T^{-2} = \nu^2$$

Here T is a time, or period of oscillation, the reciprocal of ν which is the frequency with which the system oscillates.

However, when we solve the equation of motion we shall find that the behaviour of x with time has a sinusoidal or cosinusoidal dependence, and it will prove more appropriate to consider, not ν, but the angular frequency $\omega = 2\pi\nu$ so that the period

$$T = \frac{1}{\nu} = 2\pi \sqrt{\frac{m}{s}}$$

where s/m is now written as ω^2. Thus the equation of simple harmonic motion

$$\ddot{x} + \frac{s}{m}x = 0$$

becomes

$$\boxed{\ddot{x} + \omega^2 x = 0} \qquad (1.1)$$

(Problem 1.1)

Displacement in Simple Harmonic Motion

The behaviour of a simple harmonic oscillator is expressed in terms of its displacement x from equilibrium, its velocity $\dot{x}$, and its acceleration $\ddot{x}$ at any given time. If we try the solution

$$x = A \cos \omega t$$

where A is a constant with the same dimensions as x, we shall find that it satisfies the equation of motion

$$\ddot{x} + \omega^2 x = 0$$

for

$$\dot{x} = -A\omega \sin \omega t$$

and

$$\ddot{x} = -A\omega^2 \cos \omega t = -\omega^2 x$$

Another solution

$$x = B \sin \omega t$$

is equally valid, where B has the same dimensions as A, for then

$$\dot{x} = B\omega \cos \omega t$$

and

$$\ddot{x} = -B\omega^2 \sin \omega t = -\omega^2 x$$

The complete or general solution of equation (1.1) is given by the addition or superposition of both values for x so we have

$$x = A \cos \omega t + B \sin \omega t \qquad (1.2)$$

with

$$\ddot{x} = -\omega^2(A \cos \omega t + B \sin \omega t) = -\omega^2 x$$

where A and B are determined by the values of x and $\dot{x}$ at a specified time. If we rewrite the constants as

$$A = a \sin \phi \quad \text{and} \quad B = a \cos \phi$$

where ϕ is a constant angle, then

$$A^2 + B^2 = a^2(\sin^2 \phi + \cos^2 \phi) = a^2$$

so that

$$a = \sqrt{A^2 + B^2}$$

and

$$x = a \sin \phi \cos \omega t + a \cos \phi \sin \omega t$$
$$= a \sin (\omega t + \phi)$$

The maximum value of $\sin (\omega t + \phi)$ is unity so the constant a is the maximum value of x, known as the amplitude of displacement. The limiting values of $\sin (\omega t + \phi)$ are ± 1 so the system will oscillate between the values of $x = \pm a$ and we shall see that the magnitude of a is determined by the total energy of the oscillator.

The angle ϕ is called the 'phase constant' for the following reason. Simple harmonic motion is often introduced by reference to 'circular motion' because each possible value of the displacement x can be represented by the projection of a radius vector of constant length a on the diameter of the circle traced by the tip of the vector as it rotates in a positive anticlockwise direction with a constant angular velocity ω. Each rotation, as the radius vector sweeps through a phase angle of 2π radians, therefore corresponds to a complete vibration of the oscillator. In the solution

$$x = a \sin(\omega t + \phi)$$

the phase constant ϕ, measured in radians, defines the position in the cycle of oscillation at the time $t = 0$, so that the position in the cycle from which the oscillator started to move is

$$x = a \sin \phi$$

The solution

$$x = a \sin \omega t$$

defines the displacement only of that system which starts from the origin $x = 0$ at time $t = 0$ but the inclusion of ϕ in the solution

$$x = a \sin(\omega t + \phi)$$

where ϕ may take all values between zero and 2π allows the motion to be defined from any starting point in the cycle. This is illustrated in fig. 1.2 for various values of ϕ.

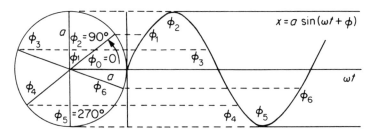

Fig. 1.2. Sinusoidal displacement of simple harmonic oscillator with time, showing variation of starting point in cycle in terms of phase angle ϕ

(Problems 1.2, 1.3, 1.4)

Velocity and Acceleration in Simple Harmonic Motion

The values of the velocity and acceleration in simple harmonic motion for

$$x = a \sin(\omega t + \phi)$$

are given by

$$\frac{dx}{dt} = \dot{x} = a\omega \cos(\omega t + \phi)$$

and

$$\frac{d^2 x}{dt^2} = \ddot{x} = -a\omega^2 \sin(\omega t + \phi)$$

The maximum value of the velocity $a\omega$ is called the *velocity amplitude* and the *acceleration amplitude* is given by $a\omega^2$.

From fig. 1.2 we see that a positive phase angle of $\pi/2$ radians converts a sine into a cosine curve. Thus the velocity

$$\dot{x} = a\omega \cos(\omega t + \phi)$$

leads the displacement

$$x = a \sin(\omega t + \phi)$$

by a phase angle of $\pi/2$ radians and its maxima and minima are always a quarter of a cycle ahead of those of the displacement; the velocity is a maximum when the displacement is zero and is zero at maximum displacement. The acceleration is 'anti-phase' (π radians) with respect to the displacement, being maximum positive when the displacement is maximum negative and vice versa. These features are shown in fig. 1.3.

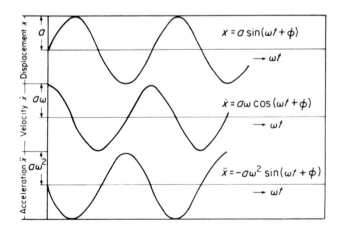

Fig. 1.3. Variation with time of displacement, velocity and acceleration in simple harmonic motion. Displacement lags velocity by $\pi/2$ radians and is π radians out of phase with the acceleration. The initial phase constant ϕ is taken as zero

Often, the relative displacement or motion between two oscillators having the same frequency may be considered in terms of their phase difference $\phi_1 - \phi_2$ which can have any value because one system may have started several cycles before the other and each complete cycle of vibration represents a change in the phase angle of $\phi = 2\pi$. When the motions of the two systems are diametrically opposed, that is, one has $x = +a$ whilst the other is at $x = -a$, the systems are 'anti-phase' and the total phase difference

$$\phi_1 - \phi_2 = n\pi \text{ radians}$$

where n is an *odd* integer. Identical systems 'in phase' have

$$\phi_1 - \phi_2 = 2n\pi \text{ radians}$$

where n is any integer. They have exactly equal values of displacement, velocity and acceleration at any instant.

(Problems 1.5, 1.6, 1.7, 1.8)

Non-linearity

As long as the displacement follows a sine or cosine behaviour with time the system is said to be linear. Non-linearity results when the stiffness s is not constant but varies with the displacement x (see Chapter 11).

Energy of a Simple Harmonic Oscillator

The fact that the velocity is zero at maximum displacement in simple harmonic motion and is a maximum at zero displacement illustrates the important concept of an exchange between kinetic and potential energy. In an ideal case the total energy remains constant but this is never realized in practice. If no energy is dissipated then all the potential energy becomes kinetic energy and vice versa, so that the values of (a) the total energy at any time, (b) the maximum potential energy and (c) the maximum kinetic energy will all be equal, that is

$$E_{\text{total}} = \text{KE} + \text{PE} = \text{KE}_{\text{max}} = \text{PE}_{\text{max}}$$

The solution $x = a \sin(\omega t + \phi)$ implies that the total energy remains constant because the amplitude of displacement $x = \pm a$ is regained every half cycle at the position of maximum potential energy; when energy is lost the amplitude gradually decays as we shall see later in this chapter. The potential energy is found by summing all the small elements of work $sx \, . \, dx$ (force sx times distance dx) *done by the system against the restoring force* over the range zero to x where $x = 0$ gives zero potential energy.

Thus the potential energy =

$$\int_0^x sx \, . \, dx = \tfrac{1}{2}sx^2$$

The kinetic energy is given by $\frac{1}{2}m\dot{x}^2$ so that the total energy

$$E = \tfrac{1}{2}m\dot{x}^2 + \tfrac{1}{2}sx^2$$

Since E is constant we have

$$\frac{\mathrm{d}E}{\mathrm{d}t} = (m\ddot{x} + sx)\dot{x} = 0$$

giving again the equation of motion

$$m\ddot{x} + sx = 0$$

The maximum potential energy occurs at $x = \pm a$ and is therefore

$$\mathrm{PE}_{\mathrm{max}} = \tfrac{1}{2}sa^2$$

The maximum kinetic energy is

$$\mathrm{KE}_{\mathrm{max}} = (\tfrac{1}{2}m\dot{x}^2)_{\mathrm{max}} = \tfrac{1}{2}ma^2\omega^2[\cos^2(\omega t + \phi)]_{\mathrm{max}}$$
$$= \tfrac{1}{2}ma^2\omega^2$$

when the cosine factor is unity.

But $m\omega^2 = s$ so the maximum values of the potential and kinetic energies are equal, showing that the energy exchange is complete.

The total energy at any instant of time or value of x is

$$E = \tfrac{1}{2}m\dot{x}^2 + \tfrac{1}{2}sx^2$$
$$= \tfrac{1}{2}ma^2\omega^2[\cos^2(\omega t + \phi) + \sin^2(\omega t + \phi)]$$
$$= \tfrac{1}{2}ma^2\omega^2$$
$$= \tfrac{1}{2}sa^2$$

as we should expect.

Fig. 1.4 shows the distribution of energy versus displacement for simple harmonic motion. Note that the potential energy curve

$$\mathrm{PE} = \tfrac{1}{2}sx^2 = \tfrac{1}{2}ma^2\omega^2\sin^2(\omega t + \phi)$$

is parabolic with respect to x and is symmetric about $x = 0$, so that energy is stored in the oscillator both when x is positive and when it is negative, e.g. a spring stores energy whether compressed or extended, as does a gas in compression or rarefaction. The kinetic energy curve

$$\mathrm{KE} = \tfrac{1}{2}m\dot{x}^2 = \tfrac{1}{2}ma^2\omega^2\cos^2(\omega t + \phi)$$

is parabolic with respect to both x and $\dot{x}$. The inversion of one curve with respect to the other displays the $\pi/2$ phase difference between the displace-

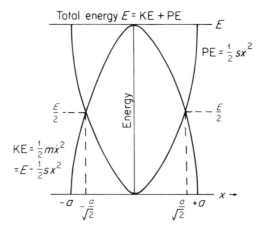

Total energy E = KE + PE

$PE = \frac{1}{2}sx^2$

$\frac{E}{2}$

$- - \frac{E}{2}$

Energy

$KE = \frac{1}{2}m\dot{x}^2$

$= E - \frac{1}{2}sx^2$

$-a$ $\quad -\frac{a}{\sqrt{2}}$ $\qquad \frac{a}{\sqrt{2}}$ $+a$ $\quad x \rightarrow$

Displacement

Fig. 1.4. Parabolic representation of potential energy and kinetic energy of simple harmonic motion versus displacement. Inversion of one curve with respect to the other shows a 90° phase difference. At any displacement value the sum of the ordinates of the curves equals the total constant energy E

ment (related to the potential energy) and the velocity (related to the kinetic energy).

For any value of the displacement x the sum of the ordinates of both curves equals the total constant energy E.

(Problems 1.9, 1.10, 1.11)

Simple Harmonic Oscillations in an Electrical System

So far we have discussed the simple harmonic motion of the mechanical and fluid systems of fig. 1.1, chiefly in terms of the inertial mass stretching the weightless spring of stiffness s. The stiffness s of a spring defines the difficulty of stretching; the reciprocal of the stiffness, the compliance C (where $s = 1/C$) defines the ease with which the spring is stretched and potential energy stored. This notation of compliance C is useful when discussing the simple harmonic oscillations of the electrical circuit of fig. 1.1(h) and fig. 1.5, where an inductance L is connected across the plates of a capacitance C. The force

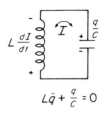

Fig. 1.5. Electrical system which oscillates simple harmonically. The sum of the voltages around the circuit is given by Kirchhoff's law as
$$L \, dI/dt + q/C = 0$$

equation of the mechanical and fluid examples now becomes the voltage equation (balance of voltages) of the electrical circuit, but the form and solution of the equations and the oscillatory behaviour of the systems are identical.

In the absence of resistance the energy of the electrical system remains constant and is exchanged between the *magnetic* field energy stored in the inductance and the *electric* field energy stored between the plates of the capacitance. At any instant, the voltage across the inductance is

$$V = -L\frac{dI}{dt} = -L\frac{d^2q}{dt^2}$$

where I is the current flowing and q is the charge, the negative sign showing that the voltage opposes the increase of current. This equals the voltage q/C across the capacitance so that

$$L\ddot{q} + q/C = 0 \qquad \text{(Kirchhoff's Law)}$$

or

$$\ddot{q} + \omega^2 q = 0$$

where

$$\omega^2 = \frac{1}{LC}$$

The energy stored in the magnetic field or inductive part of the circuit throughout the cycle, as the current increases from 0 to I, is formed by integrating the power at any instant with respect to time, that is

$$E_L = \int VI.\,dt$$

(where V is the magnitude of the voltage across the inductance).

So

$$E_L = \int VI\,dt = \int L\frac{dI}{dt}I\,dt = \int_0^I LI\,dI$$
$$= \tfrac{1}{2}LI^2 = \tfrac{1}{2}L\dot{q}^2$$

The potential energy stored mechanically by the spring is now stored electrostatically by the capacitance and equals

$$\tfrac{1}{2}CV^2 = \frac{q^2}{2C}$$

Comparison between the equations for the mechanical and electrical oscillators

$$\text{mechanical (force)} \rightarrow m\ddot{x} + sx = 0$$

$$\text{electrical (voltage)} \rightarrow L\ddot{q} + \frac{q}{C} = 0$$

$$\text{mechanical (energy)} \rightarrow \tfrac{1}{2}m\dot{x}^2 + \tfrac{1}{2}sx^2 = E$$

$$\text{electrical (energy)} \rightarrow \frac{1}{2}L\dot{q}^2 + \frac{1}{2}\frac{q^2}{C} = E$$

shows that magnetic field inertia (defined by the inductance L) controls the rate of change of current for a given voltage in a circuit in exactly the same way as the inertial mass controls the change of velocity for a given force. Magnetic inertial or inductive behaviour arises from the tendency of the magnetic flux threading a circuit to remain constant and reaction to any change in its value generates a voltage and hence a current which flows to oppose the change of flux. This is the physical basis of Fleming's right hand rule.

Superposition of Two Simple Harmonic Vibrations in One Dimension

(1) *Vibrations Having Equal Frequencies*

In the following chapters we shall meet physical situations which involve the superposition of two or more simple harmonic vibrations on the same system.

We have already seen how the displacement in simple harmonic motion may be represented in magnitude and phase by a constant length vector rotating in the positive (anticlockwise) sense with a constant angular velocity ω. To find the resulting motion of a system which moves in the x direction under the simultaneous effect of two simple harmonic oscillations of equal angular frequencies but of different amplitudes and phases, we can represent each simple harmonic motion by its appropriate vector and carry out a vector addition.

If the displacement of the first motion is given by

$$x_1 = a_1 \cos{(\omega t + \phi_1)}$$

and that of the second by

$$x_2 = a_2 \cos{(\omega t + \phi_2)}$$

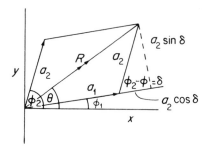

Fig. 1.6. Addition of vectors, each representing simple harmonic motion along the x
axis at angular frequency ω to give a resulting simple harmonic motion displacement
$x = R \cos(\omega t + \theta)$—here shown for $t = 0$

then fig. 1.6 shows that the resulting displacement amplitude R is given by

$$R^2 = (a_1 + a_2 \cos \delta)^2 + (a_2 \sin \delta)^2$$
$$= a_1^2 + a_2^2 + 2a_1 a_2 \cos \delta$$

where $\delta = \phi_2 - \phi_1$ is constant.

The phase constant θ of R is given by

$$\tan \theta = \frac{a_1 \sin \phi_1 + a_2 \sin \phi_2}{a_1 \cos \phi_1 + a_2 \cos \phi_2}$$

so the resulting simple harmonic motion has a displacement

$$x = R \cos(\omega t + \theta)$$

an oscillation of the same frequency ω but having an amplitude R and a phase
constant θ.

(Problem 1.12)

(2) *Vibrations Having Different Frequencies*

Suppose we now consider what happens when two vibrations of equal amplitudes but different frequencies are superposed. If we express them as

$$x_1 = a \sin \omega_1 t$$

and

$$x_2 = a \sin \omega_2 t$$

where

$$\omega_2 > \omega_1$$

then the resulting displacement is given by

$$x = x_1 + x_2 = a(\sin \omega_1 t + \sin \omega_2 t)$$

$$= 2a \sin \frac{(\omega_1 + \omega_2)t}{2} \cos \frac{(\omega_2 - \omega_1)t}{2}$$

This expression is illustrated in fig. 1.7. It represents a sinusoidal oscillation at the average frequency $(\omega_1 + \omega_2)/2$ having a displacement amplitude of $2a$

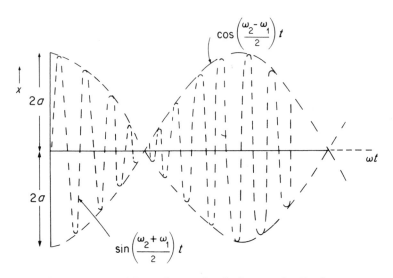

Fig. 1.7. Superposition of two simple harmonic displacements $x_1 = a \sin \omega_1 t$ and $x_2 = a \sin \omega_2 t$ when $\omega_2 > \omega_1$. The slow $\cos[\omega_2 - \omega_1)/2]t$ envelope modulates the $\sin[(\omega_2 + \omega_1)/2]t$ curve between the values $x = \pm 2a$

which modulates, that is, varies between $2a$ and zero under the influence of the cosine term of a much slower frequency equal to half the difference $(\omega_2 - \omega_1)/2$ between the original frequencies.

When ω_1 and ω_2 are almost equal the sine term has a frequency very close to both ω_1 and ω_2 whilst the cosine envelope modulates the amplitude $2a$ at a frequency $(\omega_2 - \omega_1)/2$ which is very slow.

Acoustically this growth and decay of the amplitude is registered as 'beats' of strong reinforcement when two sounds of almost equal frequency are heard. The frequency of the 'beats' is $(\omega_2 - \omega_1)$, the difference between the separate frequencies (not half the difference) because the maximum amplitude of $2a$ occurs twice in every period associated with the frequency $(\omega_2 - \omega_1)/2$. We shall meet this situation again when we consider the coupling of two oscillators in Chapter 3 and the wave group of two components in Chapter 4.

Superposition of Two Perpendicular Simple Harmonic Vibrations

(1) *Vibrations Having Equal Frequencies*

Suppose that a particle moves under the simultaneous influence of two simple harmonic vibrations of equal frequency, one along the x axis, the other along the perpendicular y axis. What is its subsequent motion?

The displacements may be written

$$x = a_1 \sin(\omega t + \phi_1)$$

$$y = a_2 \sin(\omega t + \phi_2)$$

and the path followed by the particle is formed by eliminating the time t from these equations to leave an expression involving only x and y and the constants ϕ_1 and ϕ_2.

Expanding the arguments of the sines we have

$$\frac{x}{a_1} = \sin \omega t \cos \phi_1 + \cos \omega t \sin \phi_1$$

and

$$\frac{y}{a_2} = \sin \omega t \cos \phi_2 + \cos \omega t \sin \phi_2$$

If we carry out the process

$$\left(\frac{x}{a_1} \sin \phi_2 - \frac{y}{a_2} \sin \phi_1 \right)^2 + \left(\frac{y}{a_2} \cos \phi_1 - \frac{x}{a_1} \cos \phi_2 \right)^2$$

this will yield

$$\frac{x^2}{a_1^2} + \frac{y^2}{a_2^2} - \frac{2xy}{a_1 a_2} \cos(\phi_2 - \phi_1) = \sin^2(\phi_2 - \phi_1) \tag{1.3}$$

which is the general equation for an ellipse.

In the most general case the axes of the ellipse are inclined to the x and y axes, but these become the principal axes when the phase difference

$$\phi_2 - \phi_1 = \frac{\pi}{2}$$

Equation 1.3 then takes the familiar form

$$\frac{x^2}{a_1^2} + \frac{y^2}{a_2^2} = 1$$

that is, an ellipse with semi-axes a_1 and a_2.

If $a_1 = a_2 = a$ this becomes the circle

$$x^2 + y^2 = a^2$$

When

$$\phi_2 - \phi_1 = 0, 2\pi, 4\pi \text{ etc.}$$

the equation simplifies to

$$y = \frac{a_2}{a_1} x$$

which is a straight line through the origin of slope a_2/a_1.

Again for $\phi_2 - \phi_1 = \pi, 3\pi, 5\pi$ etc., we obtain

$$y = -\frac{a_2}{a_1} x$$

a straight line through the origin of equal but opposite slope.

The paths traced out by the particle for various values of $\delta = \phi_2 - \phi_1$ are shown in fig. 1.8 and are most easily demonstrated on a cathode ray oscilloscope.

When

$$\phi_2 - \phi_1 = 0, \pi, 2\pi \text{ etc.}$$

and the ellipse degenerates into a straight line, the resulting vibration lies wholly in one plane and the oscillations are said to be *plane polarized.*

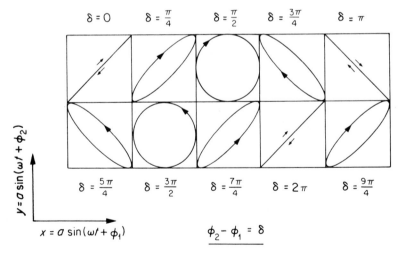

Fig. 1.8. Paths traced by a system vibrating simultaneously in two perpendicular directions with simple harmonic motions of equal frequency. The phase angle δ is the angle by which the y motion leads the x motion

Convention defines the plane of polarization as that plane perpendicular to the plane containing the vibrations. Similarly the other values of

$$\phi_2 - \phi_1$$

yield *circular* or *elliptic* polarization where the tip of the vector resultant traces out the appropriate conic section.

(Problems 1.13, 1.14, 1.15)

*Polarization

Polarization is a fundamental topic in optics and arises from the superposition of two perpendicular simple harmonic optical vibrations. We shall see in Chapter 7 that when a light wave is plane polarized its electrical field oscillation lies within a single plane and traces a sinusoidal curve along the direction of wave motion. Substances such as quartz and calcite are capable of splitting light into two waves whose planes of polarization are perpendicular to each other. Except in a specified direction, known as the optic axis, these waves have different velocities. One wave, the ordinary or O wave, travels at the same velocity in all directions and its electric field vibrations are always perpendicular to the optic axis. The extraordinary or E wave has a velocity which is

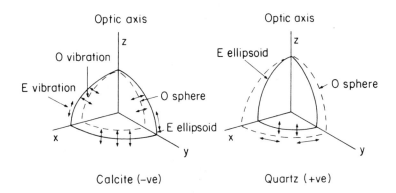

Calcite (−ve)   Quartz (+ve)

Fig. 1.9a. Ordinary (spherical) and extraordinary (ellipsoidal) wave surfaces in doubly refracting calcite and quartz. In calcite the E wave is faster than the O wave, except along the optic axis. In quartz the O wave is faster. The O vibrations are always perpendicular to the optic axis, and the O and E vibrations are always tangential to their wave surfaces

---

* This section may be omitted at a first reading.

direction-dependent. Both ordinary and extraordinary light have their own refractive indices, and thus quartz and calcite are known as doubly refracting materials. When the ordinary light is faster, as in quartz, a crystal of the substance is defined as positive, but in calcite the extraordinary light is faster and its crystal is negative. The surfaces, spheres and ellipsoids, which are the loci of the values of the wave velocities in any direction are shown in fig. 1.9(a), and for a given direction the electric field vibrations of the separate waves are

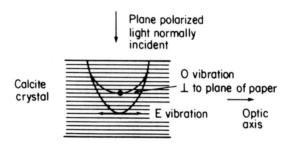

Fig. 1.9b. Plane polarized light normally incident on a calcite crystal face cut parallel to its optic axis. The advance of the E wave over the O wave is equivalent to a gain in phase

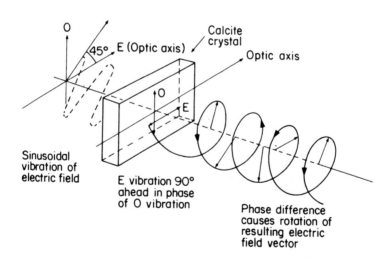

Fig. 1.9c. The crystal of fig. 1.9c is thick enough to produce a phase gain of $\pi/2$ radians in the E wave over the O wave. Wave recombination on leaving the crystal produces circularly polarized light

tangential to the surface of the sphere or ellipsoid as shown. Fig. 1.9(b) shows plane polarized light normally incident on a calcite crystal cut parallel to its optic axis. Within the crystal the faster E wave has vibrations parallel to the optic axis, while the O wave vibrations are perpendicular to the plane of the paper. The velocity difference results in a phase gain of the E vibration over the O vibration which increases with the thickness of the crystal. Fig. 1.9(c) shows plane polarized light normally incident on the crystal of fig. 1.9(b) with its vibration at an angle of 45° to the optic axis. The crystal splits the vibration into equal E and O components, and for a given thickness the E wave emerges with a phase gain of 90° over the O component. Recombination of the two vibrations produces circularly polarized light, of which the electric field vector now traces a helix in the anticlockwise direction as shown.

(2) Vibrations Having Different Frequencies (Lissajous Figures)

When the frequencies of the two perpendicular simple harmonic vibrations are not equal the resulting motion becomes more complicated. The patterns which are traced are called Lissajous figures and examples of these are shown in fig. 1.10 where the axial frequencies bear the simple ratios shown and

$$\delta = \phi_2 - \phi_1 = \frac{\pi}{2}$$

If the amplitudes of the vibrations are respectively a and b the resulting

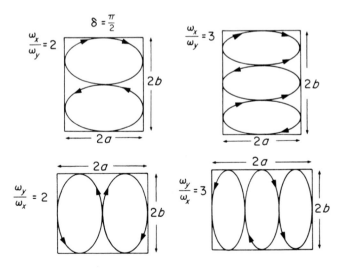

Fig. 1.10. Simple Lissajous figures produced by perpendicular simple harmonic motions of different angular frequencies

Lissajous figure will always be contained within the rectangle of sides $2a$ and $2b$. The sides of the rectangle will be tangential to the curve at a number of points and the ratio of the numbers of these tangential points along the x axis to those along the y axis is the inverse of the ratio of the corresponding frequencies (as indicated in fig. 1.10).

Superposition of a Large Number n of Simple Harmonic Vibrations of Equal Amplitude a and Equal Successive Phase Difference δ

Figure 1.11 shows the addition of n vectors of equal length a, each representing a simple harmonic vibration with a constant phase difference δ from its neighbour. Two general physical situations are characterized by such a superposition. The first is met in Chapter 4 as a wave group problem where the phase difference δ arises from a small *frequency difference*, $\delta\omega$, between consecutive components. The second appears in Chapter 10 where the intensity of optical interference and diffraction patterns are considered. There, the superposed

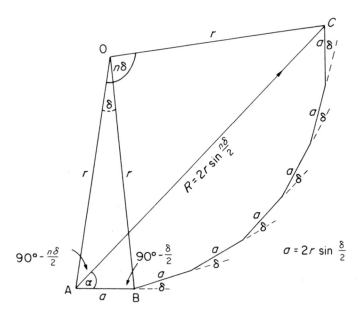

Fig. 1.11. Vector superposition of a large number n of simple harmonic vibrations of equal amplitude a and equal successive phase difference δ. The amplitude of the resultant

$$R = 2r \sin \frac{n\delta}{2} = a \frac{\sin n\delta/2}{\sin \delta/2}$$

and its phase with respect to the first contribution is given by $\alpha = (n-1)\,\delta/2$

harmonic vibrations will have the same frequency but each component will have a constant phase difference from its neighbour because of the extra *distance* it has travelled.

The figure displays the mathematical expression

$$R \cos(\omega t + \alpha) = a \cos \omega t + a \cos(\omega t + \delta) + a \cos(\omega t + 2\delta)$$
$$+ \ldots + a \cos(\omega t + [n-1]\delta)$$

where R is the magnitude of the resultant and α is its phase difference with respect to the first component $a \cos \omega t$.

Geometrically we see that each length

$$a = 2r \sin \frac{\delta}{2}$$

where r is the radius of the circle enclosing the (incomplete) polygon.

From the isosceles triangle OAC the magnitude of the resultant

$$R = 2r \sin \frac{n\delta}{2} = a \frac{\sin n\delta/2}{\sin \delta/2}$$

and its phase angle is seen to be

$$\alpha = \hat{OAB} - \hat{OAC}$$

In the isosceles triangle OAC

$$\hat{OAC} = 90° - \frac{n\delta}{2}$$

and in the isosceles triangle OAB

$$\hat{OAB} = 90° - \frac{\delta}{2}$$

so

$$\alpha = \left(90° - \frac{\delta}{2}\right) - \left(90° - \frac{n\delta}{2}\right) = (n-1)\frac{\delta}{2}$$

that is, half the phase difference between the first and the last contributions. Hence the resultant

$$R \cos(\omega t + \alpha) = a \frac{\sin n\delta/2}{\sin \delta/2} \cos\left[\omega t + (n-1)\frac{\delta}{2}\right]$$

We shall obtain the same result later in this chapter as an example on the use of exponential notation.

For the moment let us examine the behaviour of the magnitude of the resultant

$$R = a \frac{\sin n\delta/2}{\sin \delta/2}$$

which is not constant but depends on the value of δ. When n is very large δ is very small and the polygon becomes an arc of the circle centre O, of length $na = A$, with R as the chord. Then

$$\alpha = (n-1)\frac{\delta}{2} \approx \frac{n\delta}{2}$$

and

$$\sin\frac{\delta}{2} \to \frac{\delta}{2} \approx \frac{\alpha}{n}$$

Hence, in this limit,

$$R = a\frac{\sin n\delta/2}{\sin \delta/2} = a\frac{\sin \alpha}{\alpha/n} = na\frac{\sin \alpha}{\alpha} = \frac{A \sin \alpha}{\alpha}$$

The behaviour of $A \sin \alpha/\alpha$ versus α is shown in fig. 1.12. The pattern is symmetric about the value $\alpha = 0$ and is zero whenever $\sin \alpha = 0$ except at $\alpha \to 0$

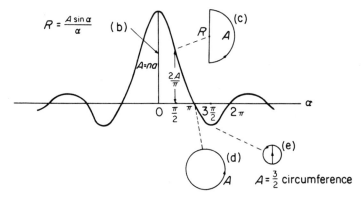

Fig. 1.12. (a) Graph of $A \sin \alpha/\alpha$ versus α, showing the magnitude of the resultants for (b) $\alpha = 0$; (c) $\alpha = \pi/2$; (d) $\alpha = \pi$ and (e) $\alpha = 3\pi/2$

that is, when $\sin \alpha/\alpha \to 1$. When $\alpha = 0$, $\delta = 0$ and the resultant of the n vectors is the straight line of length A, fig. 1.12(b). As δ increases A becomes the arc of a circle until at $\alpha = \pi/2$ the first and last contributions are out of phase ($2\alpha = \pi$) and the arc A has become a semicircle of which the diameter is the resultant R fig. 1.12(c). A further increase in δ increases α and curls the constant length A into the circumference of a circle ($\alpha = \pi$) with a zero resultant, fig. 1.12(d). At $\alpha = 3\pi/2$, fig. 1.12(e) the length A is now $3/2$ times the circumference of a circle whose diameter is the amplitude of the first minimum.

*Superposition of n Equal SHM Vectors of Length a with Random Phase

When the phase difference between the successive vectors of the last section

---

* This section may be omitted at a first reading.

may take random values ϕ between zero and 2π (measured from the x axis) the vector superposition and resultant R may be represented by fig. 1.13.

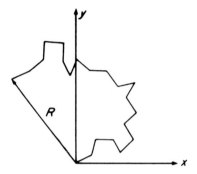

Fig. 1.13. The resultant $R = \sqrt{n}a$ of n vectors, each of length a, having random phase. This result is important in optical incoherence and in energy loss from waves from random dissipation processes

The components of R on the x and y axes are given by

$$R_x = a \cos \phi_1 + a \cos \phi_2 + a \cos \phi_3 \ldots a \cos \phi_n$$

$$= a \sum_{i=1}^{n} \cos \phi_i$$

and

$$R_y = a \sum_{i=1}^{n} \sin \phi_i$$

where

$$R^2 = R_x^2 + R_y^2$$

Now

$$R_x^2 = a^2 \left(\sum_{i=1}^{n} \cos \phi_i \right)^2 = a^2 \left[\sum_{i=1}^{n} \cos^2 \phi_i + \sum_{\substack{i=1 \\ i \neq j}}^{n} \cos \phi_i \sum_{j=1}^{n} \cos \phi_j \right]$$

In the typical term $2 \cos \phi_i \cos \phi_j$ of the double summation, $\cos \phi_i$ and $\cos \phi_j$ have random values between ± 1 and the averaged sum of sets of these products is effectively zero.

The summation

$$\sum_{i=1}^{n} \cos^2 \phi_i = n \overline{\cos^2 \phi}$$

that is, the number of terms n times the average value $\overline{\cos^2 \phi}$ which is the integrated value of $\cos^2 \phi$ over the interval zero to 2π divided by the total interval 2π, or

$$\overline{\cos^2 \phi} = \frac{1}{2\pi} \int_0^{2\pi} \cos^2 \phi \, d\phi = \frac{1}{2} = \overline{\sin^2 \phi}$$

So

$$R_x^2 = a^2 \sum_{i=1}^{n} \cos^2 \phi_i = na^2 \overline{\cos^2 \phi_i} = \frac{na^2}{2}$$

and

$$R_y^2 = a^2 \sum_{i=1}^{n} \sin^2 \phi_i = na^2 \overline{\sin^2 \phi_i} = \frac{na^2}{2}$$

giving

$$R^2 = R_x^2 + R_y^2 = na^2$$

or

$$R = \sqrt{n}a$$

Thus the amplitude R of a system subjected to n equal simple harmonic motions of amplitude a with random phases is only $\sqrt{n}a$ whereas, if the motions were all in phase R would equal na.

Such a result illustrates a very important principle of random behaviour.

(Problem 1.16)

Applications

(a) *Incoherent sources in optics.* The result above is directly applicable to the problem of coherence in optics. Light sources which are in phase are said to be coherent and this condition is essential for producing optical interference effects experimentally. If the amplitude of a light source is given by the quantity a its intensity is proportional to a^2, n coherent sources have a resulting amplitude na and a total intensity n^2a^2. Incoherent sources have random phases, n such sources each of amplitude a have a resulting amplitude $\sqrt{n}a$ and a total intensity of na^2.

(b) *Random processes and energy absorption.* From our present point of view the importance of random behaviour is the contribution it makes to energy loss or absorption from waves moving through a medium. We shall meet this in all the waves we discuss. Random processes, for example collisions between particles in Brownian motion, are of great significance in physics. Diffusion, viscosity or frictional resistance and thermal conductivity are all the result of random collision processes. These energy dissipating phenomena represent the transport of mass, momentum and energy, and change only in the direction of increasing disorder. They are known as 'thermodynamically irreversible' processes and are associated with the increase of entropy. Heat, for example, can flow only from a body at a higher temperature to one at a lower temperature. Using the earlier analysis where the length a is no longer a simple harmonic amplitude but is now the average distance a particle travels between random collisions (its mean free path), we see that after n such collisions (with, on average, equal time intervals between collisions) the particle will, on average, have travelled only a distance $\sqrt{n}a$ from its position at time $t = 0$, so that the distance travelled varies only with the square root of the time elapsed instead of being directly proportional to it. This is a feature of all random processes.

Not all the particles of the system will have travelled a distance $\sqrt{n}a$ but this distance is the most probable and represents a statistical average.

Random behaviour is described by the diffusion equation (see the last section of Chapter 6) and a constant coefficient called the diffusivity of the process will always arise. The dimensions of a diffusivity are always length^2/time and must be interpreted in terms of a characteristic distance of the process which varies only with the square root of time.

Some Useful Mathematics

The Exponential Series

By a 'natural process' of growth or decay we mean a process in which a quantity changes by a constant fraction of itself in a given interval of space or time. A 5% per annum compound interest represents a natural growth law; attenuation processes in physics usually describe natural decay.

The law is expressed differentially as

$$\frac{dN}{N} = \pm \alpha \, dx \quad \text{or} \quad \frac{dN}{N} = \pm \alpha \, dt$$

where N is the changing quantity, α is a constant and the positive and negative signs represent growth and decay respectively. The gradients dN/dx or dN/dt are therefore proportional to the value of N at which the gradient is measured.

Integration yields $N = N_0 \, e^{\pm \alpha x}$ or $N = N_0 \, e^{\pm \alpha t}$ where N_0 is the value at x or $t = 0$ and e is the exponential or the base of natural logarithms. The exponential series is defined as

$$e^x = 1 + x + \frac{x^2}{2!} + \frac{x^3}{3!} + \dots + \frac{x^n}{n!} + \dots$$

and is shown graphically for positive and negative x in fig. 1.14. It is important to note that whatever the form of the index of the logarithmic base e, it is the power to which the base is raised, and is therefore always non-dimensional. Thus $e^{\alpha x}$ is non-dimensional and α must have the dimensions of x^{-1}. Writing

$$e^{\alpha x} = 1 + \alpha x + \frac{(\alpha x)^2}{2!} + \frac{(\alpha x)^3}{3!} + \dots$$

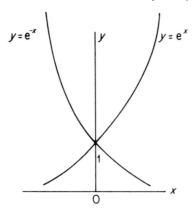

Fig. 1.14. The behaviour of the exponential series $y = e^x$ and $y = e^{-x}$

it follows immediately that

$$\frac{d}{dx}(e^{\alpha x}) = \alpha + \frac{2\alpha^2}{2!}x + \frac{3\alpha^3}{3!}x^2 + \dots$$

$$= \alpha\left[1 + \alpha x + \frac{(\alpha x)^2}{2!} + \frac{(\alpha x)^3}{3!}\right) + \dots\right]$$

$$= \alpha\, e^{\alpha x}$$

Similarly

$$\frac{d^2}{dx^2}(e^{\alpha x}) = \alpha^2\, e^{\alpha x}$$

In this chapter we shall use $d(e^{\alpha t})/dt = \alpha\, e^{\alpha t}$ and $d^2(e^{\alpha t})/dt^2 = \alpha^2\, e^{\alpha t}$ on a number of occasions.

By taking logarithms it is easily shown that $e^x e^y = e^{(x+y)}$ since $\log_e (e^x e^y) = \log_e e^x + \log_e e^y = x + y$.

The Notation $i = \sqrt{-1}$

The combination of the exponential series with the complex number notation $i = \sqrt{-1}$ is particularly convenient in physics. Here we shall show the mathematical convenience in expressing sine or cosine (oscillatory) behaviour in the form $e^{ix} = \cos x + i \sin x$.

In Chapter 2 we shall see the additional merit of i in its role of vector operator.

The series representation of $\sin x$ is written

$$\sin x = x - \frac{x^3}{3!} + \frac{x^5}{5!} - \frac{x^7}{7!} \dots$$

and that of cos x is

$$\cos x = 1 - \frac{x^2}{2!} + \frac{x^4}{4!} - \frac{x^6}{6!} \cdots$$

Since

$$i = \sqrt{-1}, i^2 = -1, i^3 = -i$$

etc., we have

$$e^{ix} = 1 + ix + \frac{(ix)^2}{2!} + \frac{(ix)^3}{3!} + \frac{(ix)^4}{4!} + \cdots$$

$$= 1 + ix - \frac{x^2}{2!} - \frac{ix^3}{3!} + \frac{x^4}{4!} + \cdots$$

$$= 1 - \frac{x^2}{2!} + \frac{x^4}{4!} + i\left(x - \frac{x^3}{3!} + \frac{x^5}{5!} + \cdots\right)$$

$$= \cos x + i \sin x$$

We also see that

$$\frac{d}{dx}(e^{ix}) = i e^{ix} = i \cos x - \sin x$$

Often we shall represent a sine or cosine oscillation by the form e^{ix} and recover the original form by taking that part of the solution preceded by i in the case of the sine, and the real part of the solution in the case of the cosine.

Examples

(1) In simple harmonic motion ($\ddot{x} + \omega^2 x = 0$) let us try the solution $x = a\, e^{i\omega t}\, e^{i\phi}$, where a is a constant length, and ϕ (and therefore $e^{i\phi}$) is a constant.

$$\frac{dx}{dt} = \dot{x} = i\omega a\, e^{i\omega t}\, e^{i\phi} = i\omega x$$

$$\frac{d^2 x}{dt^2} = \ddot{x} = i^2 \omega^2 a\, e^{i\omega t}\, e^{i\phi} = -\omega^2 x$$

Therefore

$$x = a\, e^{i\omega t}\, e^{i\phi} = a\, e^{i(\omega t + \phi)}$$

$$= a \cos(\omega t + \phi) + i\, a \sin(\omega t + \phi)$$

is a complete solution of $\ddot{x} + \omega^2 x = 0$.

On page 5 we used the sine form of the solution; the cosine form is equally valid and merely involves an advance of $\pi/2$ in the phase ϕ.

(2)

$$e^{ix} + e^{-ix} = 2\left(1 - \frac{x^2}{2!} + \frac{x^4}{4!} - \cdots\right) = 2 \cos x$$

$$e^{ix} - e^{-ix} = 2i\left(x - \frac{x^3}{3!} + \frac{x^5}{5!} - \ldots\right) = 2i\sin x$$

(3) On page 20 we used a geometrical method to show that the resultant of the superposed harmonic vibrations

$$a\cos\omega t + a\cos(\omega t + \delta) + a\cos(\omega t + 2\delta) + \ldots + a\cos(\omega t + [n-1]\delta)$$

$$= a\frac{\sin n\delta/2}{\sin \delta/2}\cos\left\{\omega t + \left(\frac{n-1}{2}\right)\delta\right\}$$

We can derive the same result using the complex exponential notation by expressing the series as the geometric progression

$$a\,e^{i\omega t} + a\,e^{i(\omega t + \delta)} + a\,e^{i(\omega t + 2\delta)} + \ldots + a\,e^{i[\omega t + (n-1)\delta]}$$

$$= a\,e^{i\omega t}(1 + e^{i\delta} + e^{i2\delta} + \ldots + e^{i(n-1)\delta})$$

$$= a\,e^{i\omega t}\frac{(1 - e^{in\delta})}{(1 - e^{i\delta})} = a\,e^{i\omega t}\frac{e^{in\delta/2}(e^{-in\delta/2} - e^{in\delta/2})}{e^{i\delta/2}(e^{-i\delta/2} - e^{i\delta/2})}$$

$$= a\,e^{i\{\omega t + [(n-1)/2]\delta\}}\frac{\sin n\delta/2}{\sin \delta/2}$$

$$= a\cos\left\{\omega t + \left(\frac{n-1}{2}\right)\delta\right\}\frac{\sin n\delta/2}{\sin \delta/2}$$

where we have recovered the original cosine term from the complex exponential notation.

(Problem 1.17)

(4) Suppose we represent a harmonic oscillation by the complex exponential form

$$z = a\,e^{i\omega t}$$

where a is the amplitude. Replacing i by $-$i defines the *complex conjugate*

$$z^* = a\,e^{-i\omega t}$$

The use of this conjugate is discussed more fully in Chapter 2 but here we can note that the product of a complex quantity and its conjugate is always equal to the square of the amplitude for

$$zz^* = a^2\,e^{i\omega t}\,e^{-i\omega t} = a^2\,e^{(i-i)\omega t} = a^2\,e^0$$

$$= a^2$$

(Problem 1.18)

Damped Simple Harmonic Motion

Initially we discussed the case of ideal simple harmonic motion where the total energy remained constant and the displacement followed a sine curve, apparently for an infinite time. In practice some energy is always dissipated by a resistive or viscous process; for example, the amplitude of a freely swinging pendulum will always decay with time as energy is lost. The presence of resistance to motion means that another force is active, which is taken as being proportional to the velocity. Like the stiffness force it always acts in a direction opposite to that of the acceleration term (see fig. 1.15) and the new balance of forces, or equation of motion, becomes

$$m\ddot{x} = -sx - r\dot{x}$$

where r is the constant of proportionality and has the dimensions of force per unit of velocity. The presence of such a term will always result in energy loss.

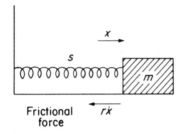

Fig. 1.15. Simple harmonic motion system with a damping or frictional force $r\dot{x}$ acting against the direction of motion. The equation of motion is
$$m\ddot{x} + r\dot{x} + sx = 0$$

The problem now is to find the behaviour of the displacement x from the equation

$$m\ddot{x} + r\dot{x} + sx = 0 \qquad (1.4)$$

where the coefficients m, r and s are constant.

When these coefficients are constant a solution of the form $x = C\,e^{\alpha t}$ can always be found. Obviously, since an exponential term is always nondimensional, C has the dimensions of x (a length, say) and α has the dimensions of inverse time, T^{-1}. We shall see that there are three possible forms of this solution, each describing a different behaviour of the displacement x with time. In two of these solutions C appears explicitly as a constant length, but in the third case it takes the form
$$C = A + Bt\ *$$

* The number of constants allowed in the general solution of a differential equation is always equal to the order (that is the highest differential coefficient) of the equation. The two values A and B are allowed because equation (1.4) is second order. The values of the constants are adjusted to satisfy the initial conditions.

where A is a length, B is a velocity and t is a time, giving C the overall dimensions of a length, as we expect. From our point of view this case is not the most important.

Taking C as a constant length gives $\dot{x} = \alpha C \, e^{\alpha t}$ and $\ddot{x} = \alpha^2 C \, e^{\alpha t}$, so that equation (1.4) may be rewritten

$$C \, e^{\alpha t}(m\alpha^2 + r\alpha + s) = 0$$

so that either

$$x = C \, e^{\alpha t} = 0 \quad \text{(which is trival)}$$

or

$$m\alpha^2 + r\alpha + s = 0$$

Solving the quadratic equation in α gives

$$\alpha = \frac{-r}{2m} \pm \sqrt{\frac{r^2}{4m^2} - \frac{s}{m}}$$

Note that $r/2m$ and $(s/m)^{1/2}$, and therefore, α, all have the dimensions of inverse time, T^{-1}, which we expect from the form of $e^{\alpha t}$.

The displacement can now be expressed as $x = C \, e^{-rt/2m \pm (r^2/4m^2 - s/m)^{1/2} t}$, where the bracket $(r^2/4m^2 - s/m)$ can be positive, zero or negative depending on the relative magnitude of the two terms inside it. Each of these conditions gives one of the three possible solutions referred to earlier and each solution describes a particular kind of behaviour. We shall discuss these solutions in order of *increasing* significance from our point of view; the third solution is the one we shall concentrate upon throughout the rest of this book.

The conditions are:

(1) *Bracket positive* $(r^2/4m^2 > s/m)$. Here the damping resistance term $r^2/4m^2$ dominates the stiffness term s/m, and heavy damping results in a *dead beat* system.

(2) *Bracket zero* $(r^2/4m^2 = s/m)$. The balance between the two terms results in a *critically damped* system.

Neither (1) nor (2) gives oscillatory behaviour.

(3) *Bracket negative* $(r^2/4m^2 < s/m)$. The system is lightly damped and gives oscillatory *damped simple harmonic motion*.

Case 1. Heavy Damping

Writing $r/2m = p$ and $(r^2/4m^2 - s/m)^{1/2} = q$, we can replace

$$x = C \, e^{-rt/2m \pm (r^2/4m^2 - s/m)^{1/2} t}$$

by

$$x = e^{-pt}(C_1 \, e^{qt} + C_2 \, e^{-qt})$$

where the C_i s are arbitrary in value but have the same dimensions as C (note that two separate values of C are allowed because the differential equation (1.4) is second order).

If now $F = C_1 + C_2$ and $G = C_1 - C_2$, the displacement is given by

$$x = e^{-pt}\left[\frac{F}{2}(e^{qt} + e^{-qt}) + \frac{G}{2}(e^{qt} - e^{-qt})\right]$$

or

$$x = e^{-pt}(F \cosh qt + G \sinh qt)$$

This represents non-oscillatory behaviour, but the actual displacement will depend upon the initial (or boundary) conditions, that is, the value of x at time $t = 0$. If $x = 0$ at $t = 0$ then $F = 0$, and

$$x = G e^{-rt/2m} \sinh \left(\frac{r^2}{4m^2} - \frac{s}{m}\right)^{1/2} t$$

Fig. 1.16 illustrates such behaviour when a heavily damped system is disturbed from equilibrium by a sudden impulse (that is, given a velocity at $t = 0$). It will

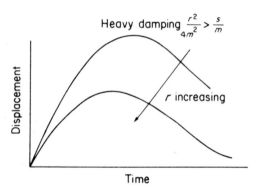

Heavy damping $\dfrac{r^2}{4m^2} > \dfrac{s}{m}$

r increasing

Displacement

Time

Fig. 1.16. Non-oscillatory behaviour of damped simple harmonic system with heavy damping (where $r^2/4m^2 > s/m$) after the system has been given an impulse from a rest position $x = 0$

return to zero displacement quite slowly without oscillating about its equilibrium position. More advanced mathematics shows that the value of the velocity dx/dt vanishes only once so that there is only one value of maximum displacement.

(Problem 1.19)

Case 2. Critical Damping $(r^2/4m^2 = s/m)$

Using the notation of Case 1, we see that $q = 0$ and that $x = C e^{-pt}$. This is, in fact, the limiting case of the behaviour of Case 1 as q changes from positive to negative. In this case the quadratic equation in α has equal roots, which, in a differential equation solution, demands that C must be written $C = A + Bt$, where A is a constant length and B a given velocity which depends on the boundary conditions. It is easily verified that the value

$$x = (A + Bt)\,e^{-rt/2m} = (A + Bt)\,e^{-pt}$$

satisfies $m\ddot{x} + r\dot{x} + sx = 0$ when $r^2/4m^2 = s/m$.

(Problem 1.20)

Application to a Ballistic Galvanometer

Critical damping is of practical importance in recording instruments such as ballistic galvanometers which experience sudden impulses and are required to return to zero displacement in the minimum time. Suppose such a galvanometer has zero displacement at $t = 0$ and receives a quantity of electric charge which gives its light spot an initial velocity V over a linear scale.

Then $x = 0$ (so that $A = 0$) and $\dot{x} = V$ at $t = 0$. However,

$$\dot{x} = B[(-pt)\,e^{-pt} + e^{-pt}] = B \text{ at } t = 0$$

so that $B = V$ and the complete solution is

$$x = Vt\,e^{-pt}$$

The maximum displacement x occurs when the instrument light spot comes to rest before returning to zero displacement. At maximum displacement

$$\dot{x} = V\,e^{-pt}(1 - pt) = 0$$

thus giving $(1 - pt) = 0$, i.e. $t = 1/p$.

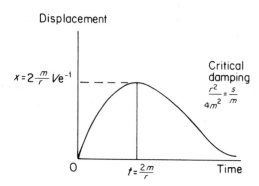

Fig. 1.17. Limiting case of non-oscillatory behaviour of damped simple harmonic system where $r^2/4m^2 = s/m$ (critical damping)

At this time the displacement is therefore

$$x = Vt\,e^{-pt} = \frac{V}{p}e^{-1}$$

$$= 0.368\,\frac{V}{p} = 0.368\,\frac{2mV}{r}$$

The curve of displacement versus time is shown in fig. 1.17; the return to zero in a critically damped system is reached in *minimum* time.

Case 3. Damped Simple Harmonic Motion

When $r^2/4m^2 < s/m$ the damping is light, and this gives from the present point of view the most important kind of behaviour, *oscillatory damped simple harmonic motion*.

The expression $(r^2/4m^2 - s/m)^{1/2}$ is an imaginary quantity, the square root of a negative number, which can be rewritten

$$\pm\left(\frac{r^2}{4m^2} - \frac{s}{m}\right)^{1/2} = \pm\sqrt{-1}\left(\frac{s}{m} - \frac{r^2}{4m^2}\right)^{1/2}$$

$$= \pm i\left(\frac{s}{m} - \frac{r^2}{4m^2}\right)^{1/2} \quad (\text{where } i = \sqrt{-1})$$

so that the displacement

$$x = C\,e^{-rt/2m}\,e^{\pm i(s/m - r^2/4m^2)^{1/2}t}$$

The bracket has the dimensions of inverse time, that is, of frequency, and can be written $(s/m - r^2/4m^2)^{1/2} = \omega'$, so that the second exponential becomes $e^{i\omega't} = \cos\omega't + i\sin\omega't$. This shows that the behaviour of the displacement x is oscillatory with a new frequency $\omega' < \omega = (s/m)^{1/2}$, the frequency of ideal simple harmonic motion. To compare the behaviour of the damped oscillator with the ideal case we should like to express the solution in a form similar to $x = A\sin(\omega't + \phi)$ as in the ideal case, where ω has been replaced by ω'.

We can do this by writing

$$x = C\,e^{-rt/2m}\,e^{\pm i\omega't}$$

$$= e^{-rt/2m}(C_1\,e^{i\omega't} + C_2\,e^{-i\omega't})$$

If we now choose

$$C_1 = \frac{A}{2i}e^{i\phi}$$

and

$$C_2 = -\frac{A}{2i}e^{-i\phi}$$

where A and ϕ (and thus $e^{i\phi}$) are constants which depend on the motion at $t = 0$, we find after substitution

$$x = A\, e^{-rt/2m} \frac{[e^{i(\omega't+\phi)} - e^{-i(\omega't+\phi)}]}{2i}$$

$$= A\, e^{-rt/2m} \sin(\omega't + \phi)$$

This procedure is equivalent to imposing the boundary condition $x = A \sin\phi$ at $t = 0$ upon the solution for x. The displacement therefore varies sinusoidally with time as in the case of simple harmonic motion, but now has a new frequency

$$\omega' = \left(\frac{s}{m} - \frac{r^2}{4m^2}\right)^{1/2}$$

and its amplitude A is modified by the exponential term $e^{-rt/2m}$, a term which decays with time.

If $x = 0$ at $t = 0$ then $\phi = 0$; fig. 1.18 shows the behaviour of x with time, its oscillations gradually decaying with the envelope of maximum amplitudes following the dotted curve $e^{-rt/2m}$. The constant A is obviously the value to which the amplitude would have risen at the first maximum if no damping were present.

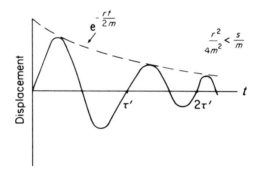

Fig. 1.18. Damped oscillatory motion where $s/m > r^2/4m^2$. The amplitude decays with $e^{-rt/2m}$, and the reduced angular frequency is given by $\omega'^2 = s/m - r^2/4m^2$

The presence of the force term $r\dot{x}$ in the equation of motion therefore introduces a loss of energy which causes the amplitude of oscillation to decay with time as $e^{-rt/2m}$.

(Problem 1.21)

Methods of Describing the Damping of an Oscillator

Earlier in this chapter we saw that the energy of an oscillator is given by

$$E = \tfrac{1}{2}ma^2\omega^2 = \tfrac{1}{2}sa^2$$

that is, proportional to the square of its amplitude.

We have just seen that, in the presence of a damping force $r\dot{x}$ the amplitude decays with time as

$$e^{-rt/2m}$$

so that the energy decay will be proportional to

$$(e^{-rt/2m})^2$$

that is $e^{-rt/m}$. The larger the value of the damping force r the more rapid the decay of the amplitude and energy. Thus we can use the exponential factor to express the rates at which the amplitude and energy are reduced.

(1) *Logarithmic Decrement*

This measures the rate at which the *amplitude* dies away. Suppose in the expression

$$x = A\, e^{-rt/2m} \sin(\omega' t + \phi)$$

we choose

$$\phi = \pi/2$$

and we write

$$x = A_0\, e^{-rt/2m} \cos \omega' t$$

with $x = A_0$ at $t = 0$. Its behaviour will follow the curve in fig. 1.19.

If the period of oscillation is τ' where $\omega' = 2\pi/\tau'$ then one period later the amplitude is given by

$$A_1 = A_0\, e^{(-r/2m)\tau'}$$

so that

$$\frac{A_0}{A_1} = e^{r\tau'/2m} = e^{\delta}$$

where

$$\delta = \frac{r}{2m}\tau' = \log_e \frac{A_0}{A_1}$$

is called the *logarithmic decrement*. It is the logarithm of the ratio of two amplitudes of oscillation which are separated by one period, the larger amplitude being the numerator since $e^{\delta} > 1$.

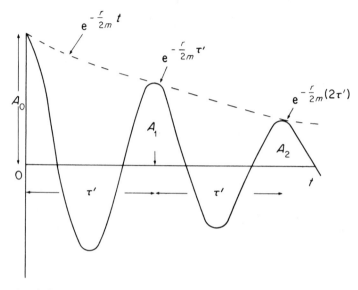

Fig. 1.19. The logarithmic ratio of any two amplitudes one period apart is the logarithmic decrement, defined as $\delta = \log_e (A_n/A_{n+1}) = r\tau'/2m$

Similarly

$$\frac{A_0}{A_2} = e^{r(2\tau')/2m} = e^{2\delta}$$

and

$$\frac{A_0}{A_n} = e^{n\delta}$$

Experimentally the value of δ is best found by comparing amplitudes of oscillations which are separated by n periods. The graph of

$$\log_e \frac{A_0}{A_n}$$

versus n for different values of n has a slope δ.

(2) *Relaxation Time or Modulus of Decay*

Another way of expressing the damping effect is by means of the time taken for the amplitude to decay to

$$e^{-1} = 0 \cdot 368$$

of its original value A_0. This time is called the *relaxation time* or *modulus of decay* and the amplitude

$$A_t = A_0 e^{-rt/2m} = A_0 e^{-1}$$

at a time $t = 2m/r$.

Measuring the natural decay in terms of the fraction e^{-1} of the original value is a very common procedure in physics. The time for a natural decay process to reach zero is, of course, theoretically infinite.

(Problem 1.22)

(3) *The Quality Factor or Q-value of a Damped Simple Harmonic Oscillator*

This measures the rate at which the *energy* decays. Since the decay of the amplitude is represented by

$$A = A_0 e^{-rt/2m}$$

the decay of energy is proportional to

$$A^2 = A_0^2 e^{(-rt/2m)^2}$$

and may be written

$$E = E_0 e^{(-r/m)t}$$

where E_0 is the energy value at $t = 0$.

The time for the energy E to decay to $E_0 e^{-1}$ is given by $t = m/r$ seconds during which time the oscillator will have vibrated through $\omega' m/r$ radians.

We define the *quality factor*

$$Q = \frac{\omega' m}{r}$$

as the *number of radians through which the damped system oscillates as its energy decays to*

$$E = E_0 e^{-1}$$

In general Q is very large so the damping resistance r is very small and

$$\frac{s}{m} \gg \frac{r^2}{4m^2}$$

so that

$$\omega' \approx \omega_0 = \left(\frac{s}{m}\right)^{\frac{1}{2}}$$

Thus we write, to a very close approximation,

$$Q = \frac{\omega_0 m}{r}$$

which is a constant of the damped system.

Since r/m now equals ω_0/Q we can write

$$E = E_0\, e^{(-r/m)t} = E_0\, e^{-\omega_0 t/Q}$$

The fact that Q is a constant $(=\omega_0 m/r)$ implies that the ratio

$$\frac{\text{energy stored in system}}{\text{energy lost per cycle}}$$

is also a constant, for

$$\frac{Q}{2\pi} = \frac{\omega_0 m}{2\pi r} = \frac{\nu_0 m}{r}$$

is the number of *cycles* (or complete oscillations) through which the system moves in decaying to

$$E = E_0\, e^{-1}$$

and if

$$E = E_0\, e^{(-r/m)t}$$

the energy lost per cycle is

$$-dE = \frac{r}{m} E\, dt = \frac{r}{m} E \frac{1}{\nu'}$$

where $dt = 1/\nu' = \tau'$, the period of oscillation.

Thus the ratio

$$\frac{\text{energy stored in system}}{\text{energy lost per cycle}} = \frac{E}{-dE} = \frac{\nu' m}{r} \approx \frac{\nu_0 m}{r}$$

$$= \frac{Q}{2\pi}$$

In the next chapter we shall meet the same quality factor Q in two other roles, the first as a measure of the power absorption bandwidth of a damped oscillator driven near its resonant frequency and again as the factor by which the displacement of the oscillator is amplified at resonance.

Example on the Q-value of a Damped Simple Harmonic Oscillator

An electron in an atom which is freely radiating power behaves as a damped simple harmonic oscillator.

If the radiated power is given by $P = q^2\omega^4 x_0^2/12\pi\epsilon_0 c^3$ watts at a wavelength of 0·6 microns (6000 Å), show that the Q-value of the atom is about 10^8 and that its free radiation lifetime is about 10^{-8} sec (the time for its energy to decay to e^{-1} of its original value).

$$q = 1\cdot6 \times 10^{-19} \; coulombs$$

$$1/4\pi\epsilon_0 = 9 \times 10^9 \; metres \; farad^{-1}$$

$$m_e = 9 \times 10^{-31} \; kilograms$$

$$c = 3 \times 10^8 \; metres \; sec^{-1}$$

$$x_0 = maximum \; amplitude \; of \; oscillation.$$

The radiated power P is $-\nu \, dE$, where $-dE$ is the energy loss per cycle, and the energy of the oscillator is given by $E = \frac{1}{2}m_e\omega^2 x_0^2$.

Thus $Q = 2\pi E/-dE = \nu\pi m_e\omega^2 x_0^2/P$, and inserting the values above with $\omega = 2\pi\nu = 2\pi c/\lambda$, where the wavelength λ is given, yields a Q value of $\sim 5 \times 10^7$.

The relation $Q = \omega t$ gives t, the radiation lifetime, a value of $\sim 10^{-8}$ sec.

Energy Dissipation

We have seen that the presence of the resistive force reduces the amplitude of oscillation with time as energy is dissipated.

The total energy remains the sum of the kinetic and potential energies

$$E = \frac{1}{2}m\dot{x}^2 + \frac{1}{2}sx^2$$

Now, however, dE/dt is not zero but negative because energy is lost, so that

$$\frac{dE}{dt} = \frac{d}{dt}(\frac{1}{2}m\dot{x}^2 + \frac{1}{2}sx^2) = \dot{x}(m\ddot{x} + sx)$$

$$= \dot{x}(-r\dot{x}) \quad for \quad m\ddot{x} + r\dot{x} + sx = 0$$

i.e. $dE/dt = -r\dot{x}^2$, which is the rate of doing work against the frictional force (dimensions of force × velocity = force × distance/time).

(Problems 1.23, 1.24)

Damped SHM in an Electrical Circuit

The force equation in the mechanical oscillator is replaced by the voltage equation in the electrical circuit of inductance, resistance and capacitance (fig. 1.20).

We have, therefore,

$$L\frac{dI}{dt} + RI + \frac{q}{C} = 0$$

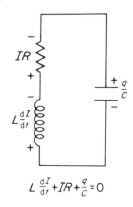

Fig. 1.20. Electrical circuit of inductance, capacitance and resistance capable of damped simple harmonic oscillations. The sum of the voltages around the circuit is given from Kirchhoff's law as

$$L\frac{\mathrm{d}I}{\mathrm{d}t} + RI + \frac{q}{C} = 0$$

or

$$L\ddot{q} + R\dot{q} + \frac{q}{C} = 0$$

and by comparison with the solutions for x in the mechanical case we know immediately that the charge

$$q = q_0\, e^{-Rt/2L \pm (R^2/4L^2 - 1/LC)^{1/2}t}$$

which, for $1/LC > R^2/4L^2$, gives oscillatory behaviour at a frequency

$$\omega^2 = \frac{1}{LC} - \frac{R^2}{4L^2}$$

From the exponential decay term we see that R/L has the dimensions of inverse time T^{-1} or ω, so that ωL has the dimensions of R, that is, ωL is measured in ohms.

Similarly, since $\omega^2 = 1/LC$, $\omega L = 1/\omega C$, so that $1/\omega C$ is also measured in ohms. We shall use these results in the next chapter.

(Problems 1.25, 1.26, 1.27)

Problem 1.1

The equation of motion

$$m\ddot{x} = -sx \quad \text{with} \quad \omega^2 = \frac{s}{m}$$

applies directly to the system in Fig. 1.1(c).
If the pendulum bob of fig. 1.1(a) is displaced a small distance x show that the stiffness (restoring force per unit distance) is mg/l and that $\omega^2 = g/l$ where g is the acceleration due to gravity. Now use the small angular displacement θ instead of x and show that ω is the same.

In fig. 1.1(b) the angular oscillations are rotational so the mass is replaced by the moment of inertia I of the disc and the stiffness by the restoring couple of the wire which is C per radian of angular displacement. Show that $\omega^2 = I/c$.

In fig. 1.1(d) show that the stiffness is $2T/l$ and that $\omega^2 = 2T/lm$.

In fig. 1.1(e) show that the stiffness of the system is $2\rho g$ and that $\omega^2 = 2g/l$ where g is the acceleration due to gravity.

In fig. 1.1(f) only the gas in the flask neck oscillates, behaving as a piston of mass ρAl. If the pressure changes are calculated from the equation of state use the adiabatic relation $pV^\gamma = $ constant and take logarithms to show that the pressure change in the flask is

$$dp = -\gamma p \frac{dV}{V} = -\gamma p \frac{Ax}{V}$$

where x is the gas displacement in the neck. Hence show that $\omega^2 = \gamma pA/l\rho V$. Note that γp is the stiffness of a gas (see Chapter 5).

In fig. 1.1(g), if the cross-sectional area of the neck is A and the hydrometer is a distance x above its normal floating level, the restoring force depends on the volume of liquid displaced, (Archimedes' principle). Show that $\omega^2 = g\rho A/m$.

Check the dimensions of ω^2 for each case.

Problem 1.2
Show by the choice of appropriate values for A and B in equation (1.2) that equally valid solutions for x are

$$x = a\cos(\omega t + \phi)$$
$$x = a\sin(\omega t - \phi)$$
$$x = a\cos(\omega t - \phi)$$

and check that these solutions satisfy the equation

$$\ddot{x} + \omega^2 x = 0$$

Problem 1.3
The pendulum in fig. 1.1(a) swings with a displacement amplitude a. If its starting point from rest is

(a)	$x = +a$
(b)	$x = -a$
(c)	$x = +a/\sqrt{2}$
(d)	$x = a/2$
(e)	$x = 0$ ·

find the different values of the phase constant ϕ for the solutions

$$x = a\sin(\omega t + \phi)$$
$$x = a\cos(\omega t + \phi)$$
$$x = a\sin(\omega t - \phi)$$
$$x = a\cos(\omega t - \phi)$$

Problem 1.4
Show that the values of ω^2 for the three simple harmonic oscillations (a), (b), (c) in the diagram are in the ratio $1 : 2 : 4$.

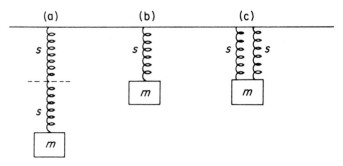

Problem 1.5

The displacement of a simple harmonic oscillator is given by

$$x = a \sin(\omega t + \phi)$$

If the oscillation started at time $t = 0$ from a position x_0 with a velocity $\dot{x} = v_0$ show that

$$\tan \phi = \omega x_0 / v_0$$

and

$$a = (x_0^2 + v_0^2/\omega^2)^{1/2}$$

Problem 1.6

A particle oscillates with simple harmonic motion along the x axis with a displacement amplitude a and spends a time dt in moving from x to $x + dx$. Show that the probability of finding it between x and $x + dx$ is given by

$$\frac{dx}{\pi(a^2 - x^2)^{1/2}}$$

(in wave mechanics such a probability is not zero for $x > a$).

Problem 1.7

Many identical simple harmonic oscillators are equally spaced along the x axis of a medium and a photograph shows that the locus of their displacements in the y direction is a sine curve. If the distance λ separates oscillators which differ in phase by 2π radians what is the phase difference between two oscillators a distance x apart.

Problem 1.8

A mass stands on a platform which vibrates simple harmonically in a vertical direction at a frequency of 5 hertz. Show that the mass loses contact with the platform when the displacement exceeds 10^{-2} meters.

Problem 1.9

A mass M is suspended at the end of a spring of length l and stiffness s. If the mass of the spring is m and the velocity of an element dy of its length is proportional to its distance y from the fixed end of the spring, show that the kinetic energy of this element is

$$\frac{1}{2}\left(\frac{m}{l}dy\right)\left(\frac{y}{l}v\right)^2$$

where v is the velocity of the suspended mass M. Hence, by integrating over the length of the spring, show that its total kinetic energy is $\frac{1}{6}mv^2$ and, from the total energy of the oscillating system, show that the frequency of oscillation is given by

$$\omega^2 = \frac{s}{M + m/3}$$

Problem 1.10

The general form for the energy of a simple harmonic oscillator is

$$E = \tfrac{1}{2} \text{ mass (velocity)}^2 + \tfrac{1}{2} \text{ stiffness (displacement)}^2$$

Set up the energy equations for the oscillators in fig. 1.1(a), (b), (c), (d), (e), (f) and (g), and use the expression

$$\frac{dE}{dt} = 0$$

to derive the equation of motion in each case.

Problem 1.11

The displacement of a simple harmonic oscillator is given by $x = a \sin \omega t$. If the values of the displacement x and the velocity $\dot{x}$ are plotted on perpendicular axes, eliminate t to show that the locus of the points $(x, \dot{x})$ is an ellipse. Show that this ellipse represents a path of constant energy.

Problem 1.12

In Chapter 10 the intensity of the pattern when light from two slits interfere (Young's experiment) will be seen to depend on the superposition of two simple harmonic oscillations of equal amplitude a and phase difference δ. Show that the intensity

$$I = R^2 = 4a^2 \cos^2 \delta/2$$

Between what values does the intensity vary?

Problem 1.13

Carry out the process indicated in the text to derive equation (1.3) on page 15.

Problem 1.14

The co-ordinates of the displacement of a particle of mass m are given by

$$x = a \sin \omega t$$
$$y = b \cos \omega t$$

Eliminate t to show that the particle follows an elliptical path and show by adding its kinetic and potential energy at any position x, y that the ellipse is a path of constant energy equal to the sum of the separate energies of the simple harmonic vibrations.

Prove that the quantity $m(x\dot{y} - y\dot{x})$ is also constant. What does this quantity represent?

Problem 1.15

Two simple harmonic motions of the same frequency vibrate in directions perpendicular to each other along the x and y axes. A phase difference

$$\delta = \phi_2 - \phi_1$$

exists between them such that the principal axes of the resulting elliptical trace are inclined at an angle to the x and y axes. Show that the measurement of two separate values of x (or y) is sufficient to determine the phase difference. (Hint: use equation (1.3) and measure y(max), and y for $x = 0$.)

Problem 1.16
There are 21 values sin ϕ_i in the set

$$\sin \phi_i = \pm 1, \pm 0.9, \pm 0.8 \ldots 0.$$

Take a random group of n values from this set and form the product

$$\sum_{\substack{i=1 \\ i \neq j}}^{n} \sin \phi_i \sum_{j=1}^{n} \sin \phi_j$$

Show that the average value obtained for several such groups is negligible with respect to $n/2$.

Problem 1.17
Use the method of example (3) (page 28) to show that

$$a \sin \omega t + a \sin (\omega t + \delta) + a \sin (\omega t + 2\delta) + \ldots + a \sin [\omega t + (n-1)\delta]$$

$$= a \sin \left[\omega t + \frac{(n-1)}{2} \delta \right] \frac{\sin n\delta/2}{\sin \delta/2}$$

Problem 1.18
If we represent the sum of the series

$$a \cos \omega t + a \cos (\omega t + \delta) + a \cos (\omega t + 2\delta) + \ldots + a \cos [\omega t + (n-1)\delta]$$

by the complex exponential form

$$z = a \, e^{i\omega t} (1 + e^{i\delta} + e^{i2\delta} + \ldots + e^{i(n-1)\delta})$$

show that

$$zz^* = a^2 \frac{\sin^2 n\delta/2}{\sin^2 \delta/2}$$

Problem 1.19
The heavily damped simple harmonic system of fig. 1.16 is displaced a distance F from its equilibrium position and released from rest. Show that its displacement is given by

$$x = F e^{-rt/2m} \cosh \left(\frac{r^2}{4m^2} - \frac{s}{m} \right) t$$

and sketch the value of its displacement versus time.

Problem 1.20
Verify that the solution

$$x = (A + Bt) \, e^{-rt/2m}$$

satisfies the equation

$$m\ddot{x} + r\dot{x} + sx = 0$$

when

$$r^2/4m^2 = s/m.$$

Problem 1.21
Show that the boundary condition $x = A \cos \phi$ at $t = 0$ imposed upon the general solution

$$x = e^{-rt/2m} (C_1 \, e^{i\omega' t} + C_2 \, e^{-i\omega' t})$$

for damped simple harmonic motion, requires

$$C_1 = \frac{A}{2} e^{i\phi} \quad \text{and} \quad C_2 = \frac{A}{2} e^{-i\phi}$$

Problem 1.22
A capacitance C with a charge q_0 at $t=0$ discharges through a resistance R. Use the voltage equation $q/C + IR = 0$ to show that the relaxation time of this process is RC seconds, that is,

$$q = q_0 e^{-t/RC}$$

(Note that t/RC is non-dimensional.)

Problem 1.23
The frequency of a damped simple harmonic oscillator is given by

$$\omega'^2 = \frac{s}{m} - \frac{r^2}{4m^2} = \omega_0^2 - \frac{r^2}{4m^2}$$

(a) If $\omega_0^2 - \omega'^2 = 10^{-6}\omega_0^2$ show that $Q = 500$ and that the logarithmic decrement $\delta = \pi/500$.
(b) If $\omega_0 = 10^6$ and $m = 10^{-10}$ kilograms show that the stiffness of the system is 100 newtons metre^{-1}, and that the resistive constant r is $2 \cdot 10^{-7}$ newton seconds metre^{-1}.
(c) If the maximum displacement at $t=0$ is 10^{-2} metres show that the energy of the system is $5 \cdot 10^{-3}$ joules and the decay to e^{-1} of this value takes $0 \cdot 5$ milliseconds.
(d) Show that the energy loss in the first cycle is $2\pi \cdot 10^{-5}$ joules.

Problem 1.24
Show that the fractional change in the resonant frequency $\omega_0(\omega_0^2 = s/m)$ of a damped simple harmonic mechanical oscillator is $\approx (8Q^2)^{-1}$ where Q is the quality factor.

Problem 1.25
Show that the quality factor of an electrical LCR series circuit is $Q = \omega_0 L/R$ where $\omega_0^2 = 1/LC$

Problem 1.26
A plasma consists of an ionized gas of ions and electrons of equal number densities $(n_i = n_e = n)$ having charges of opposite sign $\pm e$, and masses m_i and m_e respectively, where $m_i > m_e$. Relative displacement between the two species sets up a restoring

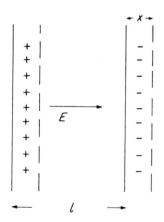

electric field which returns the electrons to equilibrium, the ions being considered stationary. In the diagram a plasma slab of thickness l has all its electrons displaced a distance x to give a restoring electric field $E = nex/\epsilon_0$, where ϵ_0 is constant. Show that the restoring force per unit area on the electrons is xn^2e^2l/ϵ_0 and that they oscillate simple harmonically with angular frequency $\omega_e{}^2 = ne^2/m_e\epsilon_0$. This frequency is called the electron plasma frequency, and only those radio waves of frequency $\omega > \omega_e$ will propagate in such an ionized medium. Hence the reflection of such waves from the ionosphere.

Problem 1.27

A simple pendulum consists of a mass m at the end of a string of length l and performs small oscillations. The length is very slowly shortened whilst the pendulum oscillates many times at a constant amplitude $l\theta$ where θ is very small. Show that if the length is changed by $-\Delta l$ the work done is $-mg\,\Delta l$ (owing to the elevation of the position of equilibrium) together with an increase in the pendulum energy

$$\Delta E = \left(mg\frac{\overline{\theta^2}}{2} - ml\overline{\dot\theta^2} \right)\Delta l$$

where $\overline{\theta^2}$ is the average value of θ^2 during the shortening. If $\theta = \theta_0 \cos \omega t$ show that the energy of the pendulum at any instant may be written

$$E = \frac{ml^2\omega^2\theta_0^2}{2} = \frac{mgl\theta_0^2}{2}$$

and hence show that

$$\frac{\Delta E}{E} = -\frac{1}{2}\frac{\Delta l}{l} = \frac{\Delta \nu}{\nu}$$

that is E/ν, the ratio of the energy of the pendulum to its frequency of oscillation remains constant during the slowly changing process. (This constant ratio under slowly varying conditions is important in quantum theory where the constant is written as a multiple of Planck's constant, h.)

Summary of Important Results

Simple Harmonic Oscillator (mass m, stiffness s, amplitude a)
Equation of motion $\ddot{x} + \omega^2 x = 0$ where $\omega^2 = s/m$
Displacement $x = a \sin (\omega t + \phi)$
Energy $= \frac{1}{2} m\dot{x}^2 + \frac{1}{2} sx^2 = \frac{1}{2} m\omega^2 a^2 = \frac{1}{2} sa^2 = $ constant

---

Superposition (Amplitude and Phase) of two SHMs
One-dimensional
Equal ω, different amplitudes, phase difference δ, resultant R where
$R^2 = a_1^2 + a_2^2 + 2a_1 a_2 \cos \delta$

Different ω, equal amplitude,

$$x = x_1 + x_2 = a(\sin \omega_1 t + \sin \omega_2 t)$$

$$= 2a \sin \frac{(\omega_1 + \omega_2)t}{2} \cos \frac{(\omega_2 - \omega_1)t}{2}$$

Two-dimensional: perpendicular axes
Equal ω, different amplitude—giving general conic section

$$\frac{x^2}{a_1^2} + \frac{y^2}{a_2^2} - \frac{2xy}{a_1 a_2} \cos (\phi_2 - \phi_1) = \sin^2(\phi_2 - \phi_1)$$

(basis of optical polarization)

Superposition of n SHM Vectors (equal amplitude a,
constant successive phase difference δ)

The resultant is $R \cos (\omega t + \alpha)$, where

$$R = a \frac{\sin n\delta/2}{\sin \delta/2}$$

and

$$\alpha = (n - 1)\delta/2$$

Important in optical diffraction and wave groups of many components

---

Damped Simple Harmonic Motion

Equation of motion $m\ddot{x} + r\dot{x} + sx = 0$
Oscillations when

$$\frac{s}{m} > \frac{r^2}{4m^2}$$

Displacement $x = A\,e^{-rt/2m} \cos (\omega' t + \phi)$ where

$$\omega'^2 = \frac{s}{m} - \frac{r^2}{4m^2}$$

Amplitude Decay

Logarithmic decrement δ—the logarithm of the ratio of two successive amplitudes one period τ' apart

$$\delta = \log_e \frac{A_n}{A_{n+1}} = \frac{r\tau'}{2m}$$

Relaxation time

Time for amplitude to decay to $A = A_0\, e^{-rt/2m} = A_0\, e^{-1}$, that is, $t = 2m/r$

Energy Decay

Quality factor Q is the number of radians during which energy decreases to $E = E_0\, e^{-1}$

$$Q = \frac{\omega_0 m}{r} = 2\pi\, \frac{\text{energy stored in system}}{\text{energy lost per cycle}}$$

$$E = E_0\, e^{-rt/m} = E_0\, e^{-1} \qquad \text{when } Q = \omega_0 t$$

In damped SHM

$$\frac{dE}{dt} = (m\ddot{x} + sx)\dot{x} = -r\dot{x}^2 \quad \text{(work rate of resistive force)}$$

For equivalent expressions in electrical oscillators replace m by L, r by R and s by $1/C$. Force equations become voltage equations.

Chapter 2

The Forced Oscillator

The Operation of i upon a Vector

We have already seen that a harmonic oscillation can be conveniently represented by the form $e^{i\omega t}$. In addition to its mathematical convenience i can also be used as a vector operator of physical significance. We say that when i precedes or operates on a vector the direction of that vector is turned through a positive angle (anticlockwise) of $\pi/2$, i.e. i acting as an operator advances the phase of a vector by 90°. The operator $-i$ rotates the vector clockwise by $\pi/2$ and retards its phase by 90°. The mathematics of i as an operator differs in no way from its use as $\sqrt{-1}$ and from now on it will play both roles.

The vector $\mathbf{r} = \mathbf{a} + i\mathbf{b}$ is shown in fig. 2.1, where the direction of $\mathbf{b}$ is perpendicular to that of $\mathbf{a}$ because it is preceded by i. The magnitude or modulus of $\mathbf{r}$ is written

$$r = |\mathbf{r}| = (a^2 + b^2)^{1/2}$$

and

$$r^2 = (a^2 + b^2) = (\mathbf{a} + i\mathbf{b})(\mathbf{a} - i\mathbf{b}) = \mathbf{r}\mathbf{r}^*,$$

where $(\mathbf{a} - i\mathbf{b}) = \mathbf{r}^*$ is defined as the complex conjugate of $(\mathbf{a} + i\mathbf{b})$, that is, the sign of i is changed.

The vector $\mathbf{r}^* = \mathbf{a} - i\mathbf{b}$ is also shown in fig. 2.1.

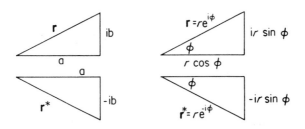

Fig. 2.1. Vector representation using i operator and exponential index. Star superscript indicates complex conjugate where $-i$ replaces i

49

The vector **r** can be written as a product of its magnitude r (scalar quantity) and its phase or direction in the form (fig. 2.1)

$$\mathbf{r} = r\,e^{i\phi} = r(\cos\phi + i\sin\phi)$$

$$= \mathbf{a} + i\mathbf{b}$$

showing that $a = r\cos\phi$ and $b = r\sin\phi$.
   It follows that

$$\cos\phi = \frac{a}{r} = \frac{a}{(a^2+b^2)^{1/2}}$$

and

$$\sin\phi = \frac{b}{r} = \frac{b}{(a^2+b^2)^{1/2}}$$

giving $\tan\phi = b/a$.
   Similarly

$$\mathbf{r}^* = r\,e^{-i\phi} = r(\cos\phi - i\sin\phi)$$

$$\cos\phi = \frac{a}{r}, \quad \sin\phi = \frac{-b}{r} \quad \text{and} \quad \tan\phi = \frac{-b}{a} \text{ (fig. 2.1)}$$

Vector form of Ohm's Law

Ohm's Law is first met as the scalar relation $V = IR$, where V is the voltage across the resistance R and I is the current through it. Its scalar form states that the voltage and current are always in phase. Both will follow a $\sin(\omega t + \phi)$ or a $\cos(\omega t + \phi)$ curve, and the value of ϕ will be the same for both voltage and current.
   However the presence of either or both of the other two electrical components, inductance L and capacitance C, will introduce a phase difference between voltage and current, and Ohm's Law takes the vector form

$$\mathbf{V} = \mathbf{I}Z_e,$$

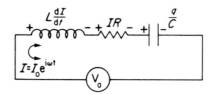

Fig. 2.2a. An electrical forced oscillator. The voltage V_a is applied to the series *LCR* circuit giving $V_a = L\,dI/dt + IR + q/C$

where $\mathbf{Z}_e$, called the *impedance*, replaces the resistance, and is the vector sum of the effective resistances of R, L, and C in the circuit.

When an alternating voltage V_a of frequency ω is applied across a resistance, inductance and condenser in series as in fig. 2.2a, the balance of voltages is given by

$$V_a = IR + L\frac{dI}{dt} + q/C$$

and the current through the circuit is given by $I = I_0 e^{i\omega t}$. The voltage across the inductance

$$V_L = L\frac{dI}{dt} = L\frac{d}{dt}I_0 e^{i\omega t} = i\omega L I_0 e^{i\omega t} = i\omega L I$$

But ωL, as we saw at the end of the last chapter, has the dimensions of ohms, being the value of the effective resistance presented by an inductance L to a current of frequency ω. The product $\omega L I$ with dimensions of ohms times current, i.e. volts, is preceded by i; this tells us that the phase of the voltage across the inductance is 90° ahead of that of the current through the circuit.

Similarly the voltage across the condenser is

$$\frac{q}{C} = \frac{1}{C}\int I\,dt = \frac{1}{C}I_0\int e^{i\omega t}\,dt = \frac{1}{i\omega C}I_0 e^{i\omega t} = -\frac{iI}{\omega C}$$

(since $1/i = -i$).

Again $1/\omega C$, measured in ohms, is the value of the effective resistance presented by the condenser to the current of frequency ω. Now however, the voltage $I/\omega C$ across the condenser is preceded by $-i$ and therefore lags the current by 90°. The voltage and current across the resistance are in phase and fig. 2.2b shows that the vector form of Ohm's Law may be written $\mathbf{V} = \mathbf{IZ}_e = I[R + i(\omega L - 1/\omega C)]$, where the impedance $\mathbf{Z}_e = R + i(\omega L - 1/\omega C)$. The quantities ωL and $1/\omega C$ are called *reactances* because they introduce a phase relationship as well as an effective resistance, and the bracket $(\omega L - 1/\omega C)$ is often written X_e, the reactive component of $\mathbf{Z}_e$.

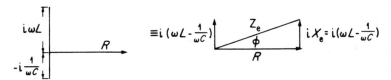

Fig. 2.2b. Vector addition of resistance and reactances to give the electrical impedance $\mathbf{Z}_e = R + i(\omega L - 1/\omega C)$

The magnitude, in ohms, i.e. the value of the impedance, is

$$Z_e = \left[R^2 + \left(\omega L - \frac{1}{\omega C}\right)^2\right]^{1/2}$$

and the vector $\mathbf{Z}_e$ may be represented by its magnitude and phase as

$$\mathbf{Z}_e = Z_e \, e^{i\phi} = Z_e \, (\cos \phi + i \sin \phi)$$

so that

$$\cos \phi = \frac{R}{Z_e}, \qquad \sin \phi = \frac{X_e}{Z_e}$$

and

$$\tan \phi = X_e / R$$

where ϕ is the phase difference between the total voltage across the circuit and the current through it.

The value of ϕ can be positive or negative depending on the relative value of ωL and $1/\omega C$: when $\omega L > 1/\omega C$, ϕ is positive, but the frequency dependence of the components shows that ϕ can change both sign and size.

The magnitude of $\mathbf{Z}_e$ is also frequency dependent and has its minimum value $Z_e = R$ when $\omega L = 1/\omega C$.

The Impedance of a Mechanical Circuit

Exactly similar arguments hold when we consider not an electrical oscillator but a mechanical circuit having mass, stiffness and resistance.

The mechanical impedance is defined as the force required to produce unit velocity in the oscillator, i.e. $\mathbf{Z}_m = \mathbf{F}/\mathbf{v}$ or $\mathbf{F} = \mathbf{v}\mathbf{Z}_m$.

Immediately we can write the mechanical impedance as

$$\mathbf{Z}_m = r + i\left(\omega m - \frac{s}{\omega}\right) = r + i X_m$$

where

$$\mathbf{Z}_m = Z_m \, e^{i\phi}$$

and

$$\tan \phi = X_m / r$$

ϕ being the phase difference between the force and the velocity. The magnitude of $Z_m = [r^2 + (\omega m - s/\omega)^2]^{1/2}$.

Mass, like inductance, produces a positive reactance, and the stiffness behaves in exactly the same way as the capacitance.

Behaviour of a Forced Oscillator

We are now in a position to discuss the physical behaviour of a mechanical oscillator of mass m, stiffness s and resistance r being driven by an alternating force $F_0 \cos \omega t$, where F_0 is the amplitude of the force (fig. 2.3). The equivalent electrical oscillator would be an alternating voltage $V_0 \cos \omega t$ applied to the circuit of inductance L, capacitance C and resistance R in fig. 2.2a.

The mechanical equation of motion, i.e., the dynamic balance of forces, is given by

$$m\ddot{x} + r\dot{x} + sx = F_0 \cos \omega t$$

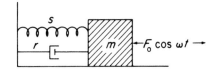

Fig. 2.3. Mechanical forced oscillator with force $F_0 \cos \omega t$ applied to damped mechanical circuit of fig. 1.15

and the voltage equation in the electrical case is

$$L\ddot{q} + R\dot{q} + q/C = V_0 \cos \omega t$$

We shall analyse the behaviour of the mechanical system but the analysis fits the electrical oscillator equally well.

The complete solution for x in the equation of motion consists of two terms:

(1) a 'transient' term which dies away with time and is, in fact, the solution to the equation $m\ddot{x} + r\dot{x} + sx = 0$ discussed in Chapter 1. This contributes the term

$$x = C e^{-rt/2m} e^{i(s/m - r^2/4m^2)^{1/2}t}$$

which decays with $e^{-rt/2m}$. The second term

(2) is called the 'steady state' term, and describes the behaviour of the oscillator after the transient term has died away.

Both terms contribute to the solution initially, but for the moment we shall concentrate on the 'steady state' term which describes the ultimate behaviour of the oscillator.

To do this we shall rewrite the force equation in vector form and represent $\cos \omega t$ by $e^{i\omega t}$ as follows:

$$m\ddot{\mathbf{x}} + r\dot{\mathbf{x}} + s\mathbf{x} = F_0 e^{i\omega t} \tag{2.1}$$

Solving for the vector $\mathbf{x}$ will give both its magnitude and phase with respect to the driving force $F_0 e^{i\omega t}$. Initially, let us try the solution $\mathbf{x} = \mathbf{A} e^{i\omega t}$, where $\mathbf{A}$ may be complex, so that it may have components in and out of phase with the driving force.

The velocity

$$\dot{\mathbf{x}} = i\omega \mathbf{A} e^{i\omega t} = i\omega \mathbf{x}$$

so that

$$\ddot{\mathbf{x}} = i^2 \omega^2 \mathbf{x} = -\omega^2 \mathbf{x}$$

and equation (2.1) becomes

$$(-\mathbf{A}\omega^2 m + i\omega \mathbf{A} r + \mathbf{A} s) e^{i\omega t} = F_0 e^{i\omega t}$$

which is true for all t when

$$\mathbf{A} = \frac{F_0}{i\omega r + (s - \omega^2 m)}$$

or, after multiplying numerator and denominator by $-i$

$$A = \frac{-iF_0}{\omega[r + i(\omega m - s/\omega)]} = \frac{-iF_0}{\omega Z_m}$$

Hence

$$x = A e^{i\omega t} = \frac{-iF_0 e^{i\omega t}}{\omega Z_m} = \frac{-iF_0 e^{i\omega t}}{\omega Z_m e^{i\phi}}$$

$$= \frac{-iF_0 e^{i(\omega t - \phi)}}{\omega Z_m}$$

where

$$Z_m = [r^2 + (\omega m - s/\omega)^2]^{1/2}$$

This vector form of the **steady state** behaviour of x gives three pieces of information and completely defines the magnitude of the displacement x and its phase with respect to the driving force after the transient term dies away. It tells us

(1) that the phase difference ϕ exists between x and the force because of the reactive part $(\omega m - s/\omega)$ of the mechanical impedance.

(2) that an extra difference is introduced by the factor $-i$ and even if ϕ were zero the displacement x would lag the force $F_0 \cos \omega t$ by 90°.

(3) that the maximum amplitude of the displacement x is $F_0/\omega Z_m$. We see that this is dimensionally correct because the velocity x/t has dimensions F_0/Z_m.

Having used $F_0 e^{i\omega t}$ to represent its real part $F_0 \cos \omega t$, we now take the real part of the solution

$$x = \frac{-iF_0 e^{i(\omega t - \phi)}}{\omega Z_m}$$

to obtain the actual value of **x**. (If the force had been $F_0 \sin \omega t$, we would now take that part of **x** preceded by i.)

Now

$$x = -\frac{iF_0}{\omega Z_m} e^{i(\omega t - \phi)}$$

$$= -\frac{iF_0}{\omega Z_m}[\cos(\omega t - \phi) + i \sin(\omega t - \phi)]$$

$$= -\frac{iF_0}{\omega Z_m} \cos(\omega t - \phi) + \frac{F_0}{\omega Z_m} \sin(\omega t - \phi)$$

The value of x resulting from $F_0 \cos \omega t$ is therefore

$$x = \frac{F_0}{\omega Z_m} \sin (\omega t - \phi)$$

[the value of x resulting from $F_0 \sin \omega t$ would be $-F_0 \cos (\omega t - \phi)/\omega Z_m$].

Note that both of these solutions satisfy the requirement that the total phase difference between displacement and force is ϕ plus the $-\pi/2$ term introduced by the $-i$ factor. When $\phi = 0$ the displacement $x = F_0 \sin \omega t/\omega Z_m$ *lags* the force $F_0 \cos \omega t$ by exactly 90°.

To find the velocity of the forced oscillation in the steady state we write

$$\mathbf{v} = \dot{\mathbf{x}} = (i\omega)\frac{(-iF_0)}{\omega Z_m} e^{i(\omega t - \phi)}$$

$$= \frac{F_0}{Z_m} e^{i(\omega t - \phi)}$$

We see immediately that

(1) there is no preceding i factor so that the velocity **v** and the force differ in phase only by ϕ, and when $\phi = 0$ the velocity and force are in phase.

(2) the amplitude of the velocity is F_0/Z_m, which we expect from the definition of mechanical impedance $\mathbf{Z}_m = \mathbf{F}/\mathbf{v}$.

Again we take the real part of the vector expression for the velocity, which will correspond to the real part of the force $F_0 e^{i\omega t}$. This is

$$v = \frac{F_0}{Z_m} \cos (\omega t - \phi)$$

Thus the *velocity is always exactly 90° ahead of the displacement in phase* and differs from the force only by a phase angle ϕ, where

$$\tan \phi = \frac{\omega m - s/\omega}{r} = \frac{X_m}{r}$$

so that a force $F_0 \cos \omega t$ gives a displacement

$$x = \frac{F_0}{\omega Z_m} \sin (\omega t - \phi)$$

and a velocity

$$v = \frac{F_0}{Z_m} \cos (\omega t - \phi)$$

(Problems 2.1, 2.2, 2.3, 2.4)

Behaviour of Velocity v in Magnitude and Phase versus Driving Force Frequency ω

The velocity amplitude is

$$\frac{F_0}{Z_m} = \frac{F_0}{[r^2 + (\omega m - s/\omega)^2]^{1/2}}$$

so that the magnitude of the velocity will vary with the frequency ω because Z_m is frequency dependent.

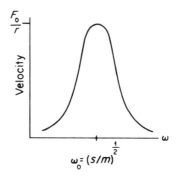

Fig. 2.4. Velocity of forced oscillator versus driving frequency ω. Maximum velocity $v_{\max} = F_0/r$ at $\omega_0^2 = s/m$

At low frequencies, the term $-s/\omega$ is the largest term in Z_m and the impedance is said to be *stiffness controlled*. At high frequencies ωm is the dominant term and the impedance is *mass controlled*. At a frequency ω_0 where $\omega_0 m = s/\omega_0$, the impedance has its minimum value $Z_m = r$ and is a real quantity with zero reactance.

The velocity F_0/Z_m then has its maximum value $v = F_0/r$, and ω_0 is said to be the frequency of *velocity resonance*. Note that $\tan \phi = 0$ at ω_0, the velocity and force being in phase.

The variation of the magnitude of the velocity with driving frequency, ω, is shown in fig. 2.4, the height and sharpness of the peak at resonance depending on r, which is the only effective term of Z_m at ω_0.

The expression

$$v = \frac{F_0}{Z_m} \cos (\omega t - \phi)$$

where

$$\tan \phi = \frac{\omega m - s/\omega}{r}$$

shows that for positive ϕ, that is, $\omega m > s/\omega$, the velocity v will lag the force because $-\phi$ appears in the argument of the cosine. When the driving force frequency ω is very high and $\omega \to \infty$, then $\phi \to 90°$ and the velocity lags the force by that amount.

When $\omega m < s/\omega$, ϕ is negative, the velocity is ahead of the force in phase, and at low driving frequencies as $\omega \to 0$ the term $s/\omega \to \infty$ and $\phi \to -90°$.

Thus at low frequencies the velocity leads the force (ϕ negative) and at high frequencies the velocity lags the force (ϕ positive).

At the frequency ω_0, however, $\omega_0 m = s/\omega_0$ and $\phi = 0$, so that velocity and force are in phase. Fig. 2.5 shows the variation of ϕ with ω for the velocity, the actual shape of the curves depending upon the value of r.

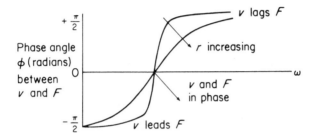

Fig. 2.5. Variation of phase angle ϕ versus driving frequency, where ϕ is the phase angle between the velocity of the forced oscillator and the driving force. $\phi = 0$ at velocity resonance. Each curve represents a fixed resistance value

(Problem 2.5)

Behaviour of Displacement versus Driving Force Frequency ω

The phase of the displacement

$$x = \frac{F_0}{\omega Z_m} \sin(\omega t - \phi)$$

is at all times exactly 90° behind that of the velocity. Whilst the graph of ϕ versus ω remains the same, the total phase difference between the displacement and the force involves the extra 90° retardation introduced by the $-i$ operator. Thus at very low frequencies, where $\phi = -\pi/2$ radians and the velocity leads the force, the displacement and the force are in phase as we should expect. At high frequencies the displacement lags the force by π radians and is exactly out of phase, so that the curve showing the phase angle between the displacement and the force is equivalent to the ϕ versus ω curve, displaced by an amount equal to $\pi/2$ radians. This is shown in fig. 2.6.

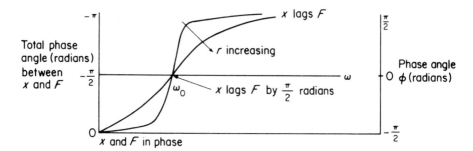

Fig. 2.6. Variation of total phase angle between displacement and driving force versus driving frequency ω. The total phase angle is $-\phi - \pi/2$ radians

The amplitude of the displacement $x = F_0/\omega Z_m$, and at low frequencies $Z_m = [r^2 + (\omega m - s/\omega)^2]^{1/2} \to s/\omega$, so that $x \approx F_0/(\omega s/\omega) = F_0/s$.

At high frequencies $Z_m \to \omega m$, so that $x \approx F_0/(\omega^2 m)$, which tends to zero as ω becomes very large. At very high frequencies, therefore, the displacement amplitude is almost zero because of the mass-controlled or inertial effect.

The velocity resonance occurs at $\omega_0^2 = s/m$, where the denominator Z_m of the velocity amplitude is a minimum, but the displacement resonance will occur, since $x = (F_0/\omega Z_m) \sin(\omega t - \phi)$, when the denominator ωZ_m is a minimum. This takes place when

$$\frac{d}{d\omega}(\omega Z_m) = \frac{d}{d\omega}\omega[r^2 + (\omega m - s/\omega)^2]^{1/2} = 0$$

i.e. when

$$2\omega r^2 + 4\omega m(\omega^2 m - s) = 0$$

or

$$2\omega[r^2 + 2m(\omega^2 m - s)] = 0$$

so that either

$$\omega = 0$$

or

$$\omega^2 = \frac{s}{m} - \frac{r^2}{2m^2} = \omega_0^2 - \frac{r^2}{2m^2}$$

Thus the *displacement resonance* occurs at a frequency slightly less than ω_0, the frequency of velocity resonance. For a small damping constant r or a large mass m these two resonances, for all practical purposes, occur at the frequency ω_0.

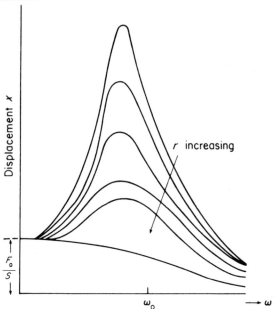

Fig. 2.7. Variation of the displacement of a forced oscillator versus driving force frequency ω for various values of r

At the displacement resonance frequency

$$\omega = \left(\frac{s}{m} - \frac{r^2}{2m^2}\right)^{1/2}$$

the maximum displacement

$$x_{max} = F_0/\omega'r$$

where

$$\omega' = \left(\frac{s}{m} - \frac{r^2}{4m^2}\right)^{1/2}$$

The maximum value of the displacement is given by

$$x_{max} = \frac{F_0}{(\omega Z_m)_{min}}$$

and inserting the value $\omega = (\omega_0^2 - r^2/2m^2)^{1/2}$ in the denominator gives the value $x_{max} = F_0/\omega'r$, where $\omega'^2 = \omega_0^2 - r^2/4m^2$.

Since $x_{max} = F_0/\omega'r$ at resonance, the amplitude at resonance is kept low by increasing r and the variation of x with ω for different values of r is shown in fig. 2.7. A negligible value of r produces a large amplification at resonance: this is the basis of high selectivity in a tuned radio circuit (see the section in this chapter on Q as an amplification factor). Keeping the resonance amplitude low is the principle of vibration insulation.

(Problems 2.6 and 2.7)

Problem on Vibration Insulation

A typical vibration insulator is shown in fig. 2.8. A heavy base is supported on a vibrating floor by a spring system of stiffness s and viscous resistance r (represented by a dashpot). The insulator will generally operate at the mass controlled end of the frequency spectrum and the resonant frequency is designed to be lower than the range of frequencies likely to be met. Suppose the vertical vibration of the floor is given by $x = A \cos \omega t$ about its equilibrium position and y is the corresponding vertical displacement of the base about its rest position. The function of the insulator is to keep the ratio y/A to a minimum.

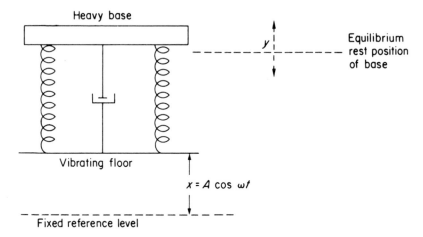

Fig. 2.8. Vibration insulator. A heavy base supported by a spring and viscous dashpot system on a vibrating floor

The equation of motion is given by

$$m\ddot{y} = -r(\dot{y} - \dot{x}) - s(y - x)$$

which, if $y - x = X$, becomes

$$m\ddot{X} + r\dot{X} + sX = -m\ddot{x} = mA\omega^2 \cos \omega t$$

$$= F_0 \cos \omega t$$

where

$$F_0 = mA\omega^2$$

Use the steady state solution of X to show that

$$y = \frac{F_0}{\omega Z_m} \sin(\omega t - \phi) + A \cos \omega t$$

and (noting that y is the superposition of two harmonic components with a constant phase difference) show that

$$\frac{y_{max}}{A} = \frac{(r^2 + s^2/\omega^2)^{1/2}}{Z_m}$$

where

$$Z_m^2 = r^2 + (\omega m - s/\omega)^2$$

Note that

$$\frac{y_{max}}{A} > 1 \quad \text{if} \quad \omega^2 < \frac{2s}{m}$$

so that s/m should be as low as possible to give protection against a given frequency ω.

(a) Show that

$$\frac{y_{max}}{A} = 1 \quad \text{for} \quad \omega^2 = \frac{2s}{m}$$

(b) Show that

$$\frac{y_{max}}{A} < 1 \quad \text{for} \quad \omega^2 > \frac{2s}{m}$$

(c) Show that if $\omega^2 = s/m$ then $y_{max}/A > 1$ but that the damping term r is helpful in keeping the motion of the base to a reasonably low level.

(d) Show that if $\omega^2 > 2s/m$ then $y_{max}/A < 1$ but damping is detrimental.

Significance of the Two Components of the Displacement Curve

Any single curve of fig. 2.7 is the superposition of the two component curves (a) and (b) in fig. 2.9, for the displacement x may be rewritten

$$x = \frac{F_0}{\omega Z_m} \sin(\omega t - \phi) = \frac{F_0}{\omega Z_m}(\sin \omega t \cos \phi - \cos \omega t \sin \phi)$$

or, since

$$\cos \phi = \frac{r}{Z_m} \quad \text{and} \quad \sin \phi = \frac{X_m}{Z_m}$$

as

$$x = \frac{F_0}{\omega Z_m} \frac{r}{Z_m} \sin \omega t - \frac{F_0}{\omega Z_m} \frac{X_m}{Z_m} \cos \omega t$$

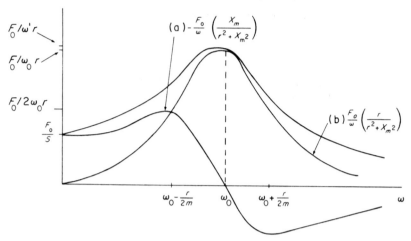

Fig. 2.9. A typical curve of fig. 2.7 resolved into its 'anti-phase' component (curve a) and its '90° out of phase' component (curve b). Curve b represents the resistive fraction of the impedance and curve a the reactive fraction. Curve b corresponds to absorption and curve a to anomalous dispersion of an electromagnetic wave in a medium having an atomic or molecular resonant frequency equal to the frequency of the wave

The cos ωt component (with a negative sign) is exactly anti phase with respect to the driving force $F_0 \cos \omega t$. Its amplitude, plotted as curve (a) may be expressed as

$$-\frac{F_0}{\omega}\frac{X_m}{Z_m^2} = \frac{F_0 m(\omega_0^2 - \omega^2)}{m^2(\omega_0^2 - \omega^2)^2 + \omega^2 r^2}$$

where $\omega_0^2 = s/m$ and ω_0 is the frequency of velocity resonance.

The sin ωt component lags the driving force $F_0 \cos \omega t$ by 90°. Its amplitude plotted as curve (b) becomes

$$\frac{F_0}{\omega}\frac{r}{r^2 + X_m^2} = \frac{F_0 \omega r}{m^2(\omega_0^2 - \omega^2)^2 + \omega^2 r^2}$$

We see immediately that curve (a) is zero and curve (b) is a maximum at ω_0 but that they combine to give a maximum at ω where

$$\omega^2 = \omega_0^2 - \frac{r^2}{2m^2}$$

the resonant frequency for amplitude displacement.

These curves are particularly familiar in the study of optical dispersion where the forced oscillator is an electron in an atom and the driving force is the oscillating field vector of an electromagnetic wave of frequency ω. When ω is the resonant frequency of the electron in the atom, the atom absorbs a large amount of energy from the electromagnetic wave and curve (b) is the shape of the characteristic absorption curve. Note that curve (b) represents the dissipating or absorbing fraction of the impedance

$$\frac{r}{(r^2 + X_m^2)^{1/2}}$$

and that part of the displacement which lags the driving force by 90°. The velocity associated with this component will therefore be in phase with the driving force and it is this part of the velocity which appears in the energy loss term $r\dot{x}^2$ due to the resistance of the oscillator and which gives rise to absorption.

On the other hand, curve (a) represents the reactive or energy storing fraction of the impedance

$$\frac{X_m}{(r^2 + X_m^2)^{1/2}}$$

and the reactive components in a medium determine the velocity of the waves in the medium which in turn governs the refractive index n. In fact, curve (a) is a graph of the value of n^2 in a region of anomalous dispersion where the ω axis represents the value $n = 1$. These regions occur at every resonant frequency of the constituent atoms of the medium. We shall return to this topic later in the book.

(Problems 2.8, 2.9, and 2.10)

Power Supplied to Oscillator by the Driving Force

In order to maintain the steady state oscillations of the system the driving force must replace the energy lost in each cycle because of the presence of the resistance. We shall now derive the most important result that:

'in the steady state the amplitude and phase of a driven oscillator adjust themselves so that the average power supplied by the driving force just equals that being dissipated by the frictional force'.

The *instantaneous power P* supplied is equal to the product of the *instantaneous driving force* and the *instantaneous velocity*, that is,

$$P = F_0 \cos \omega t \frac{F_0}{Z_m} \cos (\omega t - \phi)$$

$$= \frac{F_0^2}{Z_m} \cos \omega t \cos (\omega t - \phi)$$

The *average power*

$$P_{av} = \frac{\text{total work per oscillation}}{\text{oscillation period}}$$

$$\therefore \quad P_{av} = \int_0^T \frac{P \, dt}{T} \text{ where } T = \text{oscillation period}$$

$$= \frac{F_0^2}{Z_m T} \int_0^T \cos \omega t \cos (\omega t - \phi) \, dt$$

$$= \frac{F_0^2}{Z_m T} \int_0^T [\cos^2 \omega t \cos \phi + \cos \omega t \sin \omega t \sin \phi) \, dt$$

$$= \frac{F_0^2}{2Z_m} \cos \phi$$

because

$$\int_0^T \cos \omega t \times \sin \omega t \, dt = 0$$

and

$$\frac{1}{T} \int_0^T \cos^2 \omega t \, dt = \tfrac{1}{2}$$

The power supplied by the driving force is not stored in the system, but dissipated as work expended in moving the system against the frictional force $r\dot{x}$.

The rate of working (instantaneous power) by the frictional force is

$$(r\dot{x})\dot{x} = r\dot{x}^2 = r\frac{F_0^2}{Z_m^2} \cos^2 (\omega t - \phi)$$

and the average value of this over one period of oscillation

$$\frac{1}{2} \frac{rF_0^2}{Z_m^2} = \frac{1}{2} \frac{F_0^2}{Z_m} \cos \phi \quad \text{for} \quad \frac{r}{Z_m} = \cos \phi$$

This proves the initial statement that the power supplied equals the power dissipated.

In an electrical circuit the power is given by $VI \cos \phi$, where V and I are the instantaneous r.m.s. values of voltage and current and $\cos \phi$ is known as the *power factor*.

$$VI \cos \phi = \frac{V^2}{Z_e} \cos \phi = \frac{V_0^2}{2Z_e} \cos \phi$$

since

$$V = \frac{V_0}{\sqrt{2}}$$

(Problem 2.11)

Variation of P_{av} with ω. Absorption Resonance Curve

Returning to the mechanical case, we see that the average power supplied

$$P_{av} = (F_0^2/2Z_m) \cos \phi$$

is a maximum when $\cos \phi = 1$, that is, when $\phi = 0$ and $\omega m - s/\omega = 0$ or $\omega_0^2 = s/m$. The force and the velocity are then in phase and Z_m has its minimum value of r. Thus

$$P_{av} \text{ (maximum)} = F_0^2/2r$$

A graph of P_{av} versus ω, the frequency of the driving force, is shown in fig. 2.10. Like the curve of displacement versus ω, this graph measures the response of the oscillator; the sharpness of its peak at resonance is also determined by the value of the damping constant r, which is the only term remaining in Z_m at the resonance frequency ω_0. The peak occurs at the frequency of velocity resonance when the power absorbed by the system from the driving force is a maximum; this curve is known as the absorption curve of the oscillator (it is similar to curve (b) of fig. 2.9).

The Q-value in Terms of the Resonance Absorption Bandwidth

In the last chapter we discussed the quality factor of an oscillator system in terms of energy decay. We may derive the same parameter in terms of the curve of fig. 2.10, where the sharpness of the resonance is precisely defined by the ratio

$$Q = \frac{\omega_0}{\omega_2 - \omega_1},$$

where ω_2 and ω_1 are those frequencies at which the power supplied

$$P_{av} = \tfrac{1}{2} P_{av} \text{ (maximum)}$$

The frequency difference $\omega_2 - \omega_1$ is often called the bandwidth.

Now

$$P_{av} = r F_0^2/2Z_m^2 = \tfrac{1}{2} P_{av} \text{ (maximum)} = \tfrac{1}{2} F_0^2/2r$$

when

$$Z_m^2 = 2r^2$$

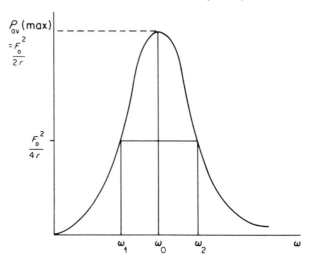

Fig. 2.10. Graph of average power versus ω supplied to an oscillator by the driving force. Bandwidth $\omega_2 - \omega_1$ of resonance curve defines response in terms of the quality factor, $Q = \omega_0/(\omega_2 - \omega_1)$, where $\omega_0^2 = s/m$

that is, when

$$r^2 + X_m^2 = 2r^2 \quad \text{or} \quad X_m = \omega m - s/\omega = \pm r.$$

If $\omega_2 > \omega_1$, then

$$\omega_2 m - s/\omega_2 = +r$$

and

$$\omega_1 m - s/\omega_1 = -r$$

Eliminating s between these equations gives

$$\omega_2 - \omega_1 = r/m$$

so that

$$Q = \omega_0 m/r$$

Note that $\omega_1 = \omega_0 - r/2m$ and $\omega_2 = \omega_0 + r/2m$ are the two significant frequencies in fig. 2.9. The quality factor of an electrical circuit is given by

$$Q = \frac{\omega_0 L}{R}$$

where

$$\omega_0^2 = (LC)^{-1}$$

Note that for high values of Q, where the damping constant r is small, the frequency ω' used in the last chapter to define $Q = \omega'm/r$ moves very close to the frequency ω_0, and the two definitions of Q become equivalent to each other and to the third definition we meet in the next section.

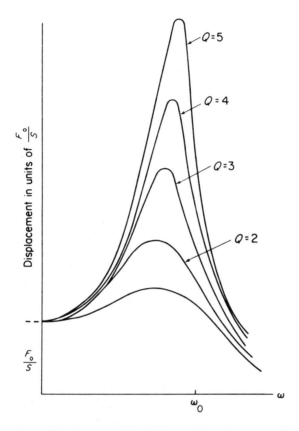

Fig. 2.11. Curves of fig. 2.7 now given in terms of the quality factor Q of the system, where Q is amplification at resonance of low frequency response $x = F_0/s$

The Q Value as an Amplification Factor

We have seen that the value of the displacement at resonance is given by

$$A_{\max} = \frac{F_0}{\omega'r} \quad \text{where} \quad \omega'^2 = \frac{s}{m} - \frac{r^2}{4m^2}$$

At low frequencies ($\omega \to 0$) the displacement has a value $A_0 = F_0/s$, so that

$$\left(\frac{A_{max}}{A_0}\right)^2 = \frac{F_0^2}{\omega'^2 r^2}\frac{s^2}{F_0^2} = \frac{m^2 \omega_0^4}{r^2[\omega_0^2 - r^2/4m^2]}$$

$$= \frac{\omega_0^2 m^2}{r^2[1 - 1/4Q^2]} = \frac{Q^2}{[1 - 1/4Q^2]}$$

Hence:

$$\frac{A_{max}}{A_0} = \frac{Q}{[1 - 1/4Q^2]^{1/2}} \approx Q\left[1 + \frac{1}{8Q^2}\right] \approx Q$$

for large Q.

Thus the displacement at low frequencies is amplified by a factor of Q at displacement resonance.

Figure 2.7 is now shown as fig. 2.11 where the Q values have been attached to each curve. In tuning radio circuits the Q-value is used as a measure of selectivity, where the sharpness of response allows a signal to be obtained free from interference from signals at nearby frequencies. In conventional radio circuits at frequencies of one megacycle, Q-values are of the order of a few hundred; at higher radio frequencies resonant copper cavities have Q-values of about 30,000 and piezo-electric crystals can produce Q-values of 500,000. Optical absorption in crystals and nuclear magnetic resonances are often described in terms of Q-values. The Mössbauer effect in nuclear physics involves a Q value of 10^{10}.

The Effect of the Transient Term

Throughout this chapter we have considered only the steady state behaviour without accounting for the transient term mentioned on page 53. This term makes an initial contribution to the total displacement but decays with time as $e^{-rt/2m}$. Its effect is best displayed by considering the vector sum of the transient and steady state components.

The steady state term may be represented by a vector of constant length rotating anticlockwise at the angular velocity ω of the driving force. The vector tip traces a circle. Upon this is superposed the transient term vector of diminishing length which rotates anti clockwise with angular velocity $\omega' = (s/m - r^2/4m^2)^{1/2}$. Its tip traces a contracting spiral.

The locus of the magnitude of the vector sum of these terms is the envelope of the varying amplitudes of the oscillator. This envelope modulates the steady state oscillations of frequency ω at a frequency which depends upon ω' and the relative phase between ωt and $\omega' t$.

Thus in fig. 2.12(a) where the total oscillator displacement is zero at time $t = 0$ we have the steady state and transient vectors equal and opposite in fig. 2.12(b) but because $\omega \neq \omega'$ the relative phase between the vectors will change

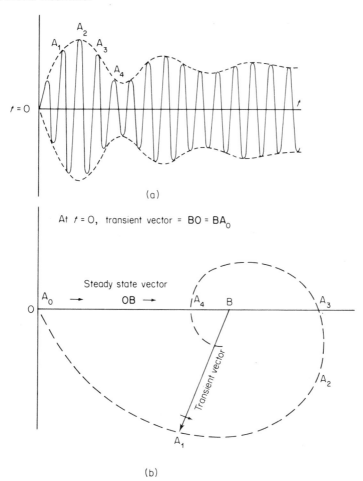

(a)

At $t = 0$, transient vector $= BO = BA_0$

Steady state vector

$OB \rightarrow$

(b)

Fig. 2.12. (a) The steady state oscillation (heavy curve) is modulated by the transient which decays exponentially with time. (b) In the vector diagram of (b) **OB** is the constant length steady state vector and $\mathbf{BA}_i$ is the transient vector. Each vector rotates anti-clockwise with its own angular velocity. At $t = 0$ the vectors **OB** and $\mathbf{BA}_0$ are equal and opposite on the horizontal axis and their vector sum is zero. At subsequent times the total amplitude is the length of OA_i which changes as A traces a contracting spiral around B. The points A_1, A_2, A_3 and A_4 indicate how the amplitude is modified in (a)

as the transient term decays. The vector tip of the transient term is shown as the dotted spiral and the total amplitude assumes the varying lengths OA_1, OA_2, OA_3, OA_4, etc.

(Problems 2.12, 2.13, 2.14, 2.15, 2.16, 2.17 and 2.18)

Problem 2.1
Show, if $F_0\,e^{i\omega t}$ represents $F_0 \sin \omega t$ in the vector form of the equation of motion for the forced oscillator that

$$x = -\frac{F_0}{\omega Z_m} \cos (\omega t - \phi)$$

and the velocity

$$v = \frac{F_0}{Z_m} \sin (\omega t - \phi)$$

Problem 2.2
The displacement of a forced oscillator is zero at time $t = 0$ and its rate of growth is governed by the rate of decay of the transient term. If this term decays to e^{-k} of its original value in a time t show that, for small damping, the average rate of growth of the oscillations is given by $x_0/t = F_0/2km\omega_0$ where x_0 is the maximum steady state displacement, F_0 is the force amplitude and $\omega_0^2 = s/m$.

Problem 2.3
The equation $m\ddot{x} + sx = F_0 \sin \omega t$ describes the motion of an undamped simple harmonic oscillator driven by a force of frequency ω. Show, by solving the equation in vector form, that the steady state solution is given by

$$x = \frac{F_0 \sin \omega t}{m(\omega_0^2 - \omega^2)} \quad \text{where} \quad \omega_0^2 = \frac{s}{m}$$

Sketch the behaviour of x versus ω and note that the change of sign as ω passes through ω_0 defines a phase change of π radians in the displacement. Now show that the general solution for the displacement is given by

$$x = \frac{F_0 \sin \omega t}{m(\omega_0^2 - \omega^2)} + A \cos \omega_0 t + B \sin \omega_0 t$$

where A and B are constant.

Problem 2.4
In problem 2.3, if $x = \dot{x} = 0$ at $t = 0$ show that

$$x = \frac{F_0}{m} \frac{1}{(\omega_0^2 - \omega^2)} \left(\sin \omega t - \frac{\omega}{\omega_0} \sin \omega_0 t \right)$$

and, by writing $\omega = \omega_0 + \Delta\omega$ where $\Delta\omega$ is small, show that, near resonance,

$$x = \frac{F_0}{2m\omega_0^2}(\sin \omega_0 t - \omega_0 t \cos \omega_0 t)$$

Sketch this behaviour, noting that the second term increases with time, allowing the oscillations to grow (resonance between free and forced oscillations).

Problem 2.5
What is the general expression for the acceleration $\dot{v}$ of a simple damped mechanical oscillator driven by a force $F_0 \cos \omega t$? Derive an expression to give the frequency of maximum acceleration and show that, if $r = \sqrt{sm}$ the acceleration amplitude at the

frequency of velocity resonance equals the limit of the acceleration amplitude at high frequencies

Problem 2.6

Prove that the **exact** amplitude at the displacement resonance of a driven mechanical oscillator may be written $x = F_0/\omega'r$ where F_0 is the driving force amplitude and

$$\omega'^2 = \frac{s}{m} - \frac{r^2}{4m^2}$$

Problem 2.7

In a forced mechanical oscillator show that the following are frequency independent (a) the displacement amplitude at low frequencies (b) the velocity amplitude at velocity resonance and (c) the acceleration amplitude at high frequencies, $(\omega \to \infty)$.

Problem 2.8

Show that curve (b) of figure 2.9 has a maximum value of $F_0/\omega_0 r$ at ω_0. Show also that for small r, the maximum value of curve (a) is $\approx F_0/2\omega_0 r$ at $\omega_1 = \omega_0 - r/2m$ and its minimum value is $\approx -F_0/2\omega_0 r$ at $\omega_2 = \omega_0 + r/2m$.

Problem 2.9

The equation $\ddot{x} + \omega_0^2 x = (-eE_0/m)\cos\omega t$ describes the motion of a free undamped electric charge $-e$ of mass m under the influence of an alternating electric field $E = E_0 \cos \omega t$. For an electron number density n show that the induced polarizability per unit volume (the dynamic susceptibility) of a medium

$$\chi_e = -\frac{n\,ex}{\epsilon_0 E} = \frac{n\,e^2}{\epsilon_0 m(\omega_0^2 - \omega^2)}$$

(The permittivity of a medium is defined as $\epsilon = \epsilon_0(1 + \chi)$ where ϵ_0 is the permittivity of free space. The relative permittivity $\epsilon_r = \epsilon/\epsilon_0$ is called the dielectric constant and is the square of the refractive index when E is the electric field of an electromagnetic wave.)

Problem 2.10

Repeat question 2.9 for the case of a damped oscillatory electron, by taking the displacement x as the component represented by curve (a) in fig. 2.9 to show that

$$\epsilon_r = 1 + \chi = 1 + \frac{n\,e^2 m(\omega_0^2 - \omega^2)}{\epsilon_0[m^2(\omega_0^2 - \omega^2)^2 + \omega^2 r^2]}$$

In fact fig. 2.9(a) plots $\epsilon_r = \epsilon/\epsilon_0$. Note that for

$$\omega \ll \omega_0, \quad \epsilon_r \approx 1 + \frac{n\,e^2}{\epsilon_0 m\omega_0^2}$$

and for

$$\omega \gg \omega_0, \quad \epsilon_r \approx 1 - \frac{n\,e^2}{\epsilon_0 m\omega^2}$$

Problem 2.11

Show that the energy dissipated per cycle by the frictional force $r\dot{x}$ at an angular frequency ω is given by $\pi r\omega x_{max}^2$.

Problem 2.12

Show that the bandwidth of the resonance absorption curve defines the phase angle range $\tan\phi = \pm 1$.

Problem 2.13

An alternating voltage, amplitude V_0 is applied across an *LCR* series circuit. Show that the voltage at current resonance across either the inductance or the condenser is QV_0.

Problem 2.14

Show that in a resonant *LCR* series circuit the maximum potential across the condenser occurs at a frequency $\omega = \omega_0(1 - 1/2Q_0^2)^{1/2}$ where $\omega_0^2 = (LC)^{-1}$ and $Q_0 = \omega_0 L/R$.

Problem 2.15

In problem 2.14 show that the maximum potential across the inductance occurs at a frequency $\omega = \omega_0(1 - 1/2Q_0^2)^{-1/2}$.

Problem 2.16

Light of wavelength 0·6 microns (6000 Å) is emitted by an electron in an atom behaving as a lightly damped simple harmonic oscillator with a Q value of 5×10^7. Show from the resonance bandwidth that the width of the spectral line from such an atom is $1·2 \times 10^{-14}$ metres.

Problem 2.17

If the Q value of problem 2·6 is high show that the width of the displacement resonance curve is approximately $\sqrt{3}r/m$ where the width is measured between those frequencies where $x = x_{max}/2$.

Problem 2.18

Show that, in problem 2.10, the mean rate of energy absorption per unit volume, that is, the power supplied is

$$P = \frac{ne^2E_0^2}{2} \frac{\omega^2 r}{m^2(\omega_0^2 - \omega^2)^2 + \omega^2 r^2}$$

Summary of Important Results

Mechanical Impedance $\mathbf{Z}_m = \mathbf{F}/\mathbf{v}$ (force per unit velocity)

$$\mathbf{Z}_m = Z_m\,e^{i\phi} = r + i(\omega m - s/\omega)$$

where $Z_m^2 = r^2 + (\omega m - s/\omega)^2$

$$\sin\phi = \frac{\omega m - s/\omega}{Z_m}, \qquad \cos\phi = \frac{r}{Z_m}, \qquad \tan\phi = \frac{\omega m - s/\omega}{r}$$

ϕ is the phase angle between the force and velocity.

Forced Oscillator

Equation of motion $m\ddot{x} + r\dot{x} + sx = F_0 \cos\omega t$
(Vector form) $m\ddot{\mathbf{x}} + r\dot{\mathbf{x}} + s\mathbf{x} = F_0\,e^{i\omega t}$
Use $\mathbf{x} = \mathbf{A}\,e^{i\omega t}$ to give steady state displacement

$$\mathbf{x} = -i\frac{F_0}{\omega Z_m}\,e^{i(\omega t - \phi)}$$

and velocity

$$\dot{\mathbf{x}} = \mathbf{v} = \frac{F_0}{Z_m} e^{i(\omega t - \phi)}$$

When $F_0 e^{i\omega t}$ represents $F_0 \cos \omega t$

$$x = \frac{F_0}{\omega Z_m} \sin (\omega t - \phi)$$

$$v = \frac{F_0}{Z_m} \cos (\omega t - \phi)$$

Maximum velocity $= \dfrac{F_0}{r}$ at **velocity** resonant frequency $\omega_0 = (s/m)^{1/2}$

Maximum displacement $= \dfrac{F_0}{\omega' r}$ where $\omega' = (s/m - r^2/4m^2)^{1/2}$ *at* **displacement** resonant frequency $\omega = (s/m - r^2/2m^2)^{1/2}$

Power absorbed by oscillator from driving force

Oscillator adjusts amplitude and phase so that power supplied equals power dissipated.

Power absorbed $= \frac{1}{2}(F_0^2/Z_m) \cos \phi$ (cos ϕ is power factor)

Maximum power absorbed $= \dfrac{F_0^2}{2r}$ at ω_0

$\dfrac{\text{Maximum power}}{2}$ absorbed $= \dfrac{F_0^2}{4r}$ at $\omega_1 = \omega_0 - \dfrac{r}{2m}$ and $\omega_2 = \omega_0 + \dfrac{r}{2m}$

Quality factor $Q = \dfrac{\omega_0 m}{r} = \dfrac{\omega_0}{\omega_2 - \omega_1}$

$Q = \dfrac{\text{maximum displacement at displacement resonance}}{\text{displacement as } \omega \to 0}$

$= \dfrac{A \text{ (max)}}{F_0/s}$

For equivalent expressions for electrical oscillators replace m by L, r by R, s by $1/C$ and F_0 by V_0 (voltage).

Chapter 3

Coupled Oscillations

The first two chapters have shown in some detail how a single vibrating system will behave. Oscillators, however, rarely exist in complete isolation; wave motion owes its existence to neighbouring vibrating systems which are able to transmit their energy to each other.

Such energy transfer takes place, in general, because two oscillators share a common component, capacitance or stiffness, inductance or mass, or resistance. Resistance coupling inevitably brings energy loss and a rapid decay in the vibration, but coupling by either of the other two parameters consumes no power, and continuous energy transfer over many oscillators is possible. This is the basis of wave motion.

We shall investigate firstly a mechanical example of stiffness coupling between two pendulums. Two atoms set in a crystal lattice experience a mutual coupling force and would be amenable to a similar treatment. Then we investigate an example of mass, or inductive, coupling, and finally we consider the coupled motion of an extended array of oscillators which leads us naturally into a discussion on wave motion.

Stiffness (or Capacitance) Coupled Oscillators

Fig. 3.1 shows two identical pendulums, each having a mass m suspended on a light rigid rod of length l. The masses are connected by a light spring of stiffness s whose natural length equals the distance between the masses when neither is displaced from equilibrium. The small oscillations we discuss are restricted to the plane of the paper.

If x and y are the respective displacements of the masses then the equations of motion are

$$m\ddot{x} = -mg\frac{x}{l} - s(x - y)$$

and

$$m\ddot{y} = -mg\frac{y}{l} + s(x - y)$$

74

These represent the normal simple harmonic motion terms of each pendulum plus a coupling term $s(x - y)$ from the spring. We see that if $x > y$ the spring is extended beyond its normal length and will act against the acceleration of x but in favour of the acceleration of y.

Fig. 3.1. Two identical pendulums, each a light rigid rod of length l supporting a mass m and coupled by a weightless spring of stiffness s and of natural length equal to the separation of the masses at zero displacement

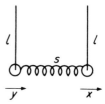

Writing $\omega_0^2 = g/l$, where ω_0 is the natural vibration frequency of each pendulum, gives

$$\ddot{x} + \omega_0^2 x = -\frac{s}{m}(x - y) \tag{3.1}$$

$$\ddot{y} + \omega_0^2 y = -\frac{s}{m}(y - x) \tag{3.2}$$

Instead of solving these equations directly for x and y we are going to choose two new coordinates

$$X = x + y$$

$$Y = x - y$$

The importance of this approach will emerge as this chapter proceeds. Adding equations (3.1) and (3.2) gives

$$\ddot{x} + \ddot{y} + \omega_0^2(x + y) = 0$$

that is

$$\ddot{X} + \omega_0^2 X = 0$$

and subtracting (3.2) from (3.1) gives

$$\ddot{Y} + (\omega_0^2 + 2s/m) Y = 0$$

The motion of the coupled system is thus described in terms of the two coordinates X and Y, each of which has an equation of motion which is simple harmonic.

If $Y = 0$, $x = y$ at all times, so that the motion is completely described by the equation

$$\ddot{X} + \omega_0^2 X = 0$$

then the frequency of oscillation is the same as that of either pendulum in isolation and the stiffness of the coupling has no effect. This is because both pendulums are always swinging in phase (fig. 3.2a) and the light spring is always at its natural length.

If $X = 0$, $x = -y$ at all times, so that the motion is completely described by

$$\ddot{Y} + (\omega_0^2 + 2s/m)Y = 0$$

The frequency of oscillation is greater because the pendulums are always out of phase (fig. 3.2b) so that the spring is either extended or compressed and the coupling is effective.

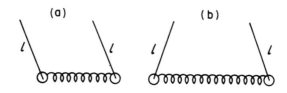

Fig. 3.2. (a) The 'in phase' mode of vibration given by $\ddot{X} + \omega_0^2 X = 0$, where X is the normal coordinate $X = x + y$ and $\omega_0^2 = g/l$. (b) 'Out of phase' mode of vibration given by $\ddot{Y} + (\omega_0^2 + 2 s/m)$ *where* Y is the normal coordinate $Y = x - y$

Normal Co-ordinates, Degrees of Freedom and Normal Modes of Vibration

The significance of choosing X and Y to describe the motion is that these parameters give a very simple illustration of normal co-ordinates.

(a) Normal co-ordinates are co-ordinates in which the equations of motion take the form of a set of linear differential equations with constant coefficients in which each equation contains *only one* dependent variable (our simple harmonic equations in X and Y).

(b) A vibration involving only one dependent variable X (or Y) is called a *normal mode of vibration* and has its own *normal frequency*. In such a *normal mode* all components of the system oscillate with the same *normal frequency*.

(c) The total energy of an undamped system may be expressed as a sum of the squares of the normal co-ordinates multiplied by constant coefficients and a sum of the squares of the first time derivatives of the co-ordinates multiplied by constant coefficients. The energy of a coupled system when the X and Y modes are both vibrating would then be expressed in terms of the squares of the velocities and displacements of X and Y (see problem).

(d) The importance of the normal modes of vibration is that they are entirely independent of each other. The energy associated with a normal mode is *never exchanged* with another mode; this is why we can add the energies of the separate modes to give the total energy. If only one mode vibrates the second mode of our system will always be at rest, acquiring no energy from the vibrating mode.

(e) Each independent way by which a system may acquire energy is called a *degree of freedom* to which is assigned its own particular normal co-ordinate. The number of such different ways in which the system can take up energy defines its number of degrees of freedom and its number of normal co-ordinates. Each harmonic oscillator has two degrees of freedom, it may take up both potential energy (normal co-ordinate X) and kinetic energy (normal co-ordinate $\dot{X}$). In our two normal modes the energies may be written

$$E_X = a\dot{X}^2 + bX^2$$

and

$$E_Y = c\dot{Y}^2 + dY^2$$

where a, b, c and d are constant.

Our system of two coupled pendulums has, then, four degrees of freedom and four normal co-ordinates.

Any configuration of our coupled system may be represented by the super-position of the two normal modes

$$X = x + y = X_0 \cos(\omega_1 t + \phi_1)$$

and

$$Y = x - y = Y_0 \cos(\omega_2 t + \phi_2)$$

where X_0 and Y_0 are the normal mode amplitudes, whilst $\omega_1^2 = g/l$ and $\omega_2^2 = (g/l + 2s/m)$ are the normal mode frequencies. To simplify the discussion let us choose

$$X_0 = Y_0 = 2a$$

and put

$$\phi_1 = \phi_2 = 0$$

The pendulum displacements are then given by

$$x = \tfrac{1}{2}(X + Y) = a \cos \omega_1 t + a \cos \omega_2 t$$

and

$$y = \tfrac{1}{2}(X - Y) = a \cos \omega_1 t - a \cos \omega_2 t$$

with velocities

$$\dot{x} = -a\omega_1 \sin \omega_1 t - a\omega_2 \sin \omega_2 t$$

and

$$\dot{y} = -a\omega_1 \sin \omega_1 t + a\omega_2 \sin \omega_2 t$$

Now let us set the system in motion by displacing the right hand mass a distance $x = 2a$ and releasing both masses from rest so that $\dot{x} = \dot{y} = 0$ at time $t = 0$.

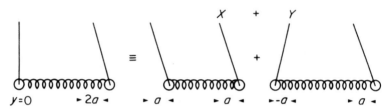

Fig. 3.3. The displacement of one pendulum
by an amount $2a$ is shown as the combination
of the two normal coordinates $X + Y$

Fig. 3.3 shows that our initial displacement $x = 2a$, $y = 0$ at $t = 0$ may be seen as a combination of the 'in phase' mode ($x = y = a$ so that $x + y = X_0 = 2a$) and of the 'out of phase' mode ($x = -y = a$ so that $Y_0 = 2a$). After release, the motion of the right hand pendulum is given by

$$x = a \cos \omega_1 t + a \cos \omega_2 t$$

$$= 2a \cos \frac{(\omega_2 - \omega_1)t}{2} \cos \frac{(\omega_1 + \omega_2)t}{2}$$

and that of the left hand pendulum is given by

$$y = a \cos \omega_1 t - a \cos \omega_2 t$$

$$= -2a \sin \frac{(\omega_1 - \omega_2)t}{2} \sin \frac{(\omega_1 + \omega_2)t}{2}$$

$$= 2a \sin \frac{(\omega_2 - \omega_1)t}{2} \sin \frac{(\omega_1 + \omega_2)t}{2}$$

If we plot the behaviour of the individual masses by showing how x and y change with time (fig. 3.4), we see that after drawing the first mass aside a distance $2a$ and releasing it x follows a cosinusoidal behaviour at a frequency which is the average of the two normal mode frequencies, but its amplitude varies cosinusoidally with a low frequency which is half the difference between the normal mode frequencies. On the other hand, y, which started at zero, vibrates sinusoidally with the average frequency but its amplitude builds up to $2a$ and then decays sinusoidally at the low frequency of half the difference

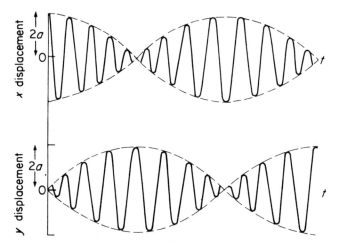

Fig. 3.4. Behaviour with time of individual pendulums, showing complete energy exchange between the pendulums as x decreases from $2a$ to zero whilst y grows from zero to $2a$

between the normal mode frequencies. In short, the y displacement mass acquires all the energy of the x displacement mass which is stationary when y is vibrating with amplitude $2a$, but the energy is then returned to the mass originally displaced. This *complete* energy exchange is only possible when the masses are identical and the ratio $(\omega_1 + \omega_2)/(\omega_2 - \omega_1)$ is an integer, otherwise neither will ever be quite stationary. The slow variation of amplitude at half the normal mode frequency difference is the phenomenon of 'beats' which occurs between two oscillations of nearly equal frequencies. We shall discuss this further in the section on wave groups in Chapter 4.

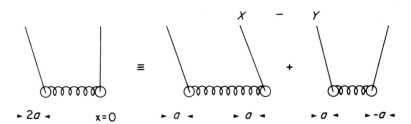

Fig. 3.5. The faster vibration of the Y mode results in a phase gain of π radians over the X mode of vibration, to give $y = 2a$, which is shown here as a combination of the normal modes $X - Y$

The important point to recognize, however, is that although the *individual* pendulums may exchange energy, there is *no* energy exchange between the normal modes. Fig. 3.3 showed the initial configuration $x = 2a$, $y = 0$, decomposed into the X and Y modes. The higher frequency of the Y mode ensures that after a number of oscillations the Y mode will have gained half a vibration (a phase of π radians) on the X mode; this is shown in fig. 3.5. The combination of the X and Y modes then gives y the value of $2a$ and $x = 0$, and the process is repeated. When Y gains another half vibration then x equals $2a$ again. The pendulums may exchange energy; the normal modes do not.

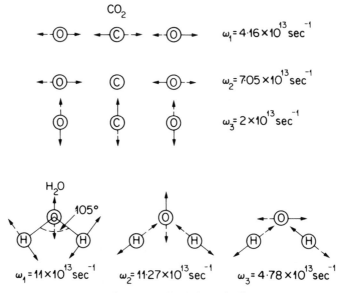

Fig. 3.6. Normal modes of vibration for triatomic molecules CO_2 and H_2O

Atoms in polyatomic molecules behave as the masses of our pendulums; the normal modes of two triatomic molecules CO_2 and H_2O are shown with their frequencies in fig. 3.6. Normal modes and their vibrations will occur frequently throughout this book.

The General Method for Finding Normal Mode Frequencies

We have just seen that when a coupled system oscillates in a *single* normal mode each component of the system will vibrate with the frequency of that mode. This allows us to adopt a method which will always yield the values of the normal mode frequencies and the relative amplitudes of the individual oscillators at each frequency.

Suppose that our system of coupled pendulums in the last section oscillates in *only one* of its normal modes of frequency ω.

Then, in the equations of motion

$$m\ddot{x} + mg(x/l) + s(x - y) = 0$$

and

$$m\ddot{y} + mg(y/l) - s(x - y) = 0$$

we may assume the solutions

$$x = A \cos \omega t$$

$$y = B \cos \omega t$$

where A and B are the displacement amplitudes of x and y at the frequency ω. Using these solutions, the equations of motion become

$$[-m\omega^2 A + (mg/l) \quad A + s(A - B)] \cos \omega t = 0$$

$$[-m\omega^2 B + (mg/l) \quad B - s(A - B)] \cos \omega t = 0$$

The sum of these expressions gives

$$(A + B)(- m\omega^2 + mg/l) = 0$$

which is satisfied when $\omega^2 = g/l$, the first normal mode frequency. The difference between the expressions gives

$$(A - B)(-m\omega^2 + mg/l + 2s) = 0$$

which is satisfied when $\omega^2 = g/l + 2s/m$, the second normal mode frequency.

Inserting the value $\omega^2 = g/l$ in the pair of equations gives $A = B$ (the 'in phase' condition), whilst $\omega^2 = g/l + 2s/m$ gives $A = -B$ (the antiphase condition).

These are the results we found in the previous section.

Because the system started from rest we have been able to assume solutions of the simple form

$$x = A \cos \omega t$$

$$y = B \cos \omega t$$

When the pendulums have an initial velocity at $t = 0$, the boundary conditions require solutions of the form

$$x = A \cos (\omega t + \alpha)$$

$$y = B \cos (\omega t + \alpha)$$

where each normal mode frequency ω has its own particular value of the phase constant α. The number of adjustable constants then allows the solutions to satisfy the arbitrary values of the initial displacements and velocities of both pendulums.

(Problems 3.1, 3.2, 3.3, 3.4, 3.5, 3.6, 3.7, 3.8, 3.9, 3.10, and 3.11)

Mass or Inductance Coupling

In a later chapter we shall discuss the propagation of voltage and current waves along a transmission line which may be considered as a series of coupled electrical oscillators having identical values of inductance and of capacitance. For the moment we shall consider the energy transfer between two electrical circuits which are inductively coupled.

A mutual inductance (shared mass) exists between two electrical circuits when the magnetic flux from the current flowing on one circuit threads the second circuit. Any change of flux induces a voltage in both circuits.

A transformer depends upon mutual inductance for its operation. The power source is connected to the transformer primary coil of n_p turns, over which is wound in the same sense a secondary coil of n_s turns. If unit current flowing in a single turn of the primary coil produces a magnetic flux ϕ, then the flux threading each primary turn (assuming no flux leakage outside the coil) is $n_p\phi$ and the total flux threading all n_p turns of the primary is

$$L_p = n_p^2\phi$$

where L_p is the self inductance of the primary coil. If unit current in a single turn of the secondary coil produces a flux ϕ, then the flux threading each secondary turn is $n_s\phi$ and the total flux threading the secondary coil is

$$L_s = n_s^2\phi$$

where L_s is the self inductance of the secondary coil.

If all the flux lines from unit current in the primary thread all the turns of the secondary, then the total flux lines threading the secondary defines the *mutual inductance*

$$M = n_s(n_p\phi) = \sqrt{L_pL_s}$$

In practice, because of flux leakage outside the coils, $M < \sqrt{L_pL_s}$ and the ratio

$$\frac{M}{\sqrt{L_pL_s}} = k, \text{ the } coefficient \text{ } of \text{ } coupling.$$

If the primary current I_p varies with $e^{i\omega t}$, a change of I_p gives an induced voltage $-L_p\, dI_p/dt = -i\omega LI_p$ in the primary and an induced voltage $-M\, dI_p/dt = -i\omega MI_p$ in the secondary.

If we consider now the two resistance-free circuits of fig. 3.7, where L_1 and L_2 are coupled by flux linkage and allowed to oscillate at some frequency ω (the voltage and current frequency of both circuits), then the voltage equations are

$$i\omega L_1 I_1 - i\frac{1}{\omega C_1}I_1 + i\omega MI_2 = 0 \tag{3.3}$$

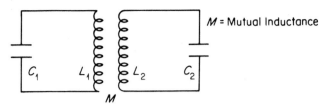

Fig. 3.7. Inductively (mass) coupled LC circuits with mutual inductance M

and

$$i\omega L_2 I_2 - i\frac{1}{\omega C_2} I_2 + i\omega M I_1 = 0 \tag{3.4}$$

where M is the mutual inductance.

Multiplying (3.3) by ω/iL_1 gives

$$\omega^2 I_1 - \frac{I_1}{L_1 C_1} + \frac{M}{L_1}\omega^2 I_2 = 0$$

and multiplying (3.4) by ω/iL_2 gives

$$\omega^2 I_2 - \frac{I_2}{L_2 C_2} + \frac{M}{L_2}\omega^2 I_1 = 0$$

where the natural frequencies of the circuit $\omega_1^2 = 1/L_1 C_1$ and $\omega_2^2 = 1/L_2 C_2$ give

$$(\omega_1^2 - \omega^2)I_1 = \frac{M}{L_1}\omega^2 I_2 \tag{3.5}$$

and

$$(\omega_2^2 - \omega^2)I_2 = \frac{M}{L_2}\omega^2 I_1 \tag{3.6}$$

The product of equations (3.5) and (3.6) gives

$$(\omega_1^2 - \omega^2)(\omega_2^2 - \omega^2) = \frac{M^2}{L_1 L_2}\omega^4 = k^2\omega^4, \tag{3.7}$$

where k is the coefficient of coupling.

Solving for ω gives the frequencies at which energy exchange between the circuits allows the circuits to resonate. If the circuits have equal natural frequencies $\omega_1 = \omega_2 = \omega_0$, say, then equation (3.7) becomes

$$(\omega_0^2 - \omega^2)^2 = k^2\omega^4$$

or

$$(\omega_0^2 - \omega^2) = \pm k\omega^2$$

that is

$$\omega = \pm \frac{\omega_0}{\sqrt{1 \pm k}}$$

The positive sign gives two frequencies

$$\omega' = \frac{\omega_0}{\sqrt{1+k}} \quad \text{and} \quad \omega'' = \frac{\omega_0}{\sqrt{1-k}}$$

at which, if we plot the current amplitude versus frequency, two maxima appear (fig. 3.8).

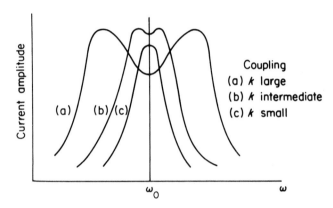

Fig. 3.8. Variation of the current amplitude in each circuit near the resonant frequency. A small resistance prevents the amplitude at resonance from reaching infinite values but this has been ignored in the simple analysis. Flattening of the response curve maximum gives 'frequency band pass' coupling

In loose coupling k and M are small, and $\omega' \approx \omega'' \approx \omega_0$, so that both systems behave almost independently. In tight coupling the frequency difference $\omega'' - \omega'$ increases, the peak values of current are displaced and the dip between the peaks is more pronounced. In this simple analysis the effect of resistance has been ignored. In practice some resistance is always present to limit the amplitude maximum.

(Problems 3.12, 3.13, 3.14, 3.15, 3.16)

Coupled Oscillations of a Loaded String

As a final example involving a large number of coupled oscillators we shall consider a light string supporting n equal masses m spaced at equal distances a along its length. The string is fixed at both ends; it has a length $(n+1)a$ and a constant tension T exists at all points and all times in the string.

Small simple harmonic oscillations of the masses are allowed in only one plane and the problem is to find the frequencies of the normal modes and the displacement of each mass in a particular normal mode.

This problem was first treated by Lagrange, its particular interest being the use it makes of normal modes and the light it throws upon the wave motion and vibration of a continuous string to which it approximates as the linear separation and the magnitude of the masses are progressively reduced.

Fig. 3.9 shows the displacement y_r of the rth mass together with those of its two neighbours. The equation of motion of this mass may be written by considering the components of the tension directed towards the equilibrium

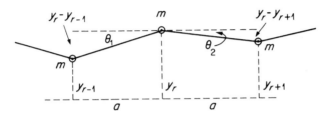

Fig. 3.9. Displacements of three masses on a loaded string under tension T giving equation of motion $m\ddot{y}_r = T(y_{r+1} - 2y_r + y_{r-1})/a$

position. The rth mass is pulled *downwards* towards the equilibrium position by a force $T \sin \theta_1$, due to the tension on its left and a force $T \sin \theta_2$ due to the tension on its right where

$$\sin \theta_1 = \frac{y_r - y_{r-1}}{a}$$

and

$$\sin \theta_2 = \frac{y_r - y_{r+1}}{a}$$

Hence the equation of motion is given by

$$m \frac{d^2 y_r}{dt^2} = -T(\sin \theta_1 + \sin \theta_2)$$

$$= -T\left(\frac{y_r - y_{r-1}}{a} + \frac{y_r - y_{r+1}}{a}\right)$$

so

$$\frac{d^2 y_r}{dt^2} = \ddot{y}_r = \frac{T}{ma}(y_{r-1} - 2y_r + y_{r+1})$$

If, in a normal mode of oscillation of frequency ω, the time variation of y_r is simple harmonic about the equilibrium axis, we may write the displacement of the rth mass in this mode as

$$y_r = A_r\, e^{i\omega t}$$

where A_r is the maximum displacement. Similarly $y_{r+1} = A_{r+1}\, e^{i\omega t}$ and $y_{r-1} = A_{r-1}\, e^{i\omega t}$. Using these values of y in the equation of motion gives

$$-\omega^2 A_r\, e^{i\omega t} = \frac{T}{ma}(A_{r-1} - 2A_r + A_{r+1})\, e^{i\omega t}$$

or

$$-A_{r-1} + \left(2 - \frac{ma\omega^2}{T}\right) A_r - A_{r+1} = 0$$

This is the fundamental equation.

The procedure now is to start with the first mass $r = 1$ and move along the string, writing out the set of similar equations as r assumes the values $r = 1, 2, 3, \ldots, n$ remembering that, because the ends are fixed

$$y_0 = A_0 = 0 \quad \text{and} \quad y_{n+1} = A_{n+1} = 0$$

Thus when $r = 1$ the equation becomes

$$\left(2 - \frac{ma\omega^2}{T}\right) A_1 - A_2 = 0 \qquad (A_0 = 0)$$

When $r = 2$ we have

$$-A_1 + \left(2 - \frac{ma\omega^2}{T}\right) A_2 - A_3 = 0$$

and when $r = n$ we have

$$-A_{n-1} + \left(2 - \frac{ma\omega^2}{T}\right) A_n = 0 \qquad (A_{n+1} = 0)$$

Thus we have a set of n equations which, when solved, will yield n different values of ω^2, each value of ω being the frequency of a normal mode, the number of normal modes being equal to the number of masses.

The formal solution of this set of n equations involves the theory of matrices. However, we may easily solve the simple cases for one or two masses on the string ($n = 1$ or 2) and, in addition, it is possible to show what the complete solution for n masses must be without using sophisticated mathematics.

Firstly, when $n = 1$, one mass on a string of length $2a$, we need only the equation for $r = 1$ where the fixed ends of the string give $A_0 = A_2 = 0$.

Hence we have

$$\left(2 - \frac{ma\omega^2}{T}\right)A_1 = 0$$

giving

$$\omega^2 = \frac{2T}{ma}$$

a single allowed frequency of vibration (fig. 3.10a).

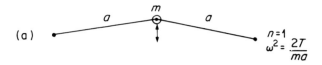

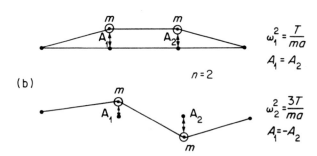

Fig. 3.10. (a) Normal vibration of a single mass m on a string of length $2a$ at a frequency $\omega^2 = 2T/ma$. (b) Normal vibrations of two masses on a string of length $3a$ showing the loose coupled 'in phase' mode of frequency $\omega_1^2 = T/ma$ and the tighter coupled 'out of phase' mode of frequency $\omega_2^2 = 3T/ma$. The number of normal modes of vibration equals the number of masses

When $n = 2$, string length $3a$ (fig. 3.10b) we need the equations for both $r = 1$ and $r = 2$, that is

$$\left(2 - \frac{ma\omega^2}{T}\right)A_1 - A_2 = 0$$

and

$$-A_1 + \left(2 - \frac{ma\omega^2}{T}\right)A_2 = 0 \qquad (A_0 = A_3 = 0)$$

Eliminating A_1 or A_2 shows that these two equations may be solved (are consistent) when

$$\left(2-\frac{ma\omega^2}{T}\right)^2 - 1 = 0$$

that is

$$\left(2-\frac{ma\omega^2}{T}-1\right)\left(2-\frac{ma\omega^2}{T}+1\right) = 0$$

Thus there are two normal mode frequencies

$$\omega_1^2 = \frac{T}{ma} \quad \text{and} \quad \omega_2^2 = \frac{3T}{ma}$$

Using the values of ω_1 in the equations for $r=1$ and $r=2$ gives $A_1 = A_2$ the slow 'in phase' oscillation of fig. 3.10b, whereas ω_2 gives $A_1 = -A_2$ the faster 'anti-phase' oscillation resulting from the increased coupling.

To find the general solution for any value of n let us rewrite the equation

$$\boxed{-A_{r-1}+\left(2-\frac{ma\omega^2}{T}\right)A_r - A_{r+1} = 0}$$

in the form

$$\frac{A_{r-1}+A_{r+1}}{A_r} = \frac{2\omega_0^2-\omega^2}{\omega_0^2} \quad \text{where} \quad \omega_0^2 = \frac{T}{ma}$$

We see that for any particular *fixed* value of the normal mode frequency ω (ω_s say) the right hand side of this equation is constant, independent of r, so the equation holds for all values of r. What values can we give to A_r which will satisfy this equation, meeting the boundary conditions $A_0 = A_{n+1} = 0$ at the end of the string?

Let us *assume* that we may express the amplitude of the rth mass at the frequency ω_s as

$$A_r = C\sin r\theta$$

where C is a constant and θ is some constant angle for a given value of ω_s. The left hand side of the equation then becomes

$$\frac{A_{r-1}+A_{r+1}}{A_r} = \frac{C[\sin(r-1)\theta+\sin(r+1)\theta]}{C\sin r\theta} = \frac{2C\sin r\theta\cos\theta}{C\sin r\theta}$$

$$= 2\cos\theta$$

which is constant and independent of r.

The value of θ_s (constant at ω_s) is easily found from the boundary conditions

$$A_0 = A_{n+1} = 0$$

for

$$A_0 = C \sin 0 = 0 \text{ (automatically at } r = 0)$$

and

$$A_{n+1} = C \sin (n+1)\theta = 0$$

when

$$(n+1)\theta_s = s\pi \quad \text{for} \quad s = 1, 2, \ldots, n$$

Hence

$$\theta_s = \frac{s\pi}{n+1}$$

and

$$A_r = C \sin r\theta = C \sin \frac{rs\pi}{n+1}$$

which is the amplitude of the rth mass at the fixed normal mode frequency ω_s.

To find the allowed values of ω_s we write

$$\frac{A_{r-1} + A_{r+1}}{A_r} = \frac{2\omega_0^2 - \omega_s^2}{\omega_0^2} = 2 \cos \theta_s = 2 \cos \frac{s\pi}{n+1}$$

giving

$$\omega_s^2 = 2\omega_0^2 \left[1 - \cos \frac{s\pi}{n+1} \right]$$

where s may take the values $s = 1, 2, \ldots, n$ and $\omega_0^2 = T/ma$.

Note that there is a maximum frequency of oscillation $\omega_s = 2\omega_0$. This is called the 'cut off' frequency and such an upper frequency limit is characteristic of all oscillating systems composed of similar elements (the masses) repeated periodically throughout the structure of the system. We shall meet this in the next chapter as a feature of wave propagation in crystals.

To summarize, we have found the normal modes of oscillation of n coupled masses on the string to have frequencies given by

$$\omega_s^2 = \frac{2T}{ma} \left[1 - \cos \frac{s\pi}{n+1} \right] \qquad (s = 1, 2, 3 \ldots n)$$

At each frequency ω_s the rth mass has an amplitude

$$A_r = C \sin \frac{rs\pi}{n+1}$$

where C is a constant.

(Problems 3.17, 3.18, 3.19, 3.20 and 3.21)

The Wave Equation

Finally, in this chapter, we show how the coupled vibrations in the periodic structure of our loaded string become waves in a continuous medium.

We found the equation of motion of the rth mass to be

$$\frac{d^2 y_r}{dt^2} = \frac{T}{ma}(y_{r+1} - 2y_r + y_{r-1})$$

Now let the separation $a = \delta x$ and consider the limit $\delta x \to 0$ as the masses merge into a continuous heavy string.

We then have

$$\frac{d^2 y_r}{dt^2} = \frac{T}{m}\left(\frac{y_{r+1} - 2y_r + y_{r-1}}{\delta x}\right) = \frac{T}{m}\left(\frac{(y_{r+1} - y_r)}{\delta x} - \frac{(y_r - y_{r-1})}{\delta x}\right)$$

$$= \frac{T}{m}\left[\left(\frac{\delta y}{\delta x}\right)_{r+1} - \left(\frac{\delta y}{\delta x}\right)_r\right]$$

Now

$$\left(\frac{dy}{dx}\right)_{x+dx} - \left(\frac{dy}{dx}\right)_x = \frac{d^2 y}{dx^2}dx$$

So in the limit $\delta x \to 0$ we may drop the subscripts and write the equation of motion for the harmonic oscillator at position x as

$$\boxed{\frac{d^2 y}{dt^2} = \frac{T}{m}\frac{d^2 y}{dx^2}dx = \frac{T}{\rho}\frac{d^2 y}{dx^2}}$$

where $\rho = m/dx$, the mass per unit length of the string (its linear density). This final expression is the WAVE EQUATION.

T/ρ has the dimensions of the square of a velocity, the velocity with which the wave (that is the phase of oscillation) is propagated. The solution for y is always that of a harmonic oscillation at some position x along the string.

Problem 3.1
Show that the potential energy of two identical simple pendulums coupled by a spring may be expressed as $aX^2 + bY^2$, where X and Y are normal co-ordinates and a and b are constant. Show that the kinetic energy may be expressed as $c\dot{X}^2 + d\dot{Y}^2$ where c and d are constants. Evaluate a, b, c and d in terms of s, l, m and g.

Problem 3.2

Express the total energy of problem 3.1 in terms of the pendulum displacements x and y as

$$E = (E_{kin} + E_{pot})_x + (E_{kin} + E_{pot})_y + (E_{pot})_{xy}$$

where the brackets give the energy of each pendulum expressed in its own co-ordinates and $(E_{pot})_{xy}$ is the coupling or interchange energy involving the product of these co-ordinates.

Problem 3.3

Figs. 3.3 and 3.5 show how the pendulum configurations $x = 2a$, $y = 0$ and $x = 0$, $y = 2a$ result from the superposition of the normal modes X and Y. Using the same initial conditions ($x = 2a$, $y = 0$, $\dot{x} = \dot{y} = 0$) draw similar sketches to show how X and Y superpose to produce $x = -2a$, $y = 0$ and $x = 0$, $y = -2a$.

Problem 3.4

In the figure two masses m_1 and m_2 are coupled by a spring of stiffness s and natural length l. If x is the extension of the spring show that equations of motion along the x axis are

$$m_1\ddot{x}_1 = sx$$

and

$$m_2\ddot{x}_2 = -sx$$

and combine these to show that the system oscillates with a frequency

$$\omega^2 = \frac{s}{\mu}$$

where

$$\mu = \frac{m_1 m_2}{m_1 + m_2}$$

is called the reduced mass.

The figure now represents a diatomic molecule as a harmonic oscillator with an effective mass equal to its reduced mass. If a sodium chloride molecule has a natural vibration frequency $= 1 \cdot 14 \times 10^{13}$ Hertz (in the infrared region of the electromagnetic spectrum) show that the interatomic force constant $s = 120$ newtons metre^{-1} (this simple model gives a higher value for s than more refined methods which account for other interactions within the salt crystal lattice)

Mass of Na atom $= 23$ a.m.u.

Mass of Cl atom $= 35$ a.m.u.

1 a.m.u. $= 1 \cdot 67 \times 10^{-27}$ kilograms

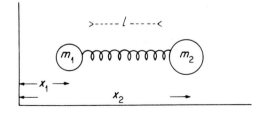

Problem 3.5

The equal masses in the figure oscillate in the vertical direction. Show that the frequencies of the normal modes of oscillation are given by

$$\omega^2 = (3 \pm \sqrt{5}) \frac{s}{2m}$$

and that in the slower mode the ratio of the amplitude of the upper mass to that of the lower mass is $\frac{1}{2}(\sqrt{5} - 1)$ whilst in the faster mode this ratio is $\frac{1}{2}(\sqrt{5} + 1)$.

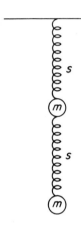

In the calculations it is not necessary to consider gravitational forces because they play no part in the forces responsible for the oscillation.

Problem 3.6

In the coupled pendulums of figure 3.3 let us write the modulated frequency $\omega_m = (\omega_2 - \omega_1)/2$ and the average frequency $\omega_a = (\omega_2 + \omega_1)/2$ and assume that the spring is so weak that it stores a negligible amount of energy. Let the modulated amplitude

$$2a \cos \omega_m t \quad \text{or} \quad 2a \sin \omega_m t$$

be constant over one cycle at the average frequency ω_a to show that the energies of the masses may be written

$$E_x = 2ma^2 \omega_a^2 \cos^2 \omega_m t$$

and

$$E_y = 2ma^2 \omega_a^2 \sin^2 \omega_m t$$

Show that the total energy E remains constant and that the energy difference at any time is

$$E_x - E_y = E \cos (\omega_2 - \omega_1)t$$

Prove that

$$E_x = \frac{E}{2}[1 + \cos (\omega_2 - \omega_1)t]$$

and

$$E_y = \frac{E}{2}[1 - \cos{(\omega_2 - \omega_1)t}]$$

to show that the constant total energy is completely exchanged between the two pendulums at the beat frequency $(\omega_2 - \omega_1)$.

Problem 3.7

When the masses of the coupled pendulums of fig. 3.1 are no longer equal the equations of motion become

$$m_1\ddot{x} = -m_1(g/l)x - s(x - y)$$

and

$$m_2\ddot{y} = -m_2(g/l)y + s(x - y)$$

Show that we may choose the normal co-ordinates

$$X = \frac{m_1x + m_2y}{m_1 + m_2}$$

with a normal mode frequency $\omega_1^2 = g/l$ and $Y = x - y$ with a normal mode frequency $\omega_2^2 = g/l + s(1/m_1 + 1/m_2)$.

Note that X is the co-ordinate of the centre of mass of the system whilst the effective mass in the Y mode is the reduced mass μ of the system where $1/\mu = 1/m_1 + 1/m_2$.

Problem 3.8

Let the system of problem 3.7 be set in motion with the initial conditions $x = A$, $y = 0$, $\dot{x} = \dot{y} = 0$ at $t = 0$. Show that the normal mode amplitudes are $X_0 = (m_1/M)A$ and $Y_0 = A$ to yield

$$x = \frac{A}{M}(m_1 \cos{\omega_1 t} + m_2 \cos{\omega_2 t})$$

and

$$y = A\frac{m_1}{M}(\cos{\omega_1 t} - \cos{\omega_2 t})$$

where $M = m_1 + m_2$.

Express these displacements as

$$x = 2A \cos{\omega_m t} \cos{\omega_a t} + \frac{2A}{M}(m_1 - m_2) \sin{\omega_m t} \sin{\omega_a t}$$

and

$$y = 2A\frac{m_1}{M} \sin{\omega_m t} \sin{\omega_a t}$$

where $\omega_m = (\omega_2 - \omega_1)/2$ and $\omega_a = (\omega_1 + \omega_2)/2$.

Problem 3.9

Apply the weak coupling conditions of problem 3.6 to the system of problem 3.8 to show that the energies

$$E_x = \frac{E}{M^2}[m_1^2 + m_2^2 + 2m_1 m_2 \cos(\omega_2 - \omega_1)t]$$

and

$$E_y = E\left(\frac{2m_1 m_2}{M^2}\right)[1 - \cos(\omega_2 - \omega_1)t]$$

Note that E_x varies between a maximum of E (at $t = 0$) and a minimum of $[(m_1 - m_2)/M]^2 E$, whilst E_y oscillates between a minimum of zero at $t = 0$ and a maximum of $4(m_1 m_2/M^2)E$ at the beat frequency of $(\omega_2 - \omega_1)$.

Problem 3.10

In the figure (see page 95) the right hand pendulum of the coupled system is driven by the horizontal force $F_0 \cos \omega t$ as shown. If a small damping constant r is included the equations of motion may be written

$$m\ddot{x} = -\frac{mg}{l}x - r\dot{x} - s(x - y) + F_0 \cos \omega t$$

and

$$m\ddot{y} = -\frac{mg}{l}y - r\dot{y} + s(x - y)$$

Show that the equations of motion for the normal co-ordinates $X = x + y$ and $Y = x - y$ are those for damped oscillators driven by a force $F_0 \cos \omega t$.

Solve these equations for X and Y and, by neglecting the effect of r, show that

$$x \approx \frac{F_0}{2m} \cos \omega t \left[\frac{1}{\omega_1^2 - \omega^2} + \frac{1}{\omega_2^2 - \omega^2}\right]$$

and

$$y \approx \frac{F_0}{2m} \cos \omega t \left[\frac{1}{\omega_1^2 - \omega^2} - \frac{1}{\omega_2^2 - \omega^2}\right]$$

where

$$\omega_1^2 = \frac{g}{l} \quad \text{and} \quad \omega_2^2 = \frac{g}{l} + \frac{2s}{m}$$

Show that

$$\frac{y}{x} \approx \frac{\omega_2^2 - \omega_1^2}{\omega_2^2 + \omega_1^2 - 2\omega^2}$$

and sketch the behaviour of the oscillator with frequency to show that outside the frequency range $\omega_2 - \omega_1$ the motion of y is attenuated.

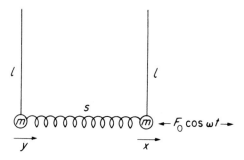

Problem 3.11

The diagram shows an oscillatory force $F_0 \cos \omega t$ acting on a mass M which is part of a simple harmonic system of stiffness k and is connected to a mass m by a spring of stiffness s. If all oscillations are along the x axis show that the condition for M to remain stationary is $\omega^2 = s/m$. (This is a simple version of small mass loading in engineering to quench undesirable oscillations).

$F_0 \cos \omega t$

Problem 3.12 (fig. on page 96)

The figure shows two identical LC circuits coupled by a common capitance C with the directions of current flow indicated by arrows. The voltage equations are

$$V_1 - V_2 = L\frac{dI_a}{dt}$$

and

$$V_2 - V_3 = L\frac{dI_b}{dt}$$

whilst the currents are given by

$$\frac{dq_1}{dt} = -I_a \qquad \frac{dq_2}{dt} = I_a - I_b$$

and

$$\frac{dq_3}{dt} = I_b$$

Solve the voltage equations for the normal co-ordinates $(I_a + I_b)$ and $(I_a - I_b)$ to show that the normal modes of oscillation are given by

$$I_a = I_b \quad \text{at } \omega_1^2 = \frac{1}{LC}$$

and

$$I_a = -I_b \quad \text{at } \omega_2^2 = \frac{3}{LC}$$

Note that when $I_a = I_b$ the coupling capacitance may be removed and $q_1 = -q_2$. When $I_a = -I_b$, $q_2 = -2q_1 = -2q_3$.

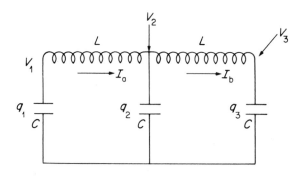

Problem 3.13

A generator of e.m.f. E is coupled to a load Z by means of an ideal transformer. From the diagram, Kirchhoff's Law gives

$$E = -e_1 = i\omega L_p I_1 - i\omega M I_2$$

and

$$I_2 Z_2 = e_2 = i\omega M I_1 - i\omega L_s I_2.$$

Show that E/I_1, the impedance of the whole system seen by the generator, is the sum of the primary impedance and a 'reflected impedance' from the secondary circuit of $\omega^2 M^2 / Z_s$ where $Z_s = Z_2 + i\omega L_s$.

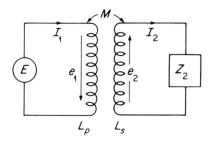

Problem 3.14

Show, for the perfect transformer of question 3.13, that the impedance seen by the generator consists of the primary impedance in parallel with an impedance $(n_p/n_s)^2 Z_2$, where n_p and n_s are the number of primary and secondary transformer coil turns respectively.

Problem 3.15

If the generator delivers maximum power when its load equals its own internal impedance show how an ideal transformer may be used as a device to match a load to a generator, e.g. a loudspeaker of a few ohms impedance to an amplifier output of 10^3 ohms impedance.

Problem 3.16

The two circuits in the diagram are coupled by a variable mutual inductance M and Kirchhoff's Law gives

$$Z_1 I_1 + Z_M I_2 = E$$

and

$$Z_M I_1 + Z_2 I_2 = 0,$$

where

$$Z_M = +i\omega M$$

M is varied at a resonant frequency where the reactance $X_1 = X_2 = 0$ to give a maximum value of I_2. Show that the condition for this maximum is $\omega M = \sqrt{R_1 R_2}$ and that this defines a 'critical coefficient of coupling' $k = (Q_1 Q_2)^{-1/2}$, where the Q's are the quality factor of the circuits.

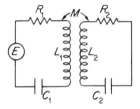

Problem 3.17

Consider the case when the number of masses on the loaded string of this chapter is $n = 3$ and show that the normal vibrations have frequencies given by $\omega_1^2 = (2 - \sqrt{2})T/ma$, $\omega_2^2 = 2T/ma$ and $\omega_3^2 = (2 + \sqrt{2})T/ma$.

Problem 3.18

Show that the relative displacements of the masses in these modes is $1 : \sqrt{2} : 1$, $1 : 0 : -1$, and $1 : -\sqrt{2} : 1$. Show by sketching these relative displacements that tighter coupling increases the mode frequency.

Problem 3.19

Taking the maximum value of

$$\omega_s^2 = \frac{2T}{ma}\left(1 - \cos\frac{s\pi}{n+1}\right)$$

at $s = n$ as that produced by the strongest coupling, deduce the relative displacements of neighbouring masses and confirm your deduction by inserting your values in consecutive difference equations relating the displacements y_{r+1}, y_r, and y_{r-1}. Why is your solution unlikely to satisfy the displacements of those masses near the ends of the string?

Problem 3.20

Expand the value of

$$\omega_s^2 = \frac{2T}{ma}\left(1 - \cos\frac{s\pi}{n+1}\right)$$

98   *The Physics of Vibrations and Waves*

when $s \ll n$ in powers of $(s/n + 1)$ to show that in the limit of very large values of n, a low frequency

$$\omega_s = \frac{s\pi}{l}\sqrt{\frac{T}{\rho}},$$

where $\rho = m/a$ and $l = (n+1)a$.

Problem 3.21
An electrical transmission line consists of equal inductances L and capacitances C arranged as shown. Using the equations

$$\frac{L\, dI_{r-1}}{dt} = V_{r-1} - V_r = \frac{q_{r-1} - q_r}{C}$$

and

$$I_{r-1} - I_r = \frac{dq_r}{dt},$$

show that an expression for I_r may be derived which is equivalent to that for y_r in the case of the mass-loaded string. (This acts as a low pass electric filter and has a cut-off frequency as in the case of the string. This cut-off frequency is a characteristic of wave propagation in periodic structures and electromagnetic wave guides.)

Summary of Important Results

In coupled systems each normal co-ordinate defines a degree of freedom, each degree of freedom defines a way in which a system may take up energy. The total energy of the system is the sum of the energies in its normal modes of oscillation because these remain separate and distinct, and energy is never exchanged between them.

A simple harmonic oscillator has two normal co-ordinates [velocity (or momentum) and displacement] and therefore two degrees of freedom, the first connected with kinetic energy, the second with potential energy.

---

n equal masses, separation a, coupled on a string under constant tension T

Equation of motion of the rth mass is

$$m\ddot{y}_r = (T/a)(y_{r-1} - 2y_r + y_{r+1})$$

which for $y_r = A_r e^{i\omega t}$ gives

$$-A_{r+1} + \left(\frac{2 - ma\omega^2}{T}\right)A_r - A_{r-1} = 0$$

There are n normal modes with frequencies ω_s given by

$$\omega_s^2 = \frac{2T}{ma}\left(1 - \cos\frac{s\pi}{n+1}\right)$$

In a normal mode of frequency ω_s the rth mass has an amplitude

$$A_r = C \sin\frac{rs\pi}{n+1}$$

where C is a constant

Wave equation

In the limit, as separation $a = \delta x \to 0$ equation of motion of rth mass on loaded string $m\ddot{y}_r = (T/a)(y_{r-1} - 2y_r + y_{r+1})$ becomes wave equation

$$\frac{d^2 y}{dt^2} = \frac{T}{\rho}\frac{d^2 y}{dx^2} = c^2\frac{d^2 y}{dx^2}$$

where ρ is mass per unit length and c is the wave velocity

Chapter 4

Transverse Wave Motion

Partial Differentiation

From this chapter onwards we shall often need to use the notation of partial differentiation.

When we are dealing with a function of only one variable, $y = f(x)$ say, we write the differential coefficient

$$\frac{\mathrm{d}y}{\mathrm{d}x} = \lim_{\delta x \to 0} \frac{f(x + \delta x) - f(x)}{\delta x}$$

but if we consider a function of two or more variables, the value of this function will vary with a change in any or all of the variables. For instance, the value of the coordinate z on the surface of a sphere centred at the origin O whose equation is $x^2 + y^2 + z^2 = a^2$, where a is the radius of the sphere, will depend on x and y so that z is a function of x and y written $z = z(x, y)$. The differential change of z which follows from a change of x and y may be written

$$\mathrm{d}z = \left(\frac{\partial z}{\partial x}\right)_y \mathrm{d}x + \left(\frac{\partial z}{\partial y}\right)_x \mathrm{d}y$$

where $(\partial z / \partial x)_y$ means differentiating z with respect to x whilst y is kept constant, so that

$$\left(\frac{\partial z}{\partial x}\right)_y = \lim_{\delta x \to 0} \frac{z(x + \delta x, y) - z(x, y)}{\delta x}$$

The total change $\mathrm{d}z$ is found by adding the separate increments due to the change of each variable in turn whilst the others are kept constant. In fig. 4.1 we can see that keeping y constant isolates a plane which cuts the spherical surface in a curved line, and the incremental contribution to $\mathrm{d}z$ along this line is exactly as though z were a function of x only. Now by keeping x constant we turn the plane through $90°$ and repeat the process with y as a variable so that the total increment of $\mathrm{d}z$ is the sum of these two processes.

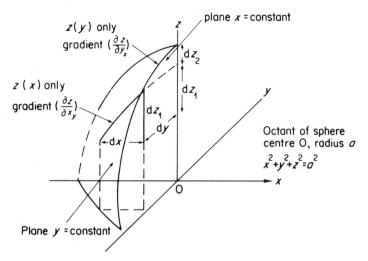

Fig. 4.1. Section of a sphere centred at the origin
showing $dz = dz_1 + dz_2 = (\partial z/\partial x)_y dx + (\partial z/\partial y)_x dy$
where each gradient is calculated with one variable
remaining constant

If only two variables are involved the subscript showing which variable is
kept constant is omitted without ambiguity.

In wave motion our functions will be those of variables of distance and time,
and we shall write $\partial/\partial x$ and $\partial^2/\partial x^2$ for the first or second derivatives with
respect to x, whilst the time t remains constant. Again, $\partial/\partial t$ and $\partial^2/\partial t^2$ will
denote first and second derivatives with respect to time, implying that x is kept
constant.

Waves

One of the simplest ways to demonstrate wave motion is to take the loose end
of a long rope which is fixed at the other end and to move the loose end quickly
up and down. Crests and troughs of the waves move down the rope, and if the
rope were infinitely long such waves would be called *progressive waves*—these
are waves travelling in an unbounded medium free from possible reflexion (fig.
4.2).

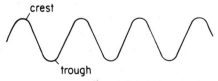

crest

trough

Progressive waves on infinitely long string

Fig. 4.2. Progressive transverse
waves moving along a string

If the medium is limited in extent, for example if the rope were reduced to a violin string, fixed at both ends, the progressive waves travelling on the string would be reflected at both ends; the vibration of the string would then be the combination of such waves moving to and fro along the string and *standing waves* would be formed.

Waves on strings are *transverse waves* where the displacements or oscillations in the medium are transverse to the direction of wave propagation. When the oscillations are parallel to the direction of wave propagation the waves are *longitudinal*. Sound waves are longitudinal waves; a gas can sustain only longitudinal waves because transverse waves require a shear force to maintain them. Both transverse and longitudinal waves can travel in a solid.

In this book we are going to discuss *plane waves* only. When we see wave motion as a series of crests and troughs we are in fact observing the vibrational motion of the individual oscillators in the medium, and in particular all of those oscillators in a plane of the medium which, at the instant of observation, have the same phase in their vibrations.

If we take a plane perpendicular to the direction of wave propagation and all oscillators lying within that plane have a common phase, we shall observe with time how that plane of common phase progresses through the medium. Over such a plane, all parameters describing the wave motion remain constant. The crests and troughs are planes of maximum amplitude of oscillation which are π radians out of phase; a crest is a plane of maximum positive amplitude, while a trough is a plane of maximum negative amplitude. In formulating such wave motion in mathematical terms we shall have to relate the phase difference between any two planes to their physical separation in space. We have, in principle, already done this in our discussion on oscillators.

Spherical waves are waves in which the surfaces of common phase are spheres and the source of waves is a central point, e.g. an explosion; each spherical surface defines a set of oscillators over which the radiating disturbance has imposed a common phase in vibration. In practice spherical waves become plane waves after travelling a very short distance. A small section of a spherical surface is a very close approximation to a plane.

Velocities in Wave Motion

At the outset we must be very clear about one point. The individual oscillators which make up the medium *do not* progress through the medium with the waves. Their motion is simple harmonic, limited to oscillations, transverse or longitudinal, about their equilibrium positions. It is their phase relationships we observe as waves, not their progressive motion through the medium.

There are three velocities in wave motion which are quite distinct although they are connected mathematically. They are

(1) *The particle velocity*, which is the simple harmonic velocity of the oscillator about its equilibrium position.

(2) *The wave or phase velocity*, the velocity with which planes of equal phase, crests or troughs, progress through the medium.

(3) *The group velocity.* A number of waves of different frequencies, wavelengths and velocities may be superposed to form a group. Waves rarely occur as single monochromatic components; a white light pulse consists of an infinitely fine spectrum of frequencies and the motion of such a pulse would be described by its group velocity. Such a group would, of course, 'disperse' with time because the wave velocity of each component would be different in all media except free space. Only in free space would it remain as white light. We shall discuss group velocity as a separate topic in a later section of this chapter. Its importance is that it is the velocity with which the energy in the wave group is transmitted. For a monochromatic wave the group velocity and the wave velocity are identical. Here we shall concentrate on particle and wave velocities.

The Wave Equation

This equation will dominate the rest of this text and we shall derive it, first of all, by considering the motion of transverse waves on a string.

We shall consider the vertical displacement y of a very short section of a uniform string. This section will perform vertical simple harmonic motions; it is our simple oscillator. The displacement y will, of course, vary with the time and also with x, the position along the string at which we choose to observe the oscillation.

The wave equation therefore will relate the displacement y of a single oscillator to distance x and time t. We shall consider oscillations only in the plane of the paper, so that our transverse waves on the string are plane polarized.

The mass of the uniform string per unit length or its linear density is ρ, and a constant tension T exists throughout the string although it is slightly extensible.

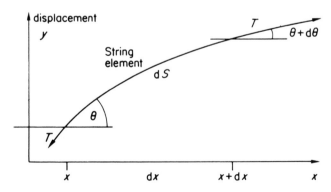

Fig. 4.3. Displaced element of string of length $ds \approx dx$ with tension T acting at an angle θ at x and at $\theta + d\theta$ at $x + dx$

This requires us to consider such a short length and such small oscillations that we may linearize our equations. The effect of gravity is neglected.

Thus in fig. 4.3 the forces acting on the curved element of length ds are T at an angle θ to the axis at one end of the element, and T at an angle $\theta + d\theta$ at the other end. The length of the curved element is

$$ds = \left[1 + \left(\frac{\partial y}{\partial x}\right)^2\right]^{1/2} dx$$

but within the limitations imposed $\partial y / \partial x$ is so small that we ignore its square and take ds = dx. The mass of the element of string is therefore $\rho ds = \rho dx$. Its equation of motion is found from Newton's Law, force equals mass times acceleration.

The perpendicular force on the element dx is $T \sin(\theta + d\theta) - T \sin \theta$ in the positive y direction, which equals the product of ρdx (mass) and $\partial^2 y / \partial t^2$ (acceleration).

Since θ is very small $\sin \theta \approx \tan \theta = \partial y / \partial x$, so that the force is given by

$$T\left[\left(\frac{\partial y}{\partial x}\right)_{x+dx} - \left(\frac{\partial y}{\partial x}\right)_x\right]$$

where the subscripts refer to the point at which the gradient is evaluated. The difference between the two terms in the bracket defines the differential coefficient of the gradient $\partial y / \partial x$ times the space interval dx, so that the force is

$$T\frac{\partial^2 y}{\partial x^2} dx$$

The equation of motion of the small element dx then becomes

$$T\frac{\partial^2 y}{\partial x^2} dx = \rho\, dx \frac{\partial^2 y}{\partial t^2}$$

or

$$\frac{\partial^2 y}{\partial x^2} = \frac{\rho}{T}\frac{\partial^2 y}{\partial t^2}$$

giving

$$\frac{\partial^2 y}{\partial x^2} = \frac{1}{c^2}\frac{\partial^2 y}{\partial t^2}$$

where T/ρ has the dimensions of c^2 and c is a velocity.

THIS IS THE WAVE EQUATION.

It relates the acceleration of a simple harmonic oscillator in a medium to the second derivative of its displacement with respect to its position, x, in the medium. The position of the term c^2 in the equation is always shown by a rapid dimensional analysis.

So far we have not explicitly stated which velocity c represents. We shall see that it is the wave or phase velocity, the velocity with which planes of common phase are propagated. In the string the velocity arises as the ratio of the tension to the inertial density of the string. We shall see, whatever the waves, that the wave velocity can always be expressed as a function of the elasticity or potential energy storing mechanism in the medium and the inertia of the medium through which its kinetic or inductive energy is stored. For longitudinal waves in a solid the elasticity is measured by Young's modulus, in a gas by γP, where γ is the specific heat ratio and P is the gas pressure.

Solution of the Wave Equation

The solution of the wave equation

$$\frac{\partial^2 y}{\partial x^2} = \frac{1}{c^2}\frac{\partial^2 y}{\partial t^2}$$

will, of course, be a function of the variables x and t. We are going to show that any function of the form $y = f_1(ct - x)$ is a solution. Moreover, any function $y = f_2(ct + x)$ will be a solution so that, generally, their superposition $y = f_1(ct - x) + f_2(ct + x)$ is the complete solution.

If f_1' represents the differentiation of the function with respect to the bracket $(ct - x)$, then

$$\frac{\partial y}{\partial x} = -f_1'(ct - x)$$

and

$$\frac{\partial^2 y}{\partial x^2} = f_1''(ct - x)$$

also

$$\frac{\partial y}{\partial t} = cf_1'(ct - x)$$

and

$$\frac{\partial^2 y}{\partial t^2} = c^2 f_1''(ct - x)$$

so that

$$\frac{\partial^2 y}{\partial x^2} = \frac{1}{c^2}\frac{\partial^2 y}{\partial t^2}$$

for $y = f_1(ct - x)$. When $y = f_2(ct + x)$ a similar result holds.

(Problems 4.1, 4.2)

If y is the simple harmonic displacement of an oscillator at position x and time t we would expect, from Chapter 1, to be able to express it in the form

$y = a \sin(\omega t - \phi)$, and in fact all of the waves we discuss in this book, because they are plane waves, will be described by sine or cosine functions.

The bracket $(ct - x)$ in the expression $y = f(ct - x)$ has the dimensions of a length and, for the function to be a sine or cosine, its argument must have the dimensions of radians so that $(ct - x)$ must be multiplied by a factor $2\pi/\lambda$, where λ is a length to be defined.

We can now write

$$y = a \sin(\omega t - \phi) = a \sin\frac{2\pi}{\lambda}(ct - x)$$

as a solution to the wave equation if $2\pi c/\lambda = \omega = 2\pi\nu$, where ν is the oscillation frequency and $\phi = 2\pi x/\lambda$.

This means that if a wave, moving to the right, passes over the oscillators in a medium and a photograph is taken at time $t = 0$, the locus of the oscillator displacements (fig. 4.4) will be given by the expression $y = a \sin(\omega t - \phi) = a \sin 2\pi(ct - x)/\lambda$. If we now observe the motion of the oscillator at the position $x = 0$ it will be given by $y = a \sin \omega t$.

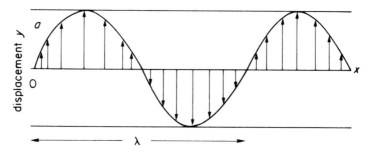

Fig. 4.4. Locus of oscillator displacements in a continuous medium as a wave passes over them travelling in the positive x-direction. The wavelength λ is defined as the distance between any two oscillators having a phase difference of 2π radians

Any oscillator to its right at some position x will be set in motion at some later time by the wave moving to the right; this motion will be given by

$$y = a \sin(\omega t - \phi) = a \sin\frac{2\pi}{\lambda}(ct - x)$$

having a phase lag of ϕ with respect to the oscillator at $x = 0$. This phase lag $\phi = 2\pi x/\lambda$, so that if $x = \lambda$ the phase lag is 2π radians, that is, equivalent to exactly one complete vibration of an oscillator.

This defines λ as the *wavelength*, the separation in space between any two oscillators with a phase difference of 2π radians. The expression $2\pi c/\lambda = \omega =$

$2\pi\nu$ gives $c = \nu\lambda$, where c, the wave or phase velocity, is the product of the frequency and the wavelength. Thus $\lambda/c = 1/\nu = \tau$, the period of oscillation, showing that the wave travels one wavelength in this time. An observer at any point would be passed by ν wavelengths per second, a distance per unit time equal to the velocity c of the wave.

If the wave is moving to the left the sign of ϕ is changed because the oscillation at x begins before that at $x = 0$. Thus the bracket

$(ct - x)$ denotes a wave moving to the right

and

$(ct + x)$ gives a wave moving in the direction of negative x.

There are several equivalent expressions for $y = f(ct - x)$ which we list here as sine functions, although cosine functions are equally valid.

They are:

$$y = a \sin \frac{2\pi}{\lambda}(ct - x)$$

$$y = a \sin 2\pi\left(\nu t - \frac{x}{\lambda}\right)$$

$$y = a \sin \omega\left(t - \frac{x}{c}\right)$$

$$y = a \sin (\omega t - kx)$$

where $k = 2\pi/\lambda = \omega/c$ is called the *wave number*; also $y = a\, e^{i(\omega t - kx)}$, the exponential representation of both sine and cosine.

Each of the expressions above is a solution to the wave equation giving the displacement of an oscillator and its phase with respect to some reference oscillator. The changes of the displacements of the oscillators and the propagation of their phases are what we observe as wave motion.

The wave or phase velocity is, of course, $\partial x/\partial t$, the rate at which the disturbance moves across the oscillators; the oscillator or particle velocity is the simple harmonic velocity $\partial y/\partial t$.

Choosing any one of the expressions above for a right going wave, e.g.

$$y = a \sin (\omega t - kx)$$

we have

$$\frac{\partial y}{\partial t} = \omega a \cos (\omega t - kx)$$

and

$$\frac{\partial y}{\partial x} = -ka \cos ((\omega t - kx)$$

so that

$$\frac{\partial y}{\partial t} = -\frac{\omega}{k}\frac{\partial y}{\partial x} = -c\frac{\partial y}{\partial x}\left(= -\frac{\partial x}{\partial t}\frac{\partial y}{\partial x}\right)$$

The particle velocity $\partial y/\partial t$ is therefore given as the product of the wave velocity

$$c = \frac{\partial x}{\partial t}$$

and the gradient of the wave profile preceded by a negative sign for a right going wave

$$y = f(ct - x)$$

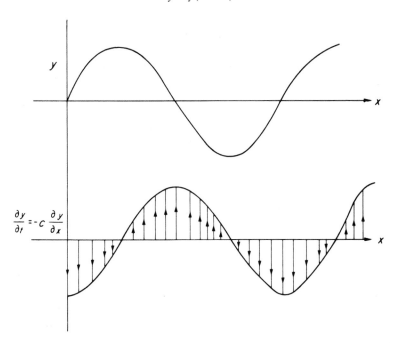

Fig. 4.5. The magnitude and direction of the particle velocity $\partial y/\partial t = -c(\partial y/\partial x)$ at any point x is shown by an arrow in the right-going sine wave above

In fig. 4.5 the arrows show the direction of the particle velocity at various points of the right going wave. It is evident that the particle velocity increases in the same direction as the transverse force in the wave and we shall see in the next section that this force is given by

$$-T\partial y/\partial x$$

where T is the tension in the string.

(Problem 4.3)

Characteristic Impedance of a String (the string as a forced oscillator)

Any medium through which waves propagate will present an impedance to those waves. If the medium is lossless, and possesses no resistive or dissipation mechanism, this impedance will be determined by the two energy storing parameters, inertia and elasticity, and it will be real. The presence of a loss mechanism will introduce a complex term into the impedance.

A string presents such an impedance to progressive waves and this is defined, because of the nature of the waves, as the transverse impedance

$$Z = \frac{\text{transverse force}}{\text{transverse velocity}} = \frac{F}{v}$$

The following analysis will emphasize the dual role of the string as a medium and as a forced oscillator.

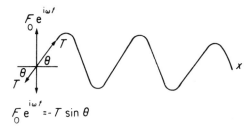

Fig. 4.6. The string as a forced oscillator with a vertical force $F_0 e^{i\omega t}$ driving it at one end

In fig. 4.6 we consider progressive waves on the string which are generated at one end by an oscillating force, $F_0 e^{i\omega t}$, which is restricted to the direction transverse to the string and operates only in the plane of the paper. The tension in the string has a constant value, T, and at the end of the string the balance of forces shows that the applied force is equal and opposite to $T \sin \theta$ at all time, so that

$$F_0 e^{i\omega t} = -T \sin \theta \approx -T \tan \theta = -T \left(\frac{\partial y}{\partial x} \right)$$

where θ is small.

The displacement of the progressive waves may be represented exponentially by

$$y = \mathbf{A}\, e^{i(\omega t - kx)}$$

where the amplitude $\mathbf{A}$ may be complex because of its phase relation with F. At the end of the string, where $x = 0$,

$$F_0 e^{i\omega t} = -T \left(\frac{\partial y}{\partial x} \right)_{x=0} = ikT\mathbf{A}\, e^{i(\omega t - k \cdot 0)}$$

giving

$$A = \frac{F_0}{ikT} = \frac{F_0}{i\omega}\left(\frac{c}{T}\right)$$

and

$$y = \frac{F_0}{i\omega}\left(\frac{c}{T}\right) e^{i(\omega t - kx)}$$

(since $c = \omega/k$).
The transverse velocity

$$v = \dot{y} = F_0\left(\frac{c}{T}\right) e^{i(\omega t - kx)}$$

where the velocity amplitude $v = F_0/Z$, gives a transverse impedance

$$Z = \frac{T}{c} = \rho c \quad (\text{since } T = \rho c^2)$$

or *Characteristic Impedance* of the string.

Since the velocity c is determined by the inertia and the elasticity, the impedance is also governed by these properties.

(We can see that the amplitude of displacement $y = F_0/\omega Z$, with the phase relationship $-i$ with respect to the force, is in complete accord with our discussion in Chapter 2.)

Reflexion and Transmission of Waves on a String at a Boundary

We have seen that a string presents a characteristic impedance ρc to waves travelling along it, and we ask how the waves will respond to a sudden change of impedance, that is, of the value ρc. We shall ask this question of all the waves we discuss, acoustic waves, voltage and current waves and electromagnetic waves, and we shall find a remarkably consistent pattern in their behaviour.

We suppose that a string consists of two sections smoothly joined at a point with a constant tension T along the whole string. The two sections have different linear densities ρ_1 and ρ_2, and therefore different wave velocities $T/\rho_1 = c_1^2$ and $T/\rho_2 = c_2^2$. The specific impedances are $\rho_1 c_1$ and $\rho_2 c_2$ respectively.

An incident wave travelling along the string meets the discontinuity in impedance at the position $x = 0$ in fig. 4.7. At this position, $x = 0$, a part of the incident wave will be reflected and part of it will be transmitted into the region of impedance $\rho_2 c_2$.

We shall denote the impedance $\rho_1 c_1$ by Z_1 and the impedance $\rho_2 c_2$ by Z_2. We write the displacement of the incident wave as $y_i = A_1 e^{i(\omega t - kx)}$, a wave of real (not complex) amplitude A_1 travelling in the positive x-direction with velocity c_1. The displacement of the reflected wave is $y_r = B_1 e^{i(\omega t + k_1 x)}$, of amplitude B_1 and travelling in the negative x-direction with velocity c_1.

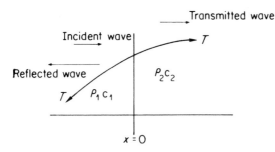

Fig. 4.7. Waves on a string of impedance $\rho_1 c_1$ reflected and transmitted at the boundary where the string changes to impedance $\rho_2 c_2$

The transmitted wave displacement is given by $y_t = A_2\, e^{i(\omega t - k_2 x)}$, of amplitude A_2 and travelling in the positive x-direction with velocity c_2.

We wish to find the reflexion and transmission amplitude coefficients, that is, the relative values of B_1 and A_2 with respect to A_1. We find these via two boundary conditions which must be satisfied at the impedance discontinuity at $x = 0$.

The boundary conditions are:

(1) A geometrical condition that the displacement is the same immediately to the left and right of $x = 0$ for all time, so that there is no discontinuity of displacement.

(2) A dynamical condition that there is a continuity of the transverse force $T(\partial y/\partial x)$ at $x = 0$, and therefore a continuous gradient. This must hold, otherwise a finite difference in the force acts on an infinitesimally small mass of the string giving an infinite acceleration; this is not permitted.

Condition (1) gives

$$y_i + y_r = y_t$$

or

$$A_1\,{}^{i(\omega t - k_1 x)} + B_1\, e^{i(\omega t + k_1 x)} = A_2\, e^{i(\omega t - k_2 x)}$$

At $x = 0$ we may cancel the exponential terms giving

$$A_1 + B_1 = A_2 \tag{4.1}$$

Condition (2) gives

$$T\frac{\partial}{\partial x}(y_i + y_r) = T\frac{\partial}{\partial x}y_t$$

at $x = 0$ for all t, so that

$$-k_1 T A_1 + k_1 T B_1 = -k_2 T A_2$$

or

$$-\omega\frac{T}{c_1}A_1 + \omega\frac{T}{c_1}B_1 = -\omega\frac{T}{c_2}A_2$$

after cancelling exponentials at $x = 0$. But $T/c_1 = \rho_1 c_1 = Z_1$ and $T/c_2 = \rho_2 c_2 = Z_2$, so that

$$Z_1(A_1 - B_1) = Z_2 A_2 \qquad (4.2)$$

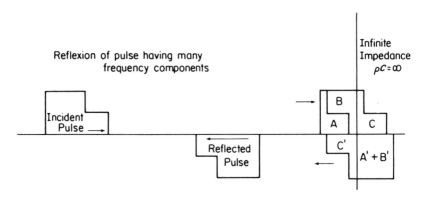

Fig. 4.8. A pulse or group of waves composed of many frequency components is reflected at an infinite impedance. Each component is perfectly reflected with a phase change of π radians, so that the reflected pulse is the inverted and reversed shape of the initial waveform. The pulse at reflexion is divided in the figure into three sections A, B, and C. At the moment of observation section C has already been reflected and suffered inversion and reversal to become C′. The actual shape of the pulse observed at this instant is A being $A+B+C'$ where $B = C'$. The displacement at the point of reflexion must be zero. The reversed and inverted section $A'+B'$ corresponds to the unreflected portion $A+B$ and will progress in the negative x-direction beyond the boundary as A and B are reflected

Equations (4.1) and (4.2) give the

$$\text{Reflexion coefficient of amplitude,} \frac{B_1}{A_1} = \frac{Z_1 - Z_2}{Z_1 + Z_2}$$

and the

$$\text{Transmission coefficient of amplitude,} \frac{A_2}{A_1} = \frac{2Z_1}{Z_1 + Z_2}$$

We see immediately that these coefficients are independent of ω and hold for waves of all frequencies; they are real and therefore free from phase changes other than that of π radians which will change the sign of a term. Moreover these ratios depend entirely upon the ratios of the impedances. (See summary on page 412). If $Z_2 = \infty$, this is equivalent to $x = 0$ being a fixed end to the string because no transmitted wave exists. This gives $B_1/A_1 = -1$, so that the incident wave is completely reflected (as we expect) with a phase change of π (phase reversal)—conditions we shall find to be necessary for standing waves to exist. A group of waves having many component frequencies will retain its shape upon reflection at $Z_2 = \infty$ but will suffer reversal (fig. 4.8). If $Z_2 = 0$, so that $x = 0$ is a free end of the string, then $B_1/A_1 = 1$ and $A_2/A_1 = 2$. This explains the 'flick' at the end of a whip or free ended string when a wave reaches it.

(Problems 4.4, 4.5, 4.6)

Reflexion and Transmission of Energy

Our interest in waves, however, is chiefly concerned with their function of transferring energy throughout a medium, and we shall now consider what happens to the energy in a wave when it meets a boundary between two media of different impedance values.

If we consider each unit length, mass ρ, of the string as a simple harmonic oscillator of maximum amplitude A, we know that its total energy will be $E = \frac{1}{2}\rho\omega^2 A^2$, where ω is the wave frequency.

The wave is travelling at a velocity c so that as each unit length of string takes up its oscillation with the passage of the wave the rate at which energy is being carried along the string is

$$(\text{energy} \times \text{velocity}) = \tfrac{1}{2}\rho\omega^2 A^2 c$$

Thus the rate of energy arriving at the boundary $x = 0$ is the energy arriving with the incident wave, that is

$$\tfrac{1}{2}\rho_1 c_1 \omega^2 A_1^2 = \tfrac{1}{2}Z_1\omega^2 A_1^2$$

The rate at which energy leaves the boundary, via the reflected and transmitted waves, is

$$\tfrac{1}{2}\rho_1 c_1 \omega^2 B_1^2 + \tfrac{1}{2}\rho_2 c_2 \omega^2 A_2^2 = \tfrac{1}{2}Z_1\omega^2 B_1^2 + \tfrac{1}{2}Z_2\omega^2 A_2^2$$

which, from the ratios B_1/A_1 and A_2/A_1,

$$= \tfrac{1}{2}\omega^2 A_1^2 \frac{Z_1(Z_1 - Z_2)^2 + 4Z_1^2 Z_2}{(Z_1 + Z_2)^2} = \tfrac{1}{2}Z_1\omega^2 A_1^2$$

Thus energy is conserved, and all energy arriving at the boundary in the incident wave leaves the boundary in the reflected and transmitted waves.

The Reflected and Transmitted Intensity Coefficients

These are given by

$$\frac{\text{Reflected Energy}}{\text{Incident Energy}} = \frac{Z_1 B_1^2}{Z_1 A_1^2} = \left(\frac{B_1}{A_1}\right)^2 = \left(\frac{Z_1 - Z_2}{Z_1 + Z_2}\right)^2$$

$$\frac{\text{Transmitted Energy}}{\text{Incident Energy}} = \frac{Z_2 A_2^2}{Z_1 A_1^2} = \frac{4 Z_1 Z_2}{(Z_1 + Z_2)^2}$$

We see that if $Z_1 = Z_2$ no energy is reflected and the *impedances are said to be matched.*

(Problems 4.7, 4.8)

The Matching of Impedances

Impedance matching represents a very important practical problem in the transfer of energy. Long distance cables carrying energy must be accurately matched at all joints to avoid wastage from energy reflection. The power transfer from any generator is a maximum when the load matches the generator impedance. A loudspeaker is matched to the impedance of the power output of an amplifier by choosing the correct turns ratio on the coupling transformer. This last example, the insertion of a coupling element between two mismatched impedances, is of fundamental importance with applications in many branches of engineering physics and optics. We shall illustrate it using waves on a string, but the results will be valid for all wave systems.

We have seen that when a smooth joint exists between two strings of different impedances, energy will be reflected at the boundary. We are now going to see that the insertion of a particular length of another string between these two mismatched strings will allow us to eliminate energy reflection and match the impedances.

In fig. 4.9 we require to match the impedances $Z_1 = \rho_1 c_1$ and $Z_3 = \rho_3 c_3$ by the smooth insertion of a string of length l and impedance $Z_2 = \rho_2 c_2$. Our problem is to find the values of l and Z_2.

The incident, reflected and transmitted displacements at the junctions $x = 0$ and $x = l$ are shown in fig. 4.9 and we seek to make the ratio

$$\frac{\text{Transmitted energy}}{\text{Incident energy}} = \frac{Z_3 A_3^2}{Z_1 A_1^2}$$

equal to unity.

The boundary conditions are that y and $T(\partial y/\partial x)$ are continuous across the junctions $x = 0$ and $x = l$.

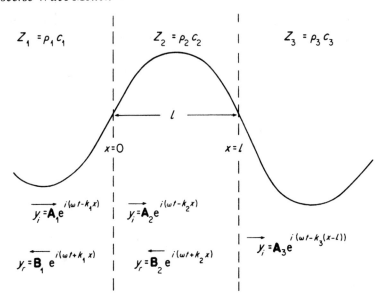

Fig. 4.9. The impedances Z_1 and Z_3 of two strings are matched by the insertion of a length l of a string of impedance Z_2. The incident and reflected waves are shown for the boundaries $x = 0$ and $x = l$. The impedances are matched when $Z_2^2 = Z_1 Z_3$ and $l = \lambda/4$ in Z_2, results which are true for waves in all media

Between Z_1 and Z_2 the continuity of y gives

$$\mathbf{A}_1 e^{i(\omega t - k_1 x)} + \mathbf{B}_1 e^{i(\omega t + k_1 x)} = \mathbf{A}_2 e^{i(\omega t - k_2 x)} + \mathbf{B}_2 e^{i(\omega t + k_2 x)}$$

or

$$\mathbf{A}_1 + \mathbf{B}_1 = \mathbf{A}_2 + \mathbf{B}_2 \quad (\text{at } x = 0) \tag{4.3}$$

Similarly the continuity of $T(\partial y/\partial x)$ at $x = 0$ gives

$$T(-ik_1\mathbf{A}_1 + ik_1\mathbf{B}_1) = T(-ik_2\mathbf{A}_2 + ik_2\mathbf{B}_2)$$

Dividing this equation by ω and remembering that $T(k/\omega) = T/c = \rho c = Z$ we have

$$Z_1(\mathbf{A}_1 - \mathbf{B}_1) = Z_2(\mathbf{A}_2 - \mathbf{B}_2) \tag{4.4}$$

Similarly at $x = l$, the continuity of y gives

$$\mathbf{A}_2 e^{-ik_2 l} + \mathbf{B}_2 e^{ik_2 l} = \mathbf{A}_3 \tag{4.5}$$

and the continuity of $T(\partial y/\partial x)$ gives

$$Z_2(\mathbf{A}_2 e^{-ik_2 l} - \mathbf{B}_2 e^{ik_2 l}) = Z_3\mathbf{A}_3 \tag{4.6}$$

From the four boundary equations (4.3), (4.4), (4.5) and (4.6) we require the ratio $\mathbf{A}_3/\mathbf{A}_1$. We use equations (4.3) and (4.4) to eliminate $\mathbf{B}_1$ and obtain $\mathbf{A}_1$ in terms of $\mathbf{A}_2$ and $\mathbf{B}_2$. We then use equations (4.5) and (4.6) to obtain both $\mathbf{A}_2$ and $\mathbf{B}_2$ in terms of $\mathbf{A}_3$. Equations (4.3) and (4.4) give

$$Z_1(\mathbf{A}_1 - \mathbf{A}_2 - \mathbf{B}_2 + \mathbf{A}_1) = Z_2(\mathbf{A}_2 - \mathbf{B}_2)$$

or

$$\mathbf{A}_1 = \frac{\mathbf{A}_2(r_{12}+1) + \mathbf{B}_2(r_{12}-1)}{2r_{12}} \tag{4.7}$$

where

$$r_{12} = \frac{Z_1}{Z_2}$$

Equations (4.5) and (4.6) give

$$\mathbf{A}_2 = \frac{r_{23}+1}{2r_{23}} \mathbf{A}_3\, e^{ik_2l}$$

and $\tag{4.8}$

$$\mathbf{B}_2 = \frac{r_{23}-1}{2r_{23}} \mathbf{A}_3\, e^{-ik_2l}$$

where

$$r_{23} = \frac{Z_2}{Z_3}$$

Equations (4.7) and (4.8) give

$$\mathbf{A}_1 = \frac{\mathbf{A}_3}{4r_{12}r_{23}}[(r_{12}+1)(r_{23}+1)\, e^{ik_2l} + (r_{12}-1)(r_{23}-1)\, e^{-ik_2l}]$$

$$= \frac{\mathbf{A}_3}{4r_{13}}[(r_{13}+1)(e^{ik_2l} + e^{-ik_2l}) + (r_{12}+r_{23})(e^{ik_2l} - e^{-ik_2l})]$$

$$= \frac{\mathbf{A}_3}{2r_{13}}[(r_{13}+1)\cos k_2l + i(r_{12}+r_{23})\sin k_2l]$$

where

$$r_{12}r_{23} = \frac{Z_1}{Z_2}\frac{Z_2}{Z_3} = \frac{Z_1}{Z_3} = r_{13}$$

Hence

$$\left(\frac{A_3}{A_1}\right)^2 = \frac{4r_{13}^2}{(r_{13}+1)^2 \cos^2 k_2l + (r_{12}+r_{23})^2 \sin^2 k_2l}$$

or

$$\frac{\text{transmitted energy}}{\text{incident energy}} = \frac{Z_3}{Z_1} \frac{A_3^2}{A_1^2} = \frac{1}{r_{13}} \frac{A_3^2}{A_1^2}$$

$$= \frac{4r_{13}}{(r_{13}+1)^2 \cos^2 k_2 l + (r_{12}+r_{23})^2 \sin^2 k_2 l}$$

If we choose $l = \lambda_2/4$, $\cos k_2 l = 0$ and $\sin k_2 l = 1$ giving

$$\frac{Z_3}{Z_1} \frac{A_3^2}{A_1^2} = \frac{4r_{13}}{(r_{12}+r_{23})^2} = 1$$

when

$$r_{12} = r_{23}$$

that is, when

$$\frac{Z_1}{Z_2} = \frac{Z_2}{Z_3} \quad \text{or} \quad Z_2 = \sqrt{Z_1 Z_3}$$

We see, therefore, that if the impedance of the coupling medium is the harmonic mean of the two impedances to be matched and the thickness of the coupling medium is

$$\frac{\lambda_2}{4} \quad \text{where} \quad \lambda_2 = \frac{2\pi}{k_2}$$

all the energy at frequency ω will be transmitted with zero reflection.

The thickness of the dielectric coating of optical lenses which eliminates reflections as light passes from air into glass is one quarter of a wavelength. The "bloomed" appearance arises because exact matching occurs at only one frequency. Transmission lines are matched to loads by inserting quarter wavelength stubs of lines with the appropriate impedance.

(Problems 4.9, 4.10)

Standing Waves on a String of Fixed Length

We have already seen that a progressive wave is completely reflected at an infinite impedance with a π phase change in amplitude. A string of fixed length l with both ends rigidly clamped presents an infinite impedance at each end; we now investigate the behaviour of waves on such a string. Let us consider the simplest case of a monochromatic wave of one frequency ω with an amplitude a travelling in the positive x-direction and an amplitude b travelling in the negative x-direction. The displacement on the string at any point would then be given by

$$y = a\, e^{i(\omega t - kx)} + b\, e^{i(\omega t + kx)}$$

with the boundary condition that $y = 0$ at $x = 0$ and $x = l$ at all times.

The condition $y = 0$ at $x = 0$ gives $0 = (a + b)\,e^{i\omega t}$ for all t, so that $a = -b$. This expresses physically the fact that a wave in either direction meeting the infinite impedance at either end is completely reflected with a π phase change in amplitude. This is a general result for all wave shapes and frequencies.
Thus

$$y = a\,e^{i\omega t}(e^{-ikx} - e^{ikx}) = (-2i)a\,e^{i\omega t} \sin kx \tag{4.9}$$

an expression for y which satisfies the *standing wave time independent form* of the wave equation

$$\partial^2 y/\partial x^2 + k^2 y = 0$$

because $(1/c^2)(\partial^2 y/\partial t^2) = (-\omega^2/c^2)y = -k^2 y$. The condition that $y = 0$ at $x = l$ for all t requires

$$\sin kl = \sin\frac{\omega l}{c} = 0 \quad \text{or} \quad \frac{\omega l}{c} = n\pi$$

limiting the values of allowed frequencies to

$$\omega_n = \frac{n\pi c}{l}$$

or

$$\nu_n = \frac{nc}{2l} = \frac{c}{\lambda_n}$$

that is

$$l = \frac{n\lambda_n}{2}$$

giving

$$\sin\frac{\omega_n x}{c} = \sin\frac{n\pi x}{l}$$

These frequencies are the *normal frequencies* or *modes of vibration* we first met in Chapter 3. They are often called *eigenfrequencies*, particularly in wave mechanics.

Such allowed frequencies define the length of the string as an exact number of half wavelengths, and fig. 4.10 shows the string displacement for the first four *harmonics* ($n = 1, 2, 3, 4$). The value for $n = 1$ is called the *fundamental*.

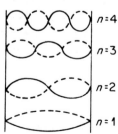

Fig. 4.10. The first four harmonics, $n = 1,2,3,4$ of the standing waves allowed between the two fixed ends of a string

As with the loaded string of Chapter 3, all normal modes may be present at the same time and the general displacement is the superposition of the displacements at each frequency. This is a more complicated problem which we discuss in Chapter 9 (Fourier Methods).

For the moment we see that for $n > 1$ there will be a number of positions along the string which are always at rest. These points occur where

$$\sin \frac{\omega_n x}{c} = \sin \frac{n\pi x}{l} = 0$$

or

$$\frac{n\pi x}{l} = r\pi \qquad (r = 0, 1, 2, 3, \ldots n)$$

The values $r = 0$ and $r = n$ give $x = 0$ and $x = l$, the ends of the string, but between the ends there are $n - 1$ positions equally spaced along the string in the nth harmonic where the displacement is always zero. These positions are called *nodes* or *nodal points*, being the positions of zero motion in a system of *standing waves*. Standing waves arise from the superposition of wave systems travelling in opposite directions. If the amplitudes of these progressive waves are equal and opposite (resulting from complete reflexion), nodal points will exist. Often, however, the reflexion is not quite complete and the waves in the opposite direction do not cancel each other to give complete nodal points. In this case we speak of a *standing wave ratio* which we shall discuss in the next section but one.

Whenever nodal points exist, however, we know that the waves travelling in opposite directions are exactly equal in all respects so that the energy carried in one direction is exactly equal to that carried in the other. This means that the total energy flux, that is, the energy carried across unit area per second in a standing wave system, is zero.

Returning to equation (4.9), we see that the complete expression for the displacement of the nth harmonic is given by

$$y_n = 2a(-\mathrm{i})(\cos \omega_n t + \mathrm{i} \sin \omega_n t) \sin \frac{\omega_n x}{c}$$

We can express this in the form

$$y_n = (A_n \cos \omega_n t + B_n \sin \omega_n t) \sin \frac{\omega_n x}{c} \tag{4.10}$$

where the amplitude of the nth mode is given by $(A_n^2 + B_n^2)^{1/2} = 2a$.

(Problem 4.11)

Energy of a Vibrating String

A vibrating string possesses both kinetic and potential energy. The kinetic energy of an element of length dx and linear density ρ is given by $\frac{1}{2}\rho \, dx \, \dot{y}^2$; the total kinetic energy is the integral of this along the length of the string.

Thus

$$E_{kin} = \tfrac{1}{2} \int_0^l \rho \dot{y}^2 \, dx$$

The potential energy is the work done by the tension T in extending an element dx to a new length ds when the string is vibrating.
Thus

$$E_{pot} = \int T(ds - dx) = \int T \left\{ \left[1 + \left(\frac{\partial y}{\partial x} \right)^2 \right]^{1/2} - 1 \right\} dx$$

$$= \tfrac{1}{2} T \int \left(\frac{\partial y}{\partial x} \right)^2 dx$$

if we neglect higher powers of $\partial y / \partial x$.

Now the change in the length of the element dx is $\tfrac{1}{2}(\partial y / \partial x)^2 \, dx$, and if the string is elastic the change in tension is proportional to the change in length so that, provided $(\partial y / \partial x)$ in the wave is of the first order of small quantities, the change in tension is of the second order and T may be considered constant.

Energy in each Normal Mode of a Vibrating String

The total displacement y in the string is the superposition of the displacements y_n of the individual harmonics and we can find the energy in each harmonic by replacing y_n for y in the results of the last section. Thus the kinetic energy in the nth harmonic is

$$E_n(\text{kinetic}) = \tfrac{1}{2} \int_0^l \rho \dot{y}_n^2 \, dx$$

and the potential energy is

$$E_n(\text{potential}) = \tfrac{1}{2} T \int_0^l \left(\frac{\partial y_n}{\partial x} \right)^2 dx$$

Since we have already shown for standing waves that

$$y_n = (A_n \cos \omega_n t + B_n \sin \omega_n t) \sin \frac{\omega_n x}{c}$$

then

$$\dot{y}_n = (-A_n \omega_n \sin \omega_n t + B_n \omega_n \cos \omega_n t) \sin \frac{\omega_n x}{c}$$

and

$$\frac{\partial y_n}{\partial x} = \frac{\omega_n}{c}(A_n \cos \omega_n t + B_n \sin \omega_n t) \cos \frac{\omega_n x}{c}$$

Thus

$$E_n(\text{kinetic}) = \tfrac{1}{2}\rho\omega_n^2[-A_n \sin \omega_n t + B_n \cos \omega_n t]^2 \int_0^l \sin^2 \frac{\omega_n x}{c}\, dx$$

and

$$E_n(\text{potential}) = \tfrac{1}{2}T\frac{\omega_n^2}{c^2}[A_n \cos \omega_n t + B_n \sin \omega_n t]^2 \int_0^l \cos^2 \frac{\omega_n x}{c}\, dx$$

Remembering that $T = \rho c^2$ we have

$$E_n(\text{kinetic} + \text{potential}) = \tfrac{1}{4}\rho l\omega_n^2(A_n^2 + B_n^2)$$
$$= \tfrac{1}{4}m\omega_n^2(A_n^2 + B_n^2)$$

where m is the mass of the string and $(A_n^2 + B_n^2)$ is the square of the maximum displacement (amplitude) of the mode. To find the exact value of the total energy E_n of the mode we would need to know the precise values of A_n and B_n and we shall evaluate these in Chapter 9 on Fourier Methods. The total energy of the vibrating string is, of course, the sum of all the E_n's of the normal modes.

(Problem 4.12)

Standing Wave Ratio

When a wave is completely reflected the superposition of the incident and reflected amplitudes will give nodal points (zero amplitude) where the incident and reflected amplitudes cancel each other, and points of maximum displacement equal to twice the incident amplitude where they reinforce.

If a progressive wave system is partially reflected from a boundary let the amplitude reflexion coefficient B_1/A_1 of the earlier section be written as r, where $r < 1$.

The maximum amplitude at reinforcement is then $A_1 + B_1$; the minimum amplitude is given by $A_1 - B_1$. In this case the ratio of maximum to minimum amplitudes in the standing wave system is called the

$$\text{Standing Wave Ratio} = \frac{A_1 + B_1}{A_1 - B_1} = \frac{1 + r}{1 - r}$$

where $r = B_1/A_1$.

Measuring the values of the maximum and minimum amplitudes gives the value of the reflexion coefficient for

$$r = B_1/A_1 = \frac{\text{SWR} - 1}{\text{SWR} + 1}$$

where SWR refers to the Standing Wave Ratio.

(Problem 4.13)

Wave Groups and Group Velocity

Our discussion so far has been limited to monochromatic waves—waves of a single frequency and wavelength. It is much more common for waves to occur as a mixture of a number or group of component frequencies; white light, for instance, is composed of a continuous visible wavelength spectrum extending from about 3000 Å in the blue to 7000 Å in the red. Examining the behaviour of such a group leads to the third kind of velocity mentioned at the beginning of this chapter, that is, the group velocity.

Superposition of Two Waves of almost Equal Frequencies

We begin by considering a group which consists of two components of equal amplitude a but frequencies ω_1 and ω_2 which differ by a small amount.

Their separate displacements are given by

$$y_1 = a \cos(\omega_1 t - k_1 x)$$

and

$$y_2 = a \cos(\omega_2 t - k_2 x)$$

Superposition of amplitude and phase gives

$$y = y_1 + y_2 = 2a \cos\left[\frac{(\omega_1 - \omega_2)t}{2} - \frac{(k_1 - k_2)x}{2}\right] \cos\left[\frac{(\omega_1 + \omega_2)t}{2} - \frac{(k_1 + k_2)x}{2}\right]$$

a wave system with a frequency $(\omega_1 + \omega_2)/2$ which is very close to the frequency of either component but with a maximum amplitude of $2a$, modulated in space and time by a very slowly varying envelope of frequency $(\omega_1 - \omega_2)/2$ and wave number $(k_1 - k_2)/2$.

This system is shown in fig. 4.11 and shows, of course, a behaviour similar to that of the equivalent coupled oscillators in Chapter 3. The velocity of the new wave is $(\omega_1 - \omega_2)/(k_1 - k_2)$ which, if the phase velocities $\omega_1/k_1 = \omega_2/k_2 = c$, gives

$$\frac{\omega_1 - \omega_2}{k_1 - k_2} = c\frac{(k_1 - k_2)}{k_1 - k_2} = c$$

so that the component frequencies and their superposition, or *group* will travel with the same velocity, the profile of their combination in fig. 4.11 remaining constant.

If the waves are sound waves the intensity is a maximum whenever the amplitude is a maximum of $2a$; this occurs twice for every period of the modulating frequency, that is, at a frequency $\nu_1 - \nu_2$.

The *beats* of maximum intensity fluctuations thus have a frequency equal to the difference $\nu_1 - \nu_2$ of the components. In the example here where the components have equal amplitudes a, superposition will produce an amplitude which varies between $2a$ and 0; this is called complete or 100% modulation.

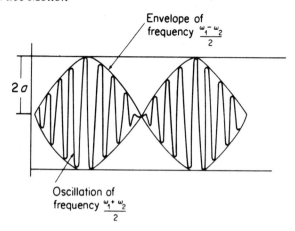

Fig. 4.11. The superposition of two waves of slightly different frequency ω_1 and ω_2 forms a group. The faster oscillation occurs at the average frequency of the two components $(\omega_1 + \omega_2)/2$ and the slowly varying group envelope has a frequency $(\omega_1 - \omega_2)/2$, half the frequency difference between the components

More generally an amplitude modulated wave may be represented by

$$y = A \cos{(\omega t - kx)}$$

where the modulated amplitude

$$A = a + b \cos{\omega' t}$$

This gives

$$y = a \cos{(\omega t - kx)} + \frac{b}{2}\{[\cos{(\omega + \omega')t - kx}] + [\cos{(\omega - \omega')t - kx}]\}$$

so that here amplitude modulation has introduced two new frequencies $\omega \pm \omega'$, known as combination tones or sidebands. Amplitude modulation of a carrier frequency is a common form of radio transmission, but its generation of sidebands has led to the crowding of radio frequencies and interference between stations.

Wave Groups and Group Velocity

Suppose now that the two frequency components of the last section have different phase velocities so that $\omega_1/k_1 \neq \omega_2/k_2$. The velocity of the maximum amplitude of the group, that is, the *group velocity*

$$\frac{\omega_1 - \omega_2}{k_1 - k_2} = \frac{\Delta\omega}{\Delta k}$$

is now different from each of these velocities; the superposition of the two waves will no longer remain constant and the group profile will change with time.

A medium in which the phase velocity is frequency dependent (ω/k not constant) is known as a dispersive medium and a *dispersion relation* expresses the variation of ω as a function of k. If a group contains a number of components of frequencies which are nearly equal the original expression for the group velocity is written

$$\frac{\Delta\omega}{\Delta k} = \frac{d\omega}{dk}$$

The group velocity is that of the maximum amplitude of the group so that it is the velocity with which the energy in the group is transmitted. Since $\omega = kv$, where v is the phase velocity, the group velocity

$$v_g = \frac{d\omega}{dk} = \frac{d}{dk}(kv) = v + k\frac{dv}{dk}$$

$$= v - \lambda\frac{dv}{d\lambda}$$

where $k = 2\pi/\lambda$.

Usually $dv/d\lambda$ is positive, so that $v_g < v$. This is called *normal dispersion*, but *anomalous dispersion* can arise when $dv/d\lambda$ is negative, so that $v_g > v$.

We shall see when we discuss electromagnetic waves that an electrical conductor is anomalously dispersive to these waves whilst a dielectric is normally dispersive except at the natural resonant frequencies of its atoms. In the chapter on forced oscillations we saw that the wave then acted as a driving force upon the atomic oscillators and that strong absorption of the wave energy was represented by the dissipation fraction of the oscillator impedance, whilst the anomalous dispersion curve followed the value of the reactive part of the impedance.

The three curves of fig. 4.12 represent

(a) a non-dispersive medium where ω/k is constant, so that $v_g = v$, for instance free space behaviour towards light waves.

(b) a normal dispersion relation $v_g < v$.

(c) an anomalous dispersion relation $v_g > v$.

Example. The electric vector of an electromagnetic wave propagates in a dielectric with a velocity $v = (\mu\epsilon)^{-1/2}$ where μ is the permeability and ϵ is the permittivity. In free space the velocity is that of light, $c = (\mu_0\epsilon_0)^{-1/2}$. The refractive index $n = c/v = \sqrt{\mu\epsilon/\mu_0\epsilon_0} = \sqrt{\mu_r\epsilon_r}$, where $\mu_r = \mu/\mu_0$ and $\epsilon_r = \epsilon/\epsilon_0$.

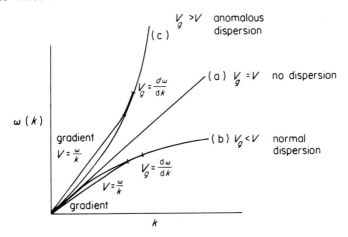

Fig. 4.12. Curves illustrating dispersion relations: (a) a straight line representing a non-dispersive medium, $v = v_g$; (b) a normal dispersion relation where the gradient $v = \omega/k > v_g = d\omega/dk$; (c) an anomalous dispersion relation where

$$v < v_g$$

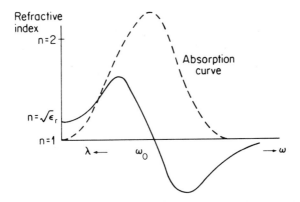

Fig. 4.13. Anomalous dispersion showing the behaviour of the refractive index $n = \sqrt{\epsilon_r}$ versus ω and λ, where ω_0 is a resonant frequency of the atoms of the medium. The absorption in such a region is also shown by the dotted line

For many substances μ_r is constant and ~ 1, but ϵ_r is frequency dependent, so that v depends on λ.

The group velocity

$$v_g = v - \lambda \, \mathrm{d}v/\mathrm{d}\lambda = v\left(1 + \frac{\lambda}{2\epsilon_r} \frac{\partial \epsilon_r}{\partial \lambda}\right)$$

so that $v_g > v$ (anomalous dispersion) when $\partial\epsilon_r/\partial\lambda$ is $+ve$. Fig. 4.13 shows the behaviour of the refractive index $n = \sqrt{\epsilon_r}$ versus ω, the frequency, and λ, the wavelength, in the region of anomalous dispersion associated with a resonant frequency. The dotted curve shows the energy absorption (compare this with fig. 2.9).

(Problems 4.14, 4.15, 4.16, 4.17, 4.18, 4.19)

Wave Group of Many Components. The Bandwidth Theorem

We have so far considered wave groups having only two frequency components. We may easily extend this to the case of a group of many frequency components, each of amplitude a, lying within the narrow frequency range $\Delta\omega$.

We have already covered the essential physics of this problem on page 20, where we found the sum of the series

$$R = \sum_{0}^{n-1} a \cos(\omega t + n\delta)$$

where δ was the constant phase difference between successive components. Here we are concerned with the constant phase difference $(\delta\omega)t$ which results from a constant frequency difference $\delta\omega$ between successive components. The frequency spectrum of this group is shown in fig. 4.14a and we wish to follow its behaviour with time.

We seek the amplitude which results from the superposition of the frequency components and write it

$$R = a \cos \omega_1 t + a \cos(\omega_1 + \delta\omega)t + a \cos(\omega_1 + 2\delta\omega)t + \ldots$$
$$+ a \cos[\omega_1 + (n-1)(\delta\omega)]t$$

The result is given on page 21 by

$$R = a\frac{\sin[n(\delta\omega)t/2]}{\sin[(\delta\omega)t/2]} \cos \bar{\omega}t$$

where the average frequency in the pulse is

$$\bar{\omega} = \omega_1 + \tfrac{1}{2}(n-1)(\delta\omega)$$

Now $n(\delta\omega) = \Delta\omega$, the pulse width, so the behaviour of the resultant R with time may be written

$$R(t) = a\frac{\sin(\Delta\omega.t/2)}{\sin(\Delta\omega.t/n2)}\cos\bar\omega t = na\frac{\sin(\Delta\omega.t/2)}{\Delta\omega.t/2}\cos\bar\omega t$$

when n is large,

or

$$R(t) = A\frac{\sin\alpha}{\alpha}\cos\bar\omega t$$

where $A = na$ and $\alpha = \Delta\omega.t/2$ is half the phase difference between the first and last components at time t.

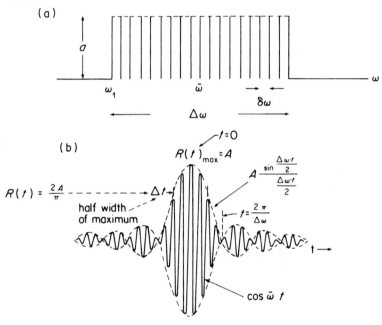

Fig. 4.14. (a) A rectangular wave pulse of width $\Delta\omega$ having n components of amplitude a with a common frequency difference $\delta\omega$. (b) Representation of the pulse on a time axis is a cosine curve at the average frequency $\bar\omega$, amplitude modulated by a $\sin\alpha/\alpha$ curve where $\alpha = \Delta\omega.t/2$. After a time $t = 2\pi/\Delta\omega$ the superposition of the components gives a zero amplitude

This expression gives us the time behaviour of the pulse and is displayed on a time axis in fig. 4.14b. We see that the amplitude $R(t)$ is given by the cosine curve of the average frequency $\bar\omega$ modified by the $A\sin\alpha/\alpha$ term.

At $t = 0$, $\sin\alpha/\alpha \to 1$ and all the components superpose with zero phase difference to give the maximum amplitude $R(t) = A = na$. After some time interval Δt when

$$\alpha = \frac{\Delta\omega\,\Delta t}{2} = \pi$$

the phases between the frequency components are such that the resulting amplitude $R(t)$ is zero.

The time Δt which is a measure of the width of the central pulse of fig. 4.14b is therefore given by

$$\frac{\Delta\omega\,\Delta t}{2} = \pi$$

or $\Delta\nu\,\Delta t = 1$ where $\Delta\omega = 2\pi\,\Delta\nu$.

The true width of the base of the central pulse is $2\Delta t$ but the interval Δt is taken as an arbitrary measure of time, centred about $t = 0$, during which the amplitude $R(t)$ remains significantly large $(>A/2)$. With this arbitrary definition the exact expression

$$\Delta\nu\,\Delta t = 1$$

becomes the approximation

$$\Delta\nu\,\Delta t \approx 1 \quad \text{or} \quad (\Delta\omega\,\Delta t \approx 2\pi)$$

and this approximation is known as the Bandwidth Theorem.

It states that the components of pulse of width $\Delta\omega$ in the frequency range will superpose to produce a significant amplitude $R(t)$ only for a time Δt before the pulse decays from random phase differences. The greater the range $\Delta\omega$ the shorter the period Δt.

Alternatively, the theorem states that a single pulse of time duration Δt is the result of the superposition of frequency components over the range $\Delta\omega$; the shorter the period Δt of the pulse the wider the range $\Delta\omega$ of the frequencies required to represent it.

When $\Delta\omega$ is zero we have a single frequency, the monochromatic wave which is therefore required (in theory) to have an infinitely long time span.

We have chosen to express our wave group in the two parameters of frequency and time (having a product of zero dimensions), but we may just as easily work in the other pair of parameters wave number k and distance x.

Replacing ω by k and t by x would define the length of the wave group as Δx in terms of the range of component wavelengths $\Delta(1/\lambda)$.

The Bandwidth Theorem then becomes

$$\Delta x\,\Delta k \approx 2\pi$$

or

$$\Delta x\,\Delta(1/\lambda) \approx 1 \quad \text{i.e.} \quad \Delta x \approx \lambda^2/\Delta\lambda$$

Note again that a monochromatic wave with $\Delta k = 0$ requires $\Delta x \to \infty$ that is, an infinitely long wavetrain.

In the wave group we have just considered the problem has been simplified by assuming all frequency components to have the same amplitude a. When this is not the case, the different values $a(\omega)$ are treated by Fourier methods as we shall see in Chapter 9.

We shall meet the ideas of this section several times in the course of this text, noting particularly that in modern physics the Bandwidth Theorem becomes Heisenberg's Uncertainty Principle.

(Problem 4.20)

Transverse Waves in a Periodic Structure

At the end of the chapter on coupled oscillations we discussed the normal transverse vibrations of n equal masses of separation a along a light string of length $(n+1)a$ under a tension T with both ends fixed. The equation of motion of the rth particle was found to be

$$m\ddot{y}_r = \frac{T}{a}(y_{r+1} + y_{r-1} - 2y_r)$$

and for n masses the frequencies of the normal modes of vibration were given by

$$\omega_s^2 = \frac{2T}{ma}\left(1 - \cos\frac{s\pi}{n+1}\right)$$

where $s = 1, 2, 3, \ldots n$. When the separation a becomes infinitesimally small ($=dx$, say) the term in the equation of motion

$$\frac{1}{a}(y_{r+1} + y_{r-1} - 2y_r) \rightarrow \frac{1}{dx}(y_{r+1} + y_{r-1} - 2y_r)$$

$$= \frac{(y_{r+1} - y_r)}{dx} - \frac{(y_r - y_{r-1})}{dx} = \left(\frac{\partial y}{\partial x}\right)_{r+1/2} - \left(\frac{\partial y}{\partial x}\right)_{r-1/2} = \left(\frac{\partial^2 y}{\partial x^2}\right)_r \cdot dx$$

so that the equation of motion becomes

$$\frac{\partial^2 y}{\partial t^2} = \frac{T}{\rho}\frac{\partial^2 y}{\partial x^2},$$

the wave equation, where $\rho = m/dx$, the linear density and

$$y \propto e^{i(\omega t - kx)}$$

We are now going to consider the propagation of transverse waves along a linear array of atoms, mass m, in a crystal lattice where the tension T now represents the elastic force between the atoms (so that T/a is the stiffness) and a, the separation between the atoms, is about 1 Å or 10^{-10} m. When the clamped ends of the string are replaced by the ends of the crystal we can express the displacement of the rth particle due to the transverse waves as

$$y_r = A_r\, e^{i(\omega t - kx)} = A_r\, e^{i(\omega t - kra)},$$

since $x = ra$. The equation of motion then becomes

$$-\omega^2 m = \frac{T}{a}(e^{ika} + e^{-ika} - 2)$$

$$= \frac{T}{a}(e^{ika/2} - e^{-ika/2})^2 = -\frac{4T}{a}\sin^2\frac{ka}{2}$$

giving the permitted frequencies

$$\omega^2 = \frac{4T}{ma}\sin^2\frac{ka}{2}$$

This expression for ω^2 is equivalent to our earlier value at the end of Chapter 3:

$$\omega_s^2 = \frac{2T}{ma}\left(1 - \cos\frac{s\pi}{n+1}\right) = \frac{4T}{ma}\sin^2\frac{s\pi}{2(n+1)}$$

if

$$\frac{ka}{2} = \frac{s\pi}{2(n+1)}$$

where $s = 1, 2, 3, \ldots n$.

But $(n+1)a = l$, the length of the string or crystal, and we have seen that wavelengths λ are allowed where $p\lambda/2 = l = (n+1)a$.

Thus

$$\frac{ka}{2} = \frac{2\pi}{\lambda}\cdot\frac{a}{2} = \frac{\pi a}{\lambda} = \frac{sa\pi}{2(n+1)a} = \frac{s}{p}\cdot\frac{\pi a}{\lambda}$$

if $s = p$. When $s = p$, a unit change in s corresponds to a change from one allowed number of half wavelengths to the next so that the minimum wavelength is $\lambda = 2a$, giving a maximum frequency $\omega_m^2 = 4T/ma$. Thus both expressions may be considered equivalent.

When $\lambda = 2a$, $\sin ka/2 = 1$ because $ka = \pi$, and neighbouring atoms are exactly π radians out of phase because

$$\frac{y_r}{y_{r+1}} \propto e^{ika} = e^{i\pi} = -1$$

The highest frequency is thus associated with maximum coupling, as we expect.

For long wavelengths or low values of the wave number k, $\sin ka/2 \to ka/2$ so that

$$\omega^2 = \frac{4T}{ma}\frac{k^2 a^2}{4}$$

and the velocity of the wave is given by

$$c^2 = \frac{\omega^2}{k^2} = \frac{Ta}{m} = \frac{T}{\rho}$$

as before, where $\rho = m/a$.
In general the phase velocity is given by

$$v = \frac{\omega}{k} = c\left[\frac{\sin ka/2}{ka/2}\right]$$

a dispersion relation which is shown in fig. 4.15. Only at very short wavelengths does the atomic spacing of the crystal structure affect the wave propagation, and here the limiting or maximum value of the wave number $k_m = \pi/a \approx 10^{10}\,\text{m}^{-1}$.

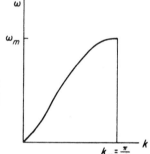

Fig. 4.15. The dispersion relation $\omega(k)$ versus k for waves travelling along a linear one-dimensional array of atoms in a periodic structure

The elastic force constant T/a for a crystal is about 15 newtons metre^{-1}; a typical 'reduced' atomic mass is about 60×10^{-27} kg. These values give a maximum frequency

$$\omega^2 = \frac{4T}{ma} \approx \frac{60}{60 \times 10^{-27}} = 10^{27}\,(\text{rad/s})^2$$

that is, a frequency $\nu \approx 5 \times 10^{12}$ Hertz.

(Note that the value of T/a used here for the crystal is a factor of 8 lower than that found in problem 3.4 for a single molecule. This is due to the interaction between neighbouring ions and the change in their equilibrium separation.)

This frequency is in the infrared region of the electromagnetic spectrum. We shall see in a later chapter that electromagnetic waves of frequency ω have a transverse electric field vector $E = E_0\,e^{i\omega t}$, where E_0 is the maximum amplitude, so that charged atoms or ions in a crystal lattice could respond as forced oscillators to radiation falling upon the crystal, which would absorb any radiation at the resonant frequency of its oscillating atoms.

Linear Array of Two Kinds of Atoms in an Ionic Crystal

We continue the discussion of this problem using a one dimensional line which contains two kinds of atoms with separation a as before, those atoms of mass M occupying the odd numbered positions, $2r-1, 2r+1$, etc., and those of mass m occupying the even numbered positions, $2r, 2r+2$, etc. The equations of motion for each type are

$$m\ddot{y}_{2r} = \frac{T}{a}(y_{2r+1} + y_{2r-1} - 2y_{2r})$$

and

$$M\ddot{y}_{2r+1} = \frac{T}{a}(y_{2r+2} + y_{2r} - 2y_{2r+1})$$

with solutions

$$y_{2r} = A_m\, e^{i(\omega t - 2rka)}$$
$$y_{2r+1} = A_M\, e^{i(\omega t - (2r+1)ka)}$$

where A_m and A_M are the amplitudes of the respective masses.
   The equations of motion thus become

$$-\omega^2 m A_m = \frac{TA_M}{a}(e^{-ika} + e^{ika}) - \frac{2TA_m}{a}$$

and

$$-\omega^2 M A_M = \frac{TA_m}{a}(e^{-ika} + e^{ika}) - \frac{2TA_M}{a}$$

equations which are consistent when

$$\omega^2 = \frac{T}{a}\left(\frac{1}{m} + \frac{1}{M}\right) \pm \frac{T}{a}\left[\left(\frac{1}{m} + \frac{1}{M}\right)^2 - \frac{4\sin^2 ka}{mM}\right]^{1/2}$$

Plotting the dispersion relation ω versus k for the positive sign and $m > M$ gives the upper curve of fig. 4.16 with

$$\omega^2 = \frac{2T}{a}\left(\frac{1}{m} + \frac{1}{M}\right) \quad \text{for} \quad k = 0$$

and

$$\omega^2 = \frac{2T}{aM} \quad \text{for} \quad k_m = \frac{\pi}{2a} \;(\text{minimum } \lambda = 4a)$$

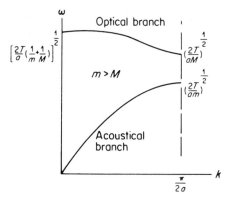

Fig. 4.16. Dispersion relations for the two modes of transverse oscillation in a crystal structure

The negative sign gives the lower curve of fig. 4.16 with

$$\omega^2 = \frac{2Tk^2a^2}{a(M+m)} \quad \text{for very small } k$$

and

$$\omega^2 = \frac{2T}{am} \quad \text{for} \quad k = \frac{\pi}{2a}$$

The upper curve is called the 'optical' branch and the lower curve is known as the 'acoustical' branch. The motions of the two types of atom for each branch are shown in fig. 4.17.

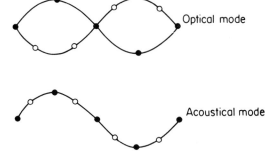

Fig. 4.17. The displacements of the different atomic species in the two modes of transverse oscillations in a crystal structure (a) the optical mode, and (b) the acoustic mode

In the optical branch for long wavelengths and small k, $A_m/A_M = -M/m$, and the atoms vibrate against each other, so that the centre of mass of the unit cell in the crystal remains fixed. This motion can be generated by the action of an electromagnetic wave when alternate atoms are ions of opposite charge; hence the name 'optical branch'. In the acoustic branch, long wavelengths and small k give $A_m = A_M$, and the atoms and their centre of mass move together (as in longitudinal sound waves). We shall see in the next chapter that the atoms may also vibrate in a longitudinal wave.

The transverse waves we have just discussed are polarized in one plane; they may also vibrate in a plane perpendicular to the plane considered here. The vibrational energy of these two transverse waves, together with that of the longitudinal wave to be discussed in the next chapter, form the basis of the theory of the specific heats of solids, a topic to which we shall return in Chapter 8.

Absorption of Infrared Radiation by Ionic Crystals

Radiation of frequency 3×10^{12} Hertz gives an infrared wavelength of 100 microns (10^{-4} m) and a wave number $k = 2\pi/\lambda \approx 6.10^4$ m^{-1}. We found the cut-off frequency in the crystal lattice to give a wave number $k_m \approx 10^{10}$ m^{-1}, so that the k value of infrared radiation is a negligible quantity relative to k_m and may be taken as zero. When the ions of opposite charge $\pm e$ move under the influence of the electric field vector $E = E_0 \, e^{i\omega t}$ of electromagnetic radiation, the equations of motion (with $k = 0$) become

$$-\omega^2 m A_m = \frac{2T}{a}(A_M - A_m) - eE_0$$

and

$$-\omega^2 M A_M = \frac{-2T}{a}(A_M - A_m) + eE_0$$

which may be solved to give

$$A_M = \frac{eE_0}{M(\omega_0^2 - \omega^2)} \quad \text{and} \quad A_m = \frac{-e}{m} \frac{E_0}{(\omega_0^2 - \omega^2)}$$

where

$$\omega_0^2 = \frac{2T}{a}\left(\frac{1}{m} + \frac{1}{M}\right)$$

the low k limit of the optical branch.

Thus when $\omega = \omega_0$ infrared radiation is strongly absorbed by ionic crystals and the ion amplitudes A_M and A_m increase. Experimentally, sodium chloride is found to absorb strongly at $\lambda = 61$ microns; potassium chloride has an absorption maximum at $\lambda = 71$ microns. (**Problem 4.21.**)

Doppler Effect

In the absence of dispersion the velocity of waves sent out by a moving source is constant but the wavelength and frequency noted by a stationary observer are altered.

In fig. 4.18 a stationary source S emits a signal of frequency ν and wavelength λ for a period t so the distance to a stationary observer O is $\nu\lambda t$. If

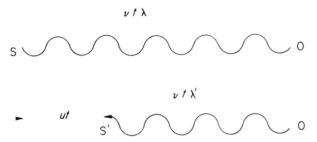

Fig. 4.18. If waves from a stationary source S are received by a stationary observer O at frequency ν and wavelength λ the frequency is observed as ν' and the wavelength as λ' at O if the source S' moves during transmission. This is the Doppler effect

the source S' moves towards O at a velocity u during the period t then O registers a new frequency ν'.

We see that

$$\nu\lambda t = ut + \nu\lambda' t$$

which, for

$$c = \nu\lambda = \nu'\lambda'$$

gives

$$\frac{c-u}{\nu} = \lambda' = \frac{c}{\nu'}$$

Hence

$$\nu' = \frac{\nu c}{c-u}$$

This observed change of frequency is called the *Doppler Effect*.

Suppose that the source S is now stationary but that an observer O' moves with a velocity v away from S. If we superimpose a velocity $-v$ on observer, source and waves, we bring the observer to rest; the source now has a velocity $-v$ and waves a velocity of $c - v$.

Using these values in the expression for ν' gives a new observed frequency

$$\nu'' = \frac{\nu(c-v)}{c}$$

(Problems 4.22, 4.23, 4.24, 4.25, 4.26, 4.27, 4.28, 4.29, 4.30, 4.31)

Problem 4.1
Show that $y = f_2(ct + x)$ is a solution of the wave equation

$$\frac{\partial^2 y}{\partial x^2} = \frac{1}{c^2}\frac{\partial^2 y}{\partial t^2}$$

Problem 4.2
Show that the wave profile, that is,

$$y = f_1(ct - x)$$

remains unchanged with time when c is the wave velocity. To do this consider the expression for y at a time $t + \Delta t$ where $\Delta t = \Delta x/c$.

Repeat the problem for $y = f_2(ct + x)$.

Problem 4.3
Show that

$$\frac{\partial y}{\partial t} = +c\frac{\partial y}{\partial x}$$

for a left going wave drawing a diagram to show the particle velocities as in fig. 4.5 (note that c is a magnitude and does not change sign).

Problem 4.4
A triangular shaped pulse of length l is reflected at the fixed end of the string on which it travels ($Z_2 = \infty$). Sketch the shape of the pulse (see fig. 4.8) after a length (a) $l/4$ (b) $l/2$ (c) $3l/4$ and (d) l of the pulse has been reflected.

Problem 4.5
A point mass M is concentrated at a point on a string of characteristic impedance ρc. A transverse wave of frequency ω moves in the positive x direction and is partially reflected and transmitted at the mass. The boundary conditions are that the string displacements just to the left and right of the mass are equal ($y_i + y_r = y_t$) and that the difference between the transverse forces just to the left and right of the mass equal the mass times its acceleration. If A_1, B_1 and A_2 are respectively the incident, reflected and transmitted wave amplitudes show that

$$\frac{B_1}{A_1} = \frac{-iq}{1+iq} \quad \text{and} \quad \frac{A_2}{A_1} = \frac{1}{1+iq}$$

where $q = \omega M/2\rho c$ and $i^2 = -1$.

Problem 4.6
In problem 4.5, writing $q = \tan \theta$, show that A_2 lags A_1 by θ and that B_1 lags A_1 by $(\pi/2 + \theta)$ for $0 < \theta < \pi/2$.

Show also that the reflected and transmitted energy coefficients are represented by $\sin^2 \theta$ and $\cos^2 \theta$ respectively.

Problem 4.7

If the wave on the string in fig. 4.6 propagates with a displacement

$$y = a \sin(\omega t - kx)$$

Show that the average rate of working by the force (average value of transverse force times transverse velocity) equals the rate of energy transfer along the string.

Problem 4.8

A transverse harmonic force of peak value 0·3 Newtons and frequency 5 Hertz initiates waves at one end of a very long string of linear density 0·01 kg/metre. Show that the rate of energy transfer along the string is $3\pi/20$ watts and that the wave velocity is $30/\pi$ metres per sec.

Problem 4.9

In the figure, media of impedances Z_1 and Z_3 are separated by a medium of intermediate impedance Z_2. A normally incident wave in the first medium has unit amplitude and the reflection and transmission coefficients for multiple reflections are shown. Show that the total reflected amplitude in medium 1 which is

$$R + tTR'(1 + rR' + r^2 R'^2 \ldots)$$

is zero if $R = -R'$ and show that this defines the condition

$$Z_2^2 = Z_1 Z_3$$

(Note that for zero total reflection in medium 1, the first reflection R is cancelled by the sum of all subsequent reflections.)

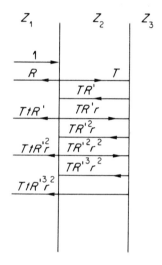

Problem 4.10

The relation between the impedance Z and the refractive index n of a dielectric is given by $Z = 1/n$. Light travelling in free space enters a glass lens which has a refractive index of 1·5 for a free space wavelength of $5·5 \times 10^{-7}$ metres. Show that reflections at this wavelength are avoided by a coating of refractive index 1·22 and thickness $1·12 \times 10^{-7}$ metres.

Problem 4.11

Prove that the displacement y_n of the standing wave expression in equation (4.10) satisfies the time independent form of the wave equation

$$\frac{\partial^2 y}{\partial x^2} + k^2 y = 0.$$

Problem 4.12

The total energy E_n of a normal mode may be found by an alternative method. Each section dx of the string is a simple harmonic oscillator with total energy equal to the maximum kinetic energy of oscillation

$$k.e._{\max} = \tfrac{1}{2}\rho \, dx (\dot{y}_n^2)_{\max} = \tfrac{1}{2}\rho \, dx \omega_n^2 (y_n^2)_{\max}$$

Now the value of $(y_n^2)_{max}$ at a point x on the string is given by

$$(y_n^2)_{max} = (A_n^2 + B_n^2) \sin^2 \frac{\omega_n x}{c}$$

Show that the sum of the energies of the oscillators along the string, that is, the integral

$$\tfrac{1}{2}\rho\omega_n^2 \int_0^l (y_n^2)_{max}\, dx$$

gives the expected result.

Problem 4.13

The displacement of a wave on a string which is fixed at both ends is given by

$$y(x, t) = A \cos (\omega t - kx) + rA \cos (\omega t + kx)$$

where r is the coefficient of amplitude reflection. Show that this may be expressed as the superposition of standing waves

$$y(x, t) = A(1 + r) \cos \omega t \cos kx + A(1 - r) \sin \omega t \sin kx.$$

Problem 4.14

A wave group consists of two wavelengths λ and $\lambda + \Delta\lambda$ where $\Delta\lambda/\lambda$ is very small.

Show that the number of wavelengths λ contained between two successive zeros of the modulating envelope is $\approx \lambda/\Delta\lambda$.

Problem 4.15

The phase velocity v of transverse waves in a crystal of atomic separation a is given by

$$v = c\left(\frac{\sin (ka/2)}{(ka/2)}\right)$$

where k is the wave number and c is constant. Show that the value of the group velocity is

$$c \cos \frac{ka}{2}$$

What is the limiting value of the group velocity for long wavelengths?

Problem 4.16

The dielectric constant of a gas at a wavelength λ is given by

$$\epsilon_r = \frac{c^2}{v^2} = A + \frac{B}{\lambda^2} - D\lambda^2$$

where A, B and D are constants, c is the velocity of light in free space and v is its phase velocity. If the group velocity is V_g show that

$$V_g \epsilon_r = v(A - 2D\lambda^2)$$

Problem 4.17

Problem 2.10 shows that the relative permittivity of an ionized gas is given by

$$\epsilon_r = \frac{c^2}{v^2} = 1 - \left(\frac{\omega_e}{\omega}\right)^2$$

where v is the phase velocity, c is the velocity of light and ω_e is the constant value of the electron plasma frequency. Show that this yields the dispersion relation $\omega^2 = \omega_e^2 + c^2 k^2$, and that as $\omega \rightarrow \omega_e$ the phase velocity exceeds that of light, c, but that the group velocity (the velocity of energy transmission) is always less than c.

Problem 4.18

The electron plasma frequency of problem 4.17 is given by

$$\omega_e^2 = \frac{n_e e^2}{m_e \epsilon_0}.$$

Show that for an electron number density $n_e \sim 10^{20}$ (10^{-5} of an atmosphere), electromagnetic waves must have wavelengths $\lambda < 3 \,.\, 10^{-3}$ m (in the microwave region) to propagate. These are typical wavelengths for probing thermonuclear plasmas at high temperatures.

$$\epsilon_0 = 8 \cdot 8 \times 10^{-12} \,\text{farads/metre}$$

$$m_e = 9 \cdot 1 \times 10^{-31} \,\text{kg}$$

$$e = 1 \cdot 6 \times 10^{-19} \,\text{coulombs}$$

Problem 4.19

In wave mechanics the dispersion relation for an electron of velocity $v = \hbar k/m$ is given by $\omega^2/c^2 = k^2 + m^2 c^2/\hbar^2$, where c is the velocity of light, m is the electron mass (considered constant at a given velocity) $\hbar = h/2\pi$ and h is Planck's constant. Show that the product of the group and particle velocities is c^2.

Problem 4.20

The figure shows a pulse of length Δt given by $y = A \cos \omega_0 t$.
    Show that the frequency representation

$$y(\omega) = a \cos \omega_1 t + a \cos (\omega_1 + \delta\omega)t \ldots + a \cos [\omega_1 + (n-1)(\delta\omega)]t$$

is centred on the average frequency ω_0 and that the range of frequencies making significant contributions to the pulse satisfy the criterion

$$\Delta\omega \, \Delta t \approx 2\pi$$

Repeat this process for a pulse of length Δx with $y = A \cos k_0 x$ to show that in k space the pulse is centred at k_0 with the significant range of wave numbers Δk satisfying the criterion $\Delta x \, \Delta k \approx 2\pi$.

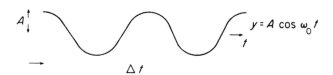

Problem 4.21

The elastic force constant for an ionic crystal is ~ 15 newtons/metre. Show that the experimental values for the frequencies of infrared absorption quoted at the end of this chapter for NaCl and KCl are in reasonable agreement with calculated values.

$$1 \text{ a.m.u.} = 1 \cdot 66 \times 10^{-27} \,\text{kg}$$

$$\text{Na mass} = 23 \text{ a.m.u.}$$

$$\text{K mass} = 39 \text{ a.m.u.}$$

$$\text{Cl mass} = 35 \text{ a.m.u.}$$

Problem 4.22

Show that, in the Doppler effect, the change of frequency noted by a stationary observer O as a moving source S' passes him is given by

$$\Delta \nu = \frac{2\nu c u}{(c^2 - u^2)}$$

where $c = \nu\lambda$, the signal velocity and u is the velocity of S'.

Problem 4.23

Suppose, in the Doppler effect, that a source S' and an observer O' move in the same direction with velocities u and v respectively. Bring the observer to rest by superimposing a velocity $-v$ on the system to show that O' now registers a frequency

$$\nu''' = \frac{\nu(c - v)}{(c - u)}$$

Problem 4.24

Light from a star of wavelength 6×10^{-7} metres is found to be shifted 10^{-11} metres towards the red when compared with the same wavelength from a laboratory source. If the velocity of light is 3×10^8 metres sec^{-1} show that the earth and the star are separating at a velocity of 5 kilometres per second.

Problem 4.25

An aircraft flying on a level course transmits a signal of 3×10^9 Hertz which is reflected from a distant point ahead on the flight path and received by the aircraft with a frequency difference of 15 kHertz. What is the aircraft speed.

Problem 4.26

Light from hot sodium atoms is centred about a wavelength of 6×10^{-7} metres but spreads 2×10^{-12} metres on either side of this wavelength due to the Doppler effect as radiating atoms move towards and away from the observer. Calculate the thermal velocity of the atoms to show that the gas temperature is ~ 900 K.

Problem 4.27

Show that in the Doppler effect when the source and observer are not moving in the same direction that the frequencies

$$\nu' = \frac{\nu c}{c - u}, \qquad \nu'' = \frac{\nu(c - v)}{c}$$

and

$$\nu''' = \nu\left(\frac{c - v}{c - u}\right)$$

are valid if u and v are not the actual velocities but the components of these velocities along the direction in which the waves reach the observer.

Problem 4.28

In extending the Doppler principle consider the accompanying figure where O is a stationary observer at the origin of the coordinate system $O(x, t)$ and O' is an observer situated at the origin of the system $O'(x', t')$ which moves with a constant velocity v in the x direction relative to the system O. When O and O' are coincident at $t = t' = 0$ a light source sends waves in the x direction with constant velocity c. These waves obey the relation

$$0 \equiv x^2 - c^2 t^2 \text{ (seen by } O) \equiv x'^2 - c^2 t'^2 \text{ (seen by } O'). \tag{1}$$

Since there is only one relative velocity v, the transformation

$$x' = k(x - vt) \tag{2}$$

and

$$x = k'(x' + vt') \tag{3}$$

must also hold. Use (2) and (3) to eliminate x' and t' from (1) and show that this identity is satisfied only by $k = k' = 1/(1 - \beta^2)^{1/2}$, where $\beta = v/c$. (Hint—in the identity of equation (1) equate coefficients of the variables to zero.).

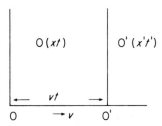

This is the Lorentz transformation in the theory of relativity giving

$$x' = \frac{(x - vt)}{(1 - \beta^2)^{1/2}}, \qquad x = \frac{x' + vt'}{(1 - \beta^2)^{1/2}}$$

$$t' = \frac{(t - (v/c^2)x)}{(1 - \beta^2)^{1/2}}, \qquad t = \frac{(t' + (v/c^2)x')}{(1 - \beta^2)^{1/2}}$$

Problem 4.29

Show that the interval $\Delta t = t_2 - t_1$ seen by O in problem 4.28 is seen as $\Delta t' = k \Delta t$ by O' and that the length $l = x_2 - x_1$ seen by O is seen by O' as $l' = l/k$.

Problem 4.30

Show that two simultaneous events at x_2 and $x_1 (t_2 = t_1)$ seen by O in the previous problems are not simultaneous when seen by O' (that is $t'_1 \neq t'_2$).

Problem 4.31

Show that the order of events seen by $O(t_2 > t_1)$ of the previous problems will not be reversed when seen by O' (that is $t'_2 > t'_1$) as long as the velocity of light c is the greatest velocity attainable.

Summary of Important Results

Wave Equation $\dfrac{\partial^2 y}{\partial x^2} = \dfrac{1}{c^2} \dfrac{\partial^2 y}{\partial t^2}$

Wave (phase) velocity $= c = \dfrac{\omega}{k} = \dfrac{\partial x}{\partial t}$

k = wave number $= \dfrac{2\pi}{\lambda}$

where the wavelength λ defines separation between two oscillations with phase difference of 2π radians.

Particle velocity $\dfrac{\partial y}{\partial t} = -c\dfrac{\partial y}{\partial x}$

Displacement $y = a\, e^{i(\omega t - kx)}$

where a is wave amplitude.

Characteristic Impedance of a string

$$Z = \frac{\text{transverse force}}{\text{transverse velocity}} = -T\frac{\partial y}{\partial x}\Big/\frac{\partial y}{\partial t} = \rho c$$

Reflexion and Transmission Coefficients

$$\frac{\text{Reflected Amplitude}}{\text{Incident Amplitude}} = \frac{Z_1 - Z_2}{Z_1 + Z_2}$$

$$\frac{\text{Transmitted Amplitude}}{\text{Incident Amplitude}} = \frac{2Z_1}{Z_1 + Z_2}$$

$$\frac{\text{Reflected Energy}}{\text{Incident Energy}} = \left(\frac{Z_1 - Z_2}{Z_1 + Z_2}\right)^2$$

$$\frac{\text{Transmitted Energy}}{\text{Incident Energy}} = \frac{4Z_1 Z_2}{(Z_1 + Z_2)^2}$$

Impedance Matching
Impedances Z_1 and Z_3 are matched by insertion of impedance Z_2 where $Z_2^2 = Z_1 Z_3$
Thickness of Z_2 is $\lambda/4$ measured in Z_2.

Standing waves. Normal Modes. Harmonics.

Solution of wave equation separates time and space dependence to satisfy time independent wave equation

$$\frac{\partial^2 y}{\partial x^2} + k^2 y = 0 \qquad (\text{cancel } e^{i\omega t})$$

Standing waves on string of length l have wavelength λ_n where

$$n\frac{\lambda_n}{2} = l$$

Displacement of nth harmonic is

$$y_n = (A_n \cos \omega_n t + B_n \sin \omega_n t) \sin \frac{\omega_n x}{c}$$

Energy of nth harmonic (string mass m)

$$E_n = KE_n + PE_n = \tfrac{1}{4} m \omega_n^2 (A_n^2 + B_n^2)$$

Group Velocity

In a dispersive medium the wave velocity v varies with frequency ω (wave number k). The energy of a group of such waves travels with the group velocity

$$v_g = \frac{d\omega}{dk} = v + \frac{k}{dk}\frac{dv}{dk} = v - \lambda \frac{dv}{d\lambda}$$

Rectangular Wave Group of n components amplitude a, width $\Delta\omega$, represented in time by

$$R(t) = a \cdot \frac{\sin(\Delta\omega \cdot t/2)}{\sin (\Delta\omega \cdot t/n \cdot 2)} \cos \bar{\omega} t$$

where $\bar{\omega}$ is average frequency. $R(t)$ is zero when

$$\frac{\Delta\omega \cdot t}{2} = \pi$$

i.e. *Bandwidth Theorem* gives

$$\Delta\omega \cdot \Delta t = 2\pi$$

or

$$\Delta x \Delta k = 2\pi$$

A pulse of duration Δt requires a frequency width $\Delta\omega$ to define it in frequency space and vice versa.

Doppler Effect

Signal of frequency ν and velocity c transmitted by a stationary source S and received by a stationary observer O becomes

$$\nu' = \frac{\nu c}{c - u}$$

when source is no longer stationary but moves towards O with a velocity u.

Chapter 5

Longitudinal Waves

In deriving the wave equation

$$\frac{\partial^2 y}{\partial x^2} = \frac{1}{c^2}\frac{\partial^2 y}{\partial t^2}$$

in Chapter 4, we used the example of a transverse wave and continued to discuss waves of this type on a vibrating string. In this chapter we consider longitudinal waves, waves in which the particle or oscillator motion is in the same direction as the wave propagation. Longitudinal waves propagate as sound waves in all phases of matter, plasmas, gases, liquids and solids, but we shall concentrate on gases and solids. In the case of gases, limitations of thermodynamic interest are imposed; in solids the propagation will depend on the dimensions of the medium. Neither a gas nor a liquid can sustain the transverse shear necessary for transverse waves, but a solid can maintain both longitudinal and transverse oscillations.

Sound Waves in Gases

Let us consider a fixed mass of gas, which at a pressure P_0 occupies a volume V_0 with a density ρ_0. These values define the equilibrium state of the gas which is disturbed, or deformed, by the compressions and rarefactions of the sound waves. Under the influence of the sound waves

the pressure P_0 becomes $P = P_0 + p$

the volume V_0 becomes $V = V_0 + v$

and

the density ρ_0 becomes $\rho = \rho_0 + \rho_d$.

The excess pressure p is the maximum pressure amplitude of the sound wave and is an alternating component superimposed on the equilibrium gas pressure P_0.

The fractional change in volume is called the *dilatation*, written $v/V_0 = \delta$, and the fractional change of density is called the *condensation*, written $\rho_d/\rho_0 =$

144

s. The values of δ and *s* are $\approx 10^{-3}$ for ordinary sound waves, and a value of $p = 2 \times 10^{-5} \, \text{n/m}^2$ (about 10^{-10} of an atmosphere) gives a sound wave which is still audible at 1000 Hertz. Thus the changes in the medium due to sound waves are of an extremely small order and define limitations within which the wave equation is appropriate.

The fixed mass of gas is equal to

$$\rho_0 V_0 = \rho V = \rho_0 V_0 (1 + \delta)(1 + s)$$

so that $(1 + \delta)(1 + s) = 1$, giving $s = -\delta$ to a very close approximation. The elastic property of the gas, a measure of its compressibility, is defined in terms of its *bulk modulus*

$$B = -\frac{dP}{dV/V} = -V\frac{dP}{dV}$$

the difference in pressure for a fractional change in volume, a volume increase with fall in pressure giving the negative sign. The value of B depends on whether the changes in the gas arising from the wave motion are adiabatic or isothermal. They must be thermodynamically reversible in order to avoid the energy loss mechanisms of diffusion, viscosity and thermal conductivity. The complete absence of these random, entropy generating processes defines an adiabatic process, a thermodynamic cycle with a 100% efficiency in the sense that none of the energy in the wave, potential or kinetic, is lost. In a sound wave such thermodynamic concepts restrict the excess pressure amplitude; too great an amplitude raises the local temperature in the gas at the amplitude peaks and thermal conductivity removes energy from the wave system. Local particle velocity gradients will also develop, leading to diffusion and viscosity.

Using a constant value of the adiabatic bulk modulus limits sound waves to small oscillations since the total pressure $P = P_0 + p$ is taken as constant; larger amplitudes lead to non-linear effects and shock waves, which we shall discuss separately in Chapter 11.

All adiabatic changes in the gas obey the relation $PV^\gamma = \text{constant}$, where γ is the ratio of the specific heats at constant pressure and volume respectively.

Differentiation gives

$$V^\gamma \, dP + \gamma P V^{\gamma - 1} \, dV = 0$$

or

$$-V\frac{dP}{dV} = \gamma P = B_a \text{ (where the subscript } a \text{ denotes adiabatic)}$$

so that the eleastic property of the gas is γP, considered to be constant. Since $P = P_0 + p$, then $dP = p$, the excess pressure, giving

$$B_a = -\frac{p}{v/V_0} \quad \text{or} \quad p = -B_a\delta = B_a s$$

In a sound wave the particle displacements and velocities are along the x-axis and we choose the co-ordinate η to define the displacement.

In obtaining the wave equation we consider the motion of an element of the gas of infinitesimal thickness dx and unit cross section. Under the influence of the sound wave the behaviour of this element is shown in fig. 5.1. The particles in the layer x are displaced a distance η and those at $x + dx$ are displaced a distance $\eta + d\eta$, so that the increase in the thickness dx of the element of unit

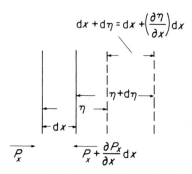

Fig. 5.1. Thin element of gas of unit cross-section and thickness dx displaced an amount η and expanded by an amount $(\partial\eta/\partial x)\,dx$ under the influence of a pressure difference $-(\partial P_x/\partial x)\,dx$

cross section (which therefore measures the increase in volume) is

$$d\eta = \frac{\partial \eta}{\partial x}dx$$

and

$$\delta = \frac{v}{V_0} = \left(\frac{\partial \eta}{\partial x}\right) dx/dx = \frac{\partial \eta}{\partial x} = -s$$

where $\partial\eta/\partial x$ is called the *strain.*

The medium is deformed because the pressures along the x-axis on either side of the thin element are not in balance (fig. 5.1). The net force acting on the element is given by

$$P_x - P_{x+dx} = \left[P_x - \left(P_x + \frac{\partial P_x}{\partial x}dx \right) \right]$$

$$= -\frac{\partial P_x}{\partial x}dx = -\frac{\partial}{\partial x}(P_0 + p)\,dx = -\frac{\partial p}{\partial x}dx$$

The mass of the element is $\rho_0\,dx$ and its acceleration is given, to a close approximation, by $\partial^2\eta/dt^2$.

From Newton's Law we have

$$-\frac{\partial p}{\partial x}dx = \rho_0\,dx\frac{\partial^2\eta}{\partial t^2}$$

where

$$p = -B_a\delta = -B_a\frac{\partial\eta}{\partial x}$$

so that

$$-\frac{\partial p}{\partial x} = B_a\frac{\partial^2\eta}{\partial x^2}, \quad\text{giving}\quad B_a\frac{\partial^2\eta}{\partial x^2} = \rho_0\frac{\partial^2\eta}{\partial t^2}$$

But $B_a/\rho_0 = \gamma P/\rho_0$ is the ratio of the elasticity to the inertia or density of the gas, and this ratio has the dimensions

$$\frac{\text{force}}{\text{area}}\cdot\frac{\text{volume}}{\text{mass}} = (\text{velocity})^2, \quad\text{so}\quad \frac{\gamma P}{\rho_0} = c^2$$

where c is the sound wave velocity.

Thus

$$\frac{\partial^2\eta}{\partial x^2} = \frac{1}{c^2}\frac{\partial^2\eta}{\partial t^2}$$

is the wave equation. Writing η_m as the maximum amplitude of displacement we have the following expressions for a wave in the *positive x-direction*:

$$\eta = \eta_m\,e^{i(\omega t - kx)} \qquad \dot{\eta} = \frac{\partial\eta}{\partial t} = i\omega\eta$$

$$\delta = \frac{\partial\eta}{\partial x} = -ik\eta = -s \quad (\text{so } s = ik\eta)$$

$$p = B_a s = iB_a k\eta$$

The phase relationships between these parameters (fig. 5.2a) show that when the wave is in the positive x-direction, the excess pressure p, the fractional density increase s and the particle velocity $\dot{\eta}$ are all $\pi/2$ radians in phase ahead of the displacement η, whilst the volume change (π radians out of phase with the density change) is $\pi/2$ radians behind the displacement. These relationships no longer hold when the wave direction is reversed (fig. 5.2b); for a *wave in the negative x-direction*

$$\eta = \eta_m\,e^{i(\omega t + kx)} \qquad \dot{\eta} = \frac{\partial\eta}{\partial t} = i\omega\eta$$

$$\delta = \frac{\partial\eta}{\partial x} = ik\eta = -s \quad (\text{so } s = -ik\eta)$$

$$p = B_a s = -iB_a k\eta$$

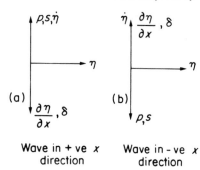

Wave in + ve *x*   Wave in - ve *x*
direction   direction

Fig. 5.2. Phase relationships between the particle displacement η, particle velocity $\dot{\eta}$, excess pressure p and condensation $s = -\delta$ (the dilatation) for waves travelling in the positive and negative x directions. The displacement η is taken in the positive x direction for both waves

In both waves the particle displacement η is measured in the positive x-direction and the thin element dx of the gas oscillates about the value $\eta = 0$, which defines its central position. For a wave in the positive x-direction the value $\eta = 0$, with $\dot{\eta}$ a maximum in the positive x-direction, gives a *maximum positive excess pressure (compression)* with a maximum condensation s (maximum density) and a minimum volume. For a wave in the negative x-direction, the same value $\eta = 0$, with $\dot{\eta}$ a maximum in the positive x-direction, gives a *maximum negative excess pressure (rarefaction)*, a maximum volume and a minimum density. To produce a compression in a wave moving in the negative x-direction the particle velocity $\dot{\eta}$ must be a maximum in the negative x-direction at $\eta = 0$. This distinction is significant when we are defining the impedance of the medium to the waves. A change of sign is involved with a change of direction—a convention we shall also have to follow when discussing the waves of Chapters 6 and 7.

Energy distribution in Sound Waves

The kinetic energy in the sound wave is found by considering the motion of the individual gas elements of thickness dx.

Each element will have a kinetic energy

$$\Delta E_{kin} = \tfrac{1}{2}\rho_0 \, dx \, \dot{\eta}^2$$

where $\dot{\eta}$ will depend upon the position x of the element. The average value of the kinetic energy density is found by taking the value of $\dot{\eta}^2$ averaged over a region of n wavelengths.

Now

$$\dot{\eta} = \dot{\eta}_m \sin \frac{2\pi}{\lambda}(ct - x)$$

so that

$$\overline{\dot{\eta}^2} = \frac{\dot{\eta}_m^2 \int_0^{n\lambda} \sin^2 2\pi(ct - x)/\lambda \ dx}{n\lambda} = \tfrac{1}{2}\dot{\eta}_m^2$$

so that the *average kinetic energy density* in the medium is

$$\overline{\Delta E}_{\text{kin}} = \tfrac{1}{4}\rho_0\dot{\eta}_m^2 = \tfrac{1}{4}\rho_0\omega^2\eta_m^2$$

(a simple harmonic oscillator of maximum amplitude a has an average kinetic energy over one cycle of $\tfrac{1}{4}m\omega^2 a^2$).

The potential energy density is found by considering the work $p \, dV$ done on the fixed mass of gas of volume V_0 during the adiabatic changes in the sound wave. This work is expressed as

$$\Delta E_{\text{pot}} = -\int p \, dV$$

where the negative sign shows that the potential energy change is positive in both a compression (p positive, dV negative) and a rarefaction (p negative, dV positive) fig. 5.3.

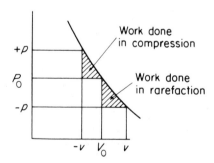

Fig. 5.3. Shaded triangles show that potential energy $pv/2$ gained by gas in compression equals that gained in rarefaction when both p and v change sign

The condensation

$$s = -\frac{\int dV}{V_0}$$

where $\int \mathrm{d}V = v$ the incremental change in the original volume V_0 so

$$\mathrm{d}V = - V_0 \,\mathrm{d}s$$

which, with

$$p = B_a s$$

gives

$$\Delta E_{\mathrm{pot}} = - \int p \,\mathrm{d}V = \int_0^s B_a s V_0 \,\mathrm{d}s$$

$$= \tfrac{1}{2} B_a s^2 \,\mathrm{d}x = \tfrac{1}{2} B_a \delta^2 \,\mathrm{d}x$$

where $s = -\delta$ and the thickness $\mathrm{d}x$ of the element of unit cross section represents its volume V_0.

Now

$$\eta = \eta_m \, \mathrm{e}^{\mathrm{i}(\omega t \pm kx)}$$

so that

$$\delta = \frac{\partial \eta}{\partial x} = \pm \frac{1}{c} \frac{\partial \eta}{\partial t}, \quad \text{where } c = \frac{\omega}{k}$$

Thus

$$\Delta E_{\mathrm{pot}} = \tfrac{1}{2} \frac{B_a}{c^2} \dot{\eta}^2 \,\mathrm{d}x = \tfrac{1}{2} \rho_0 \dot{\eta}^2 \,\mathrm{d}x$$

and its average value over $n\lambda$ gives the potential energy density

$$\overline{\Delta E_{\mathrm{pot}}} = \tfrac{1}{4} \rho_0 \dot{\eta}_m^2$$

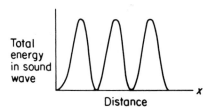

Total
energy
in sound
wave

Distance

Fig. 5.4. Energy distribution is space for a sound wave in a gas. Both potential and kinetic energies are at a maximum when the particle velocity $\dot{\eta}$ is a maximum and zero at $\dot{\eta} = 0$

We see that the average values of the kinetic and potential energy density in the sound wave are equal, but more important, since the value of each for the element $\mathrm{d}x$ is $\tfrac{1}{2} \rho_0 \dot{\eta}^2 \,\mathrm{d}x$, we observe that the element possesses maximum (or minimum) potential and kinetic energy at the same time. A compression or rarefaction produces a maximum in the energy of the element since the value $\dot{\eta}$ governs the energy content. Thus the energy in the wave is distributed in the wave system with distance as shown in fig. 5.4.

Intensity of Sound Waves

This is a measure of the energy flux, the rate at which energy crosses unit area, so that it is the product of the energy density (kinetic plus potential) and the wave velocity c. Normal sound waves range in intensity between 10^{-12} and 1 watt per square metre, extremely low levels which testify to the sensitivity of the ear. The roar of a large football crowd greeting a goal will just about heat a cup of coffee.

The intensity may be written

$$I = \tfrac{1}{2}\rho_0 c \dot{\eta}_m^2 = \tfrac{1}{2}\rho_0 c \omega^2 \eta_m^2 = \rho_0 c \dot{\eta}_{rms}^2 = p_{rms}^2 / \rho_0 c = p_{rms} \dot{\eta}_{rms}$$

A commonly used standard of sound intensity is given by

$$I_0 = 10^{-2} \text{ watts/metre}^2$$

which is about the level of the average conversational tone between two people standing next to each other. Shouting at this range raises the intensity by a factor of 100 and in the range $100 I_0$ to $1000 I_0$ (10 watts/m^2) the sound is painful.

Whenever the sound intensity increases by a factor of 10 it is said to have increased by 1 bel so the dynamic range of the ear is about 12 bels. An intensity increase by a factor of

$$10^{0.1} = 1 \cdot 26$$

increases the intensity by 1 decibel (1 db), a change of loudness which is just detected by a person with good hearing.

We see that the product $\rho_0 c$ appears in most of the expressions for the intensity; its significance becomes apparent when we define the impedance of the medium to the waves as the

$$\textit{Specific Acoustic Impedance} = \frac{\text{excess pressure}}{\text{particle velocity}} = \frac{p}{\dot{\eta}}$$

(the ratio of a force per unit area to a velocity).

Now, for a wave in the positive x-direction,

$$p = B_a s = i B_a k \eta \quad \text{and} \quad \dot{\eta} = i\omega\eta, \text{ so that,}$$

$$\frac{p}{\dot{\eta}} = \frac{B_a k}{\omega} = \frac{B_a}{c} = \rho_0 c$$

Thus the acoustic impedance presented by the medium to these waves, as in the case of the transverse waves on the string, is given by the product of the density and the wave velocity and is governed by the elasticity and inertia of the medium. For a wave in the negative x-direction, the specific acoustic impedance

$$\frac{p}{\dot{\eta}} = -\frac{iB_a k\eta}{i\omega\eta} = -\rho_0 c$$

with a change of sign because of the changed phase relationship.

The units of $\rho_0 c$ are normally stated as $\text{kg m}^{-2} \text{sec}^{-1}$ in books on practical acoustics; in these units air has a specific acoustic impedance value of ~ 400, water a value of $1\cdot45 \times 10^6$ and steel a value of $3\cdot9 \times 10^7$. These values will become more significant when we use them later in examples on the reflexion and transmission of sound waves.

Although the specific acoustic impedance $\rho_0 c$ is a real quantity for plane sound waves, it has an added reactive component ik/r for spherical waves, where r is the distance travelled by the wavefront. This component tends to zero with increasing r as the spherical wave becomes effectively plane.

(Problems 5.1, 5.2, 5.3, 5.4, 5.5, 5.6, 5.7, 5.8)

Longitudinal Waves in a Solid

The velocity of longitudinal waves in a solid depends upon the dimensions of the specimen in which the waves are travelling. If the solid is a thin bar of finite cross section the analysis for longitudinal waves in a gas is equally valid, except that the bulk modulus B_a is replaced by Young's modulus Y, the ratio of the longitudinal stress in the bar to its longitudinal strain.

The wave equation is then

$$\frac{\partial^2 \eta}{\partial x^2} = \frac{1}{c^2}\frac{\partial^2 \eta}{\partial t^2}, \quad \text{with } c^2 = \frac{Y}{\rho}$$

A longitudinal wave in a medium compresses the medium and distorts it laterally. Because a solid can develop a shear force in any direction, such a lateral distortion is accompanied by a transverse shear. The effect of this upon the wave motion in solids of finite cross section is quite complicated and has been ignored in the very thin specimen above. In bulk solids, however, the longitudinal and transverse modes may be considered separately.

We have seen that the longitudinal compression produces a strain $\partial\eta/\partial x$; the accompanying lateral distortion produces a strain $\partial\beta/\partial y$ (of opposite sign to $\partial\eta/\partial x$ and perpendicular to the x-direction).

Here β is the displacement in the y-direction and is a function of both x and y. The ratio of these strains

$$-\frac{\partial\beta}{\partial y} \bigg/ \frac{\partial\eta}{\partial x} = \sigma$$

is known as Poisson's ratio and is expressed in terms of Lamé's elastic constants λ and μ for a solid as

$$\sigma = \frac{\lambda}{2(\lambda + \mu)} \quad \text{where} \quad \lambda = \frac{\sigma Y}{(1+\sigma)(1-2\sigma)}$$

These constants are always positive, so that $\sigma < \frac{1}{2}$, and is commonly $\approx \frac{1}{3}$. In terms of these constants Young's modulus becomes

$$Y = (\lambda + 2\mu - 2\lambda\sigma)$$

The constant μ is the transverse coefficient of rigidity, that is, the ratio of the transverse stress to the transverse strain. It plays the role of the elasticity in the propagation of pure transverse waves in a bulk solid which Young's modulus plays for longitudinal waves in a thin specimen. Fig 5.5 illustrates the shear in a

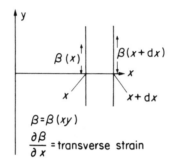

Fig. 5.5. Shear in a bulk solid producing a transverse wave. The transverse shear strain is $\partial\beta/\partial x$ and the transverse shear stress is $\mu \ \partial\beta/\partial x$, where μ is the shear modulus of rigidity

transverse plane wave, where the transverse strain is defined by $\partial\beta/\partial x$. The transverse stress at x is therefore $T_x = \mu \ \partial\beta/\partial x$. The equation of transverse motion of the thin element dx is then given by

$$T_{x+dx} - T_{dx} = \rho \ dx \ \ddot{y}$$

where ρ is the density, or

$$\frac{\partial}{\partial x}\left(\mu \frac{\partial\beta}{\partial x}\right) = \rho\ddot{y}$$

but $\ddot{y} = \partial^2\beta/\partial t^2$, hence

$$\frac{\partial^2\beta}{\partial x^2} = \frac{\rho}{\mu}\frac{\partial^2\beta}{\partial t^2}$$

the wave equation with a velocity given by $c^2 = \mu/\rho$.

The effect of the transverse rigidity μ is to stiffen the solid and increase the elastic constant governing the propagation of longitudinal waves. In a bulk solid the velocity of these waves is no longer given by $c^2 = Y/\rho$, but becomes

$$c^2 = \frac{\lambda + 2\mu}{\rho}$$

Since Young's modulus $Y = \lambda + 2\mu - 2\lambda\sigma$, the elasticity is increased by the amount $2\lambda\sigma \approx \lambda$, so that longitudinal waves in a bulk solid have a higher velocity than the same waves along a thin specimen.

In an isotropic solid, where the velocity of propagation is the same in all directions, the concept of a bulk modulus, used in the discussion on waves in gases, holds equally well. Expressed in terms of Lamé's elastic constants the bulk modulus for a solid is written

$$B = \lambda + \tfrac{2}{3}\mu = Y[3(1 - 2\sigma)]^{-1}$$

the longitudinal wave velocity for a bulk solid becomes

$$c_L = \left(\frac{B + (4/3)\mu}{\rho} \right)^{1/2}$$

whilst the transverse velocity remains as

$$c_T = \left(\frac{\mu}{\rho} \right)^{1/2}$$

Application to Earthquakes

The values of these velocities are well known for seismic waves generated by earthquakes. Near the surface of the earth the longitudinal waves have a velocity of 8 kilometres per second and the transverse waves travel at 4·45 kilometres per second. The velocity of the longitudinal waves increases with depth until, at a depth of about 1800 miles, no waves are transmitted because of a discontinuity and severe mismatch of impedances associated with the fluid core.

At the surface of the earth the transverse wave velocity is affected by the fact that stress components directed through the surface are zero there and these waves, known as Rayleigh Waves, travel with a velocity given by

$$c = f(\sigma) \left(\frac{\mu}{\rho} \right)^{1/2}$$

where

$$f(\sigma) = 0\cdot9194 \quad \text{when } \sigma = 0\cdot25$$

and

$$f(\sigma) = 0\cdot9553 \quad \text{when } \sigma = 0\cdot5$$

The energy of the Rayleigh Waves is confined to two dimensions; their amplitude is often much higher than that of the three dimensional longitudinal waves and therefore they are potentially more damaging.

In an earthquake the arrival of the fast longitudinal waves is followed by the Rayleigh Waves and then by a complicated pattern of reflected waves including those affected by the stratification of the earth's structure, known as Love Waves.

(Problem 5.9)

Longitudinal Waves in a Periodic Structure

Lamé's elastic constants, λ and μ, which are used to define such macroscopic quantities as Young's modulus and the bulk modulus, are themselves determined by forces which operate over interatomic distances. The discussion on transverse waves in a periodic structure has already shown that in a one-

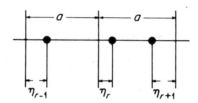

Fig. 5.6. Displacement of atoms in a linear array
due to a longitudinal wave in a crystal structure

dimensional array representing a crystal lattice a stiffness $s = T/a$ dyne cm^{-1} can exist between two atoms separated by a distance a.

When the waves along such a lattice are longitudinal the atomic displacements from equilibrium are represented by η (fig. 5.6). An increase in the separation between two atoms from a to $a + \eta$ gives a strain $\epsilon = \eta/a$, and a stress normal to the face area a^2 of a unit cell in a crystal equal to $s\eta/a^2 = s\epsilon/a$, a force per unit area.

Now Young's modulus is the ratio of this longitudinal stress to the longitudinal strain, so that $Y = s\epsilon/\epsilon a$ or $s = Ya$. The longitudinal vibration frequency of the atoms of mass m connected by stiffness constants s is given, very approximately by

$$\nu = \frac{\omega}{2\pi} = \frac{1}{2\pi}\sqrt{\frac{s}{m}} \approx \frac{1}{2\pi a}\sqrt{\frac{Y}{\rho}} \approx \frac{c_0}{2\pi a}$$

where $m = \rho a^3$ and c_0 is the velocity of sound in a solid. The value of $c_0 \approx 5 \times 10^3$ m sec^{-1}, and $a \approx 2 \times 10^{-10}$ m, so that $\nu \approx 3 \times 10^{12}$ Hertz, which is almost the same value as the frequency of the transverse wave in the infrared region of the electromagnetic spectrum. The highest ultrasonic frequency generated so far is about a factor of 10 lower than $v = c_0/2\pi a$. At frequencies $\approx 5 \times 10^{12}$ to 10^{13} Hertz many interesting experimental results must be expected. A more precise mathematical treatment yields the same equation of motion for the rth particle as in the transverse wave, namely

$$m\ddot{\eta}_r = s(\eta_{r+1} + \eta_{r-1} - 2\eta_r)$$

where $s = T/a$ and

$$\eta_r = \eta_{max}\, e^{i(\omega t - kra)}$$

The results are precisely the same as in the case of transverse waves and the shape of the dispersion curve is also similar. The maximum value of the cut-off frequency ω_m is, however, higher for the longitudinal than for the transverse waves. This is because the longitudinal elastic constant Y is greater than the transverse constant μ, that is, the force required for a given displacement in the longitudinal direction is greater than that for the same displacement in the transverse direction.

Reflexion and Transmission of Sound Waves at Boundaries

When a sound wave meets a boundary separating two media of different acoustic impedances two boundary conditions must be met in considering the reflexion and transmission of the wave. They are that

(i) the particle velocity $\dot{\eta}$

and

(ii) the acoustic excess pressure p

are both continuous across the boundary. Physically this ensures that the two media are in complete contact everywhere across the boundary.

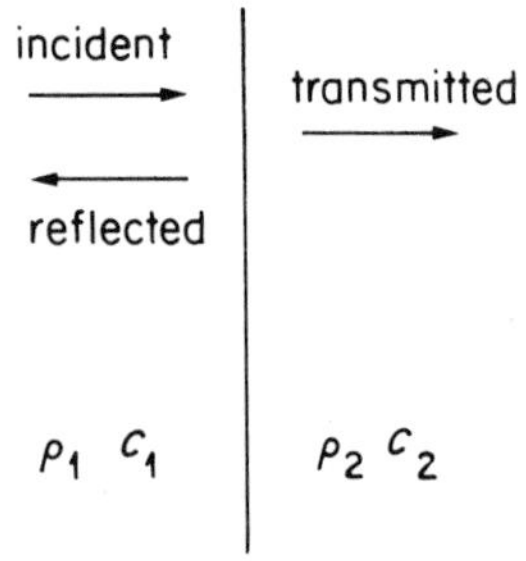

Fig. 5.7. Incident, reflected and transmitted sound waves at a plane boundary between media of specific acoustic impedances $\rho_1 c_1$ and $\rho_2 c_2$

Fig. 5.7 shows that we are considering a plane sound wave travelling in a medium of specific acoustic impedance $Z_1 = \rho_1 c_1$ and meeting, at normal incidence, an infinite plane boundary separating the first medium from another of specific acoustic impedance $Z_2 = \rho_2 c_2$. If the subscripts i, r and t denote incident, reflected and transmitted respectively, then the boundary conditions give

$$\dot{\eta}_i + \dot{\eta}_r = \dot{\eta}_t \tag{5.1}$$

and

$$p_i + p_r = p_t \tag{5.2}$$

For the incident wave $p_i = \rho_1 c_1 \dot{\eta}_i$ and for the reflected wave $p_r = -\rho_1 c_1 \dot{\eta}_r$, so equation (5.2) becomes

$$\rho_1 c_1 \dot{\eta}_i - \rho_1 c_1 \dot{\eta}_r = \rho_2 c_2 \dot{\eta}_t$$

or

$$Z_1 \dot{\eta}_i - Z_1 \dot{\eta}_r = Z_2 \dot{\eta}_t \tag{5.3}$$

Eliminating $\dot{\eta}_t$ from (5.1) and (5.3) gives

$$\frac{\dot{\eta}_r}{\dot{\eta}_i} = \frac{\omega \eta_r}{\omega \eta_i} = \frac{\eta_r}{\eta_i} = \frac{Z_1 - Z_2}{Z_1 + Z_2}$$

Eliminating $\dot{\eta}_r$ from (5.1) and (5.3) gives

$$\frac{\dot{\eta}_t}{\dot{\eta}_i} = \frac{\eta_t}{\eta_i} = \frac{2Z_1}{Z_1 + Z_2}$$

Now

$$\frac{p_r}{p_i} = -\frac{Z_1 \dot{\eta}_r}{Z_1 \dot{\eta}_i} = \frac{Z_2 - Z_1}{Z_1 + Z_2} = -\frac{\dot{\eta}_r}{\dot{\eta}_i}$$

and

$$\frac{p_t}{p_i} = \frac{Z_2 \dot{\eta}_t}{Z_1 \dot{\eta}_i} = \frac{2Z_2}{Z_1 + Z_2}$$

We see that if $Z_1 > Z_2$ the incident and reflected particle velocities are in phase, whilst the incident and reflected acoustic pressures are out of phase. The superposition of incident and reflected velocities which are in phase leads to a cancellation of pressure (a pressure node in a standing wave system). If $Z_1 < Z_2$ the pressures are in phase and the velocities are out of phase.

The transmitted particle velocity and acoustic pressure are always in phase with their incident counterparts.

At a rigid wall, where Z_2 is infinite, the velocity $\dot{\eta} = 0 = \dot{\eta}_i + \dot{\eta}_r$, which leads to a doubling of pressure at the boundary. (See Summary on page 412.)

Reflexion and Transmission of Sound Intensity

The intensity coefficients of reflexion and transmission are given by

$$\frac{I_r}{I_i} = \frac{Z_1(\dot{\eta}_r^2)_{rms}}{Z_1(\dot{\eta}_i^2)_{rms}} = \left(\frac{Z_1 - Z_2}{Z_1 + Z_2}\right)^2$$

and

$$\frac{I_t}{I_i} = \frac{Z_2(\dot{\eta}_t^2)_{rms}}{Z_1(\dot{\eta}_i^2)_{rms}} = \frac{Z_2}{Z_1}\left(\frac{2Z_1}{Z_1 + Z_2}\right)^2 = \frac{4Z_1 Z_2}{(Z_1 + Z_2)^2}$$

The conservation of energy gives

$$\frac{I_r}{I_i} + \frac{I_t}{I_i} = 1 \quad \text{or} \quad I_i = I_t + I_r$$

The great disparity between the specific acoustic impedance of air on the one hand and water or steel on the other leads to an extreme mismatch of impedances when the transmission of acoustic energy between these media is attempted.•

There is an almost total reflexion of sound wave energy at an air–water interface, independent of the side from which the wave approaches the boundary. Only 14% of acoustic energy can be transmitted at a steel–water interface, a limitation which has severe implications for underwater transmission and detection devices which rely on acoustics.

(Problems 5.10, 5.11, 5.12, 5.13, 5.14, 5.15, 5.16, 5.17)

Problem 5.1
Show that in a gas at temperature T the average thermal velocity of a molecule is approximately equal to the velocity of sound.

Problem 5.2
The velocity of sound in air of density $1 \cdot 29 \text{ kg m}^{-3}$ may be taken as 330 metres sec^{-1}. Show that the acoustic pressure for the painful sound of 10 watts metre$^{-2} \approx 6 \cdot 5 \times 10^{-4}$ of an atmosphere.

Problem 5.3
Show that the displacement amplitude of an air molecule at a painful sound level of 10 watts metre^{-2} at 500 Hertz $\approx 3 \cdot 10^{-4}$ metre.

Problem 5.4
Barely audible sound in air has an intensity of $10^{-10} I_0$. Show that the displacement amplitude of an air molecule for sound at this level at 500 Hertz is $\approx 10^{-10}$ metre, that is, about the size of the molecular diameter.

Problem 5.5
Hi-fi equipment is played very loudly at an intensity of $100 I_0$ in a small room of cross section 3 metres $\times$ 3 metres. Show that this audio output is about 10 watts.

Problem 5.6

Two sound waves, one in water and one in air, have the same intensity. Show that the ratio of their pressure amplitudes (p water/p air) is about 60. When the pressure amplitudes are equal show that the intensity ratio is $\approx 3 \times 10^{-2}$.

Problem 5.7

A spring of mass m, stiffness s and length L is stretched to a length $L + l$. When longitudinal waves propagate along the spring the equation of motion of a length dx may be written

$$m \, dx \frac{\partial^2 \eta}{\partial t^2} = \frac{\partial F}{\partial x} \, dx$$

where η is the longitudinal displacement and F is the restoring force. Derive the wave equation to show that the wave velocity v is given by $v^2 = s/\rho$ where ρ is the mass per unit length of the spring.

Problem 5.8

In problem 1.9 we showed that a mass M suspended by a spring of stiffness s and mass m oscillated simple harmonically at a frequency given by

$$\omega^2 = \frac{s}{M + m/3}$$

We may consider the same problem in terms of standing waves along the vertical spring with displacement

$$\eta = (A \cos kx + B \sin kx) \sin \omega t$$

where $k = \omega/v$ is the wave number. The boundary conditions are that $\eta = 0$ at $x = 0$ (the top of the spring) and

$$M \frac{\partial^2 \eta}{\partial t^2} = -sL \frac{\partial \eta}{\partial x} \quad \text{at } x = L$$

(the bottom of the spring). Show that these lead to the expression

$$kL \tan kL = \frac{m}{M}$$

and expand tan kL in powers of kL to show that, in the second order approximation

$$\omega^2 = \frac{s}{M + m/3}$$

The value of v is given in problem 5.7.

Problem 5.9

A solid has a Poissons ratio $\sigma = 0.25$. Show that the ratio of the longitudinal wave velocity to the transverse wave velocity is $\sqrt{3}$. Use the values of these velocities given in the text to derive an appropriate value of σ for the earth.

Problem 5.10

Show that when sound waves are normally incident on a plane steel water interface 86% of the energy is deflected. If the waves are travelling in water and are normally incident on a plane water–ice interface show that $82 \cdot 3\%$ of the energy is transmitted.

$$(\rho c \text{ values in } \mathrm{kg\,m^{-2}\,sec^{-1}})$$

$$\text{water} = 1 \cdot 43 \times 10^6$$
$$\text{ice}\ \ \ = 3 \cdot 49 \times 10^6$$
$$\text{steel}\ = 3 \cdot 9\ \times 10^7$$

Problem 5.11

Use the boundary conditions for standing acoustic waves in a tube to confirm the following:

	Particle displacement		Pressure	
	closed end	open end	closed end	open end
Phase change on reflexion	180°	0	0	180°
	node	antinode	antinode	node

Problem 5.12

Standing acoustic waves are formed in a tube of length l with (a) both ends open and (b) one end open and the other closed. If the particle displacement

$$\eta = (A \cos kx + B \sin kx) \sin \omega t$$

and the boundary conditions are as shown in the diagrams, show that for

(a) $\eta = A \cos kx \sin \omega t$ with $\lambda = 2l/n$

and for

(b) $\eta = A \cos kx \sin \omega t$ with $\lambda = 4l/(2n+1)$

Sketch the first three harmonics for each case.

Problem 5.13

On page 114 we discussed the problem of matching two strings of impedances Z_1 and Z_3 by the insertion of a quarter wave element of impedance

$$Z_2 = (Z_1 Z_3)^{1/2}$$

Repeat this problem for the acoustic case where the expressions for the string displacements

$$y_i, y_r, y_t$$

now represent the appropriate acoustic pressures p_i, p_r and p_t.

Show that the boundary condition for pressure continuity at $x = 0$ *is*

$$\mathbf{A}_1 + \mathbf{B}_1 = \mathbf{A}_2 + \mathbf{B}_2$$

and that for continuity of particle velocity is

$$Z_2(\mathbf{A}_1 - \mathbf{B}_1) = Z_1(\mathbf{A}_2 - \mathbf{B}_2)$$

Similarly, at $x = l$, show that the boundary conditions are

$$\mathbf{A}_2\, e^{-ik_2 l} + \mathbf{B}_2\, e^{ik_2 l} = \mathbf{A}_3$$

and

$$Z_3(\mathbf{A}_2\, e^{-ik_2 l} - \mathbf{B}_2\, e^{ik_2 l}) = Z_2\mathbf{A}_3$$

Hence prove that the coefficient of sound transmission

$$\frac{Z_1}{Z_3}\frac{A_3^2}{A_1^2} = 1$$

when

$$Z_2^2 = Z_1 Z_3 \quad \text{and} \quad l = \frac{\lambda_2}{4}$$

(Note that the expressions for both boundary conditions and transmission coefficient differ from those in the case of the string.)

Problem 5.14
For sound waves of high amplitude the adiabatic bulk modulus may no longer be considered as a constant. Use the adiabatic condition that

$$\frac{P}{P_0} = \left[\frac{V_0}{V_0(1+\delta)}\right]^\gamma$$

in deriving the wave equation to show that each part of the high amplitude wave has its own sound velocity $c_0(1+s)^{(\gamma+1)/2}$, where $c_0^2 = \gamma P_0/\rho_0$, δ is the dilatation, s the condensation and γ the ratio of the specific heats at constant pressure and volume.

Problem 5.15
Some longitudinal waves in a plasma exhibit a combination of electrical and acoustical phenomena. They obey a dispersion relation at temperature T of $\omega^2 = \omega_e^2 + 3aTk^2$, where ω_e is the constant electron plasma frequency (see problem 4.18) and the Boltzmann constant is written as a to avoid confusion with the wave number k. Show that the product of the phase and group velocities is related to the average thermal energy of an electron (found from $pV = RT$).

Problem 5.16
It is possible to obtain the wave equation for tidal waves (long waves in shallow water) by the method used in deriving the acoustic wave equation. In the figure a constant mass of fluid in an element of unit width, height h and length Δx moves a distance η and assumes

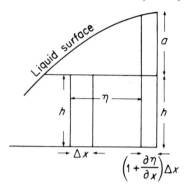

a new height $h + \alpha$ and length $(1 + \partial\eta/\partial x)\,\Delta x$, but retains unit width. Show that, to a first approximation,

$$\alpha = -h\frac{\partial\eta}{\partial x}$$

Neglecting surface tension, the force on the element face of height $h + \alpha$ arises from the product of the height and the mean hydrostatic pressure. Show that the net force on the liquid element is given by

$$-\frac{\partial F}{\partial x}\Delta x = -\rho g h\frac{\partial\alpha}{\partial x}\Delta x$$

Continue the derivation using the acoustic case as a model to show that these waves are non-dispersive with a phase velocity given by $v^2 = gh$.

Problem 5.17

Waves near the surface of a non-viscous incompressible liquid of density ρ have a phase velocity given by

$$v^2(k) = \left[\frac{g}{k} + \frac{Tk}{\rho}\right]\tanh kh$$

where g is the acceleration due to gravity, T is the surface tension, k is the wave number and h is the liquid depth. When $h \ll \lambda$ the liquid is shallow; when $h \gg \lambda$ the liquid is deep.

(a) Show that, when gravity and surface tension are equally important and $h \gg \lambda$, the wave velocity is a minimum at $v^4 = 4gT/\rho$, and show that this occurs for a 'critical' wavelength $\lambda_c = 2\pi(T/\rho g)^{1/2}$.

(b) The condition $\lambda \gg \lambda_c$ defines a *gravity* wave, and surface tension is negligible. Show that gravity waves in a shallow liquid are non dispersive with a velocity $v = \sqrt{gh}$ (see problem 5.16).

(c) Show that gravity waves in a deep liquid have a phase velocity $v = \sqrt{g/k}$ and a group velocity of half this value.

(d) The condition $\lambda < \lambda_c$ defines a ripple (dominated by surface tension). Show that short ripples in a deep liquid have a phase velocity $v = \sqrt{Tk/\rho}$ and a group velocity of $\frac{3}{2}v$. (Note the anomalous dispersion.)

Summary of Important Results

Wave velocity

$$c^2 = \frac{\text{Bulk Modulus}}{\rho} = \frac{\gamma P}{\rho}$$

Specific Acoustic Impedance

$$Z = \frac{\text{acoustic pressure}}{\text{particle velocity}}$$

$Z = \rho c$ (for right going wave)

$\quad = -\rho c$ (for left going wave because pressure and particle velocity become anti-phase)

$$\text{Intensity} = \tfrac{1}{2}\rho c \dot{\eta}_m^2 = \frac{p_{\text{rms}}^2}{\rho c} = p_{\text{rms}} \dot{\eta}_{\text{rms}}$$

Reflexion and Transmission Coefficients

$$\frac{\text{Reflected Amplitude}}{\text{Incident Amplitude}} \left\{ \begin{array}{l} \text{displacement} \\ \text{and velocity} \end{array} \right\} = \frac{Z_1 - Z_2}{Z_1 + Z_2} = -\frac{\text{Reflected pressure}}{\text{Incident pressure}}$$

$$\frac{\text{Transmitted Amplitude}}{\text{Incident Amplitude}} \begin{array}{l} \text{(displacement)} \\ \text{(and velocity)} \end{array} = \frac{2Z_1}{Z_1 + Z_2}$$

$$= \frac{Z_1}{Z_2} \times \frac{\text{Transmitted pressure}}{\text{Incident pressure}}$$

$$\frac{\text{Reflected Intensity}}{\text{Incident Intensity}} \text{ (energy)} = \left(\frac{Z_1 - Z_2}{Z_1 + Z_2}\right)^2$$

$$\frac{\text{Transmitted Intensity}}{\text{Incident Intensity}} \text{ (energy)} = \frac{4Z_1 Z_2}{(Z_1 + Z_2)^2}$$

Chapter 6

Waves on Transmission Lines

In the wave motion discussed so far four major points have emerged. They are

(i) Individual particles in the medium oscillate about their equilibrium positions with simple harmonic motion but do not propagate through the medium.

(ii) Crests and troughs and all planes of equal phase are transmitted through the medium to give the wave motion.

(iii) The wave or phase velocity is governed by the product of the inertia of the medium and its capacity to store potential energy, that is, its elasticity.

(iv) The impedance of the medium to this wave motion is governed by ratio of the inertia to the elasticity (see table on p. xiv).

In this chapter we wish to investigate the wave propagation of voltages and currents and we shall see that the same physical features are predominant. Voltage and current waves are usually sent along a geometrical configuration of wires and cables known as transmission lines. The physical scale or order of magnitude of these lines can vary from that of an oscilloscope cable on a laboratory bench to the electric power distribution lines supported on pylons over hundreds of miles or the submarine telecommunication cables lying on an ocean bed.

Any transmission line can be simply represented by a pair of parallel wires into one end of which power is fed by an a.c. generator. Fig. 6.1a shows such a line at the instant when the generator terminal A is positive with respect to terminal B, with current flowing out of the terminal A and into terminal B as the generator is doing work. A half cycle later the position is reversed and B is the positive terminal, the net result being that along each of the two wires there will be a distribution of charge as shown, reversing in sign at each half cycle due to the oscillatory simple harmonic motion of the charge carriers (fig. 6.1b). These carriers move a distance equal to a fraction of a wavelength on either side of their equilibrium positions. As the charge moves current flows, having a maximum value where the product of charge density and velocity is greatest.

The existence along the cable of maximum and minimum current values varying simple harmonically in space and time describes a current wave along

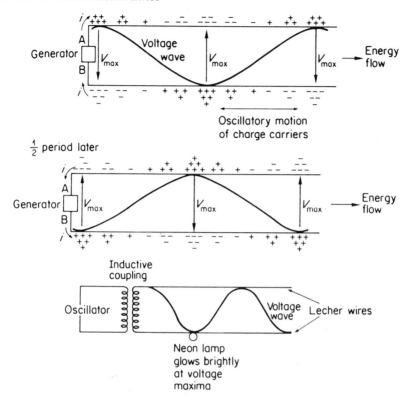

Fig. 6.1. Power fed continuously by a generator into an infinitely long transmission line. Charge distribution and voltage waves for (a) generator terminal positive at A and (b) a half period later, generator terminal positive at B. Laboratory demonstration (c) of voltage maxima along a Lecher wire system. The neon lamp glows when held near a position of V_{max}

the cable. Associated with these currents there are voltage waves (fig. 6.1a), and if the voltage and current at the generator are always in phase then power is continuously fed into the transmission line and the waves will always be carrying energy away from the generator. In a laboratory the voltage and current waves may be shown on a Lecher Wire system (fig. 6.1c).

In deriving the wave equation for both voltage and current to obtain the velocity of wave propagation we shall concentrate our attention on a short element of the line having a length very much less than that of the waves. Over this element we may consider the variables to change linearly to the first order and we can use differentials.

The currents which flow will generate magnetic flux lines which thread the region between the cables, giving rise to a self inductance L_0 per unit length measured in henries per metre. Between the lines, which form a condenser, there is an electrical capacitance C_0 per unit length measured in farads per metre. In the absence of any resistance in the line these two parameters completely decribe the line, which is known as *ideal* or *lossless*.

Ideal or Lossless Transmission Line

Fig. 6.2 represents a short element of zero resistance of an ideal transmission line length $dx \ll \lambda$ (the voltage or current wavelength). The self inductance of the element is $L_0\, dx$ henries and its capacitance is $C_0\, dx$ farads.

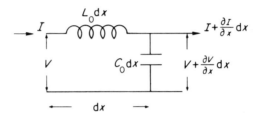

Fig. 6.2. Representation of element of an ideal transmission line of inductance L_0 henries per unit length and capacitance C_0 farads per unit length. The element length $\ll \lambda$, the voltage and current wavelength

If the rate of change of voltage per unit length at constant time is $\partial V/\partial x$ then the voltage difference between the ends of the element dx is $\partial V/\partial x\, dx$, which equals the voltage drop from the self inductance $-(L_0\, dx)\, \partial I/\partial t$.
    Thus

$$\frac{\partial V}{\partial x}\, dx = -(L_0\, dx)\frac{\partial I}{\partial t}$$

or

$$\frac{\partial V}{\partial x} = -L_0\frac{\partial I}{\partial t} \qquad (6.1)$$

If the rate of change of current per unit length at constant time is $\partial I/\partial x$ there is a loss of current along the length dx of $-\partial I/\partial x\, dx$ because some current has charged the capacitance $C_0\, dx$ of the line to a voltage V.
    If the amount of charge is $q = (C_0\, dx)\, V$,

$$dI = \frac{dq}{dt} = \frac{\partial}{\partial t}(C_0\, dx)\, V$$

so that

$$\frac{-\partial I}{\partial x}\,\mathrm{d}x = \frac{\partial}{\partial t}(C_0\,\mathrm{d}x)\,V$$

or

$$\frac{-\partial I}{\partial x} = C_0\frac{\partial V}{\partial t} \tag{6.2}$$

Since $\partial^2/\partial x\partial t = \partial^2/\partial t\partial x$ it follows, by taking $\partial/\partial x$ of equation (6.1) and $\partial/\partial t$ of equation (6.2) that

$$\frac{\partial^2 V}{\partial x^2} = L_0 C_0 \frac{\partial^2 V}{\partial t^2} \tag{6.3}$$

a pure wave equation for the voltage with a velocity of propagation given by $v^2 = 1/L_0 C_0$.

Similarly $\partial/\partial t$ of (6.1) and $\partial/\partial x$ of (6.2) gives

$$\frac{\partial^2 I}{\partial x^2} = L_0 C_0 \frac{\partial^2 I}{\partial t^2} \tag{6.4}$$

showing that the current waves propagate with the same velocity $v^2 = 1/L_0 C_0$. We must remember here, in checking dimensions, that L_0 and C_0 are defined per unit length.

So far then, the oscillatory motion of the charge carriers (our particles in a medium) has led to the propagation of voltage and current waves with a velocity governed by the product of the magnetic inertia or inductance of the medium and its capacity to store potential energy.

Coaxial Cables

Many transmission lines are made in the form of coaxial cables, e.g. a cylinder of dielectric material such as polythene having one conductor along its axis and the other surrounding its outer surface. This configuration has an inductance per unit length of

$$L_0 = \frac{\mu}{2\pi}\log_e\frac{r_2}{r_1}\text{ henries}$$

where r_1 and r_2 are the radii of the inner and outer conductors respectively and μ is the magnetic permeability of the dielectric (henries per metre). Its capacitance per unit length

$$C_0 = \frac{2\pi\epsilon}{\log_e r_2/r_1}\text{ farads}$$

where ϵ is the permittivity of the dielectric (farads per metre) so that $v^2 = 1/L_0C_0 = 1/\mu\epsilon$.

The velocity of the voltage and current waves along such a cable is wholly determined by the properties of the dielectric medium. We shall see in the next chapter on electromagnetic waves that μ and ϵ represent the inertial and elastic properties of any medium in which such waves are propagating; the velocity of these waves will be given by $v^2 = 1/\mu\epsilon$. In free space these parameters have the values

$$\mu_0 = 4\pi \times 10^{-7} \text{ henries per metre}$$

$$\epsilon_0 = (36\pi \times 10^9)^{-1} \text{ farads per metre}$$

and v^2 becomes $c^2 = (\mu_0\epsilon_0)^{-1}$ where c is the velocity of light, equal to 3×10^8 metres per second.

Coaxial cables can be made to a very high degree of precision and the time for an electrical signal to travel a given length can be accurately calculated because the velocity is known.

Such a cable can be used as a 'delay line' in order to separate the arrival of signals at a given point by very small intervals of time.

Characteristic Impedance of a Transmission Line

The solutions to equations (6.3) and (6.4) are, of course,

$$V_+ = V_{0+} \sin \frac{2\pi}{\lambda}(vt - x)$$

and

$$I_+ = I_{0+} \sin \frac{2\pi}{\lambda}(vt - x)$$

where V_0 and I_0 are the maximum values and where the subscript + refers to a wave moving in the positive x-direction. Equation (6.1), $\partial V/\partial x = -L_0 \partial I/\partial t$, therefore gives $-V'_+ = -vL_0I'_+$, where the superscript refers to differentiation with respect to the bracket $(vt - x)$.

Integration of this equation gives

$$V_+ = vL_0I_+$$

where the constant of integration has no significance because we are considering only oscillatory values of voltage and current whilst the constant will change merely the d.c. level.

The ratio

$$\frac{V_+}{I_+} = vL_0 = \sqrt{\frac{L_0}{C_0}} \text{ ohms}$$

and the value of $\sqrt{L_0/C_0}$, written as Z_0, is a constant for a transmission line of given properties and is called the *characteristic impedance*. Note that it is a pure resistance (no dimensions of length are involved) and it is the impedance seen by the wave system propagating along an infinitely long line, just as an acoustic wave experiences a specific acoustic impedance ρc. The physical correspondence between ρc and $L_0 v = \sqrt{L_0/C_0} = Z_0$ is immediately evident.

The value of Z_0 for the coaxial cable considered earlier can be shown to be

$$Z_0 = \frac{1}{2\pi}\sqrt{\frac{\mu}{\epsilon}}\log_e\frac{r_2}{r_1}$$

Electromagnetic waves in free space experience an impedance $Z_0 = \sqrt{\mu_0/\epsilon_0} = 376\cdot6$ ohms.

So far we have considered waves travelling only in the x-direction. Waves which travel in the negative x-direction will be represented (from solving the wave equation) by

$$V_- = V_{0-}\sin\frac{2\pi}{\lambda}(vt+x)$$

and

$$I_- = I_{0-}\sin\frac{2\pi}{\lambda}(vt+x)$$

where the negative subscript denotes the negative x-direction of propagation.

Equation (6.1) then yields the result that

$$\frac{V_-}{I_-} = -vL_0 = -Z_0$$

so that, in common with the specific acoustic impedance, a negative sign is introduced into the ratio when the waves are travelling in the negative x-direction.

When waves are travelling in both directions along the transmission line the total voltage and current at any point will be given by

$$V = V_+ + V_-$$

and

$$I = I_+ + I_-$$

When a transmission line has waves only in the positive direction the voltage and current waves are always in phase, energy is propagated and power is being fed into the line by the generator at all times. This situation is destroyed when waves travel in both directions; waves in the negative x-direction are produced by reflexion at a boundary when a line is terminated or mismatched; we shall now consider such reflexions.

(Problems 6.1 and 6.2)

The Physics of Vibrations and Waves

Reflexions from the End of a Transmission line

Suppose that a transmission line of characteristic impedance Z_0 has a finite length and that the end opposite that of the generator is terminated by a load of impedance Z_L as shown in fig. 6.3.

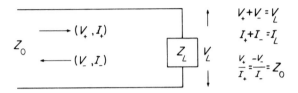

Fig. 6.3. Transmission line termi-nated by impedance Z_L to produce reflected waves unless $Z_L = Z_0$, the characteristic impedance

A wave travelling to the right (V_+, I_+) may be reflected to produce a wave (V_-, I_-).

The boundary conditions at Z_L must be $V_+ + V_- = V_L$, where V_L is the voltage across the load and $I_+ + I_- = I_L$. In addition $V_+/I_+ = Z_0$, $V_-/I_- = -Z_0$ and $V_L/I_L = Z_L$. It is easily shown that these equations yield

$$\frac{V_-}{V_+} = \frac{Z_L - Z_0}{Z_L + Z_0}$$

(the voltage amplitude reflexion coefficient),

$$\frac{I_-}{I_+} = \frac{Z_0 - Z_L}{Z_L + Z_0}$$

(the current amplitude reflexion coefficient),

$$\frac{V_L}{V_+} = \frac{2Z_L}{Z_L + Z_0}$$

and

$$\frac{I_L}{I_+} = \frac{2Z_0}{Z_L + Z_0}$$

in complete correspondence with the reflexion and transmission coefficients we have met so far. (See summary on page 412.)

We see that if the line is terminated by a load $Z_L = Z_0$, its characteristic impedance, the line is matched, all the energy propagating down the line is absorbed and there is no reflected wave. When $Z_L = Z_0$, therefore, the wave in the positive direction continues to behave as though the transmission line were infinitely long.

Short Circuited Transmission Line ($Z_L = 0$)

If the ends of the transmission line are short circuited (fig. 6.4), $Z_L = 0$, and we have

$$V_L = V_+ + V_- = 0$$

so that $V_+ = -V_-$, and there is total reflexion with a phase change of π. But this

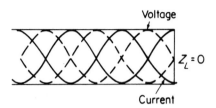

Fig. 6.4. Short circuited transmission line of length $(2n+1)\lambda/4$ produces a standing wave with a current maximum and zero voltage at end of line

is the condition, as we saw in an earlier chapter, for the existence of standing waves; we shall see that such waves exist on the transmission line.

At any position x on the line we may express the two voltage waves by

$$V_+ = Z_0 I_+ = V_{0+}\, e^{i(\omega t - kx)}$$

and

$$V_- = -Z_0 I_- = V_{0-}\, e^{i(\omega t + kx)}$$

where, with total reflexion and π phase change, $V_{0+} = -V_{0-}$. The total voltage at x is

$$V_x = (V_+ + V_-) = V_{0+}(e^{-ikx} - e^{ikx})\, e^{i\omega t} = (-i)2V_{0+} \sin kx\ e^{i\omega t}$$

and the total current at x is

$$I_x = (I_+ + I_-) = \frac{V_{0+}}{Z_0}(e^{-ikx} + e^{ikx})\, e^{i\omega t} = \frac{2V_{0+}}{Z_0} \cos kx\ e^{i\omega t}$$

We see then that at any point x along the line the voltage V_x varies as $\sin kx$ and the current I_x varies as $\cos kx$, so that voltage and current are 90° out of phase in space. In addition the $-i$ factor in the voltage expression shows that the voltage lags the current 90° in time, so that if we take the voltage to vary with $\cos \omega t$ from the $e^{i\omega t}$ term, then the current will vary with $-\sin \omega t$. If we take the time variation of voltage to be as $\sin \omega t$ the current will change with $\cos \omega t$.

Voltage and current at all points are 90° out of phase in space and time, and the power factor $\cos \phi = \cos 90° = 0$, so that no power is consumed. A standing wave system exists with equal energy propagated in each direction and the total energy propagation equal to zero. Nodes of voltage and current are spaced along the transmission line as shown in fig. 6.4, with I always a maximum where $V = 0$ and vice versa.

If the current I varies with $\cos \omega t$ it will be at a maximum when $V = 0$; when V is a maximum the current is zero. The energy of the system is therefore completely exchanged each quarter cycle between the magnetic inertial energy $\frac{1}{2}L_0 I^2$ and the electric potential energy $\frac{1}{2}C_0 V^2$.

(Problems 6.3, 6.4, 6.5, 6.6, 6.7, 6.8, 6.9, 6.10, 6.11)

Effect of Resistance in a Transmission Line

The discussion so far has concentrated on a transmission line having only inductance and capacitance, i.e. wattless components which consume no power. In practice, of course, no such line exists: there is always some resistance in the wires which will be responsible for energy losses. We shall take this resistance into account by supposing that the transmission line has a series

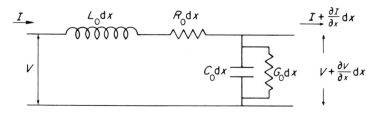

Fig. 6.5. Real transmission line element includes a series resistance R_0 ohms per unit length and a shunt conductance G_0 siemens per unit length

resistance R_0 ohms per unit length and a short circuiting or shunting resistance between the wires, which we express as a shunt conductance (inverse of resistance) written as G_0, where G_0 has the dimensions of siemens per metre. Our model of the short element of length dx of the transmission line now appears in fig. 6.5, with a resistance $R_0\,dx$ in series with $L_0\,dx$ and the conductance $G_0\,dx$ shunting the capacitance $C_0\,dx$. Current will now leak across the transmission line because the dielectric is not perfect. We have seen that the time-dependence of the voltage and current variations along a transmission line may be written

$$V = V_0\,e^{i\omega t} \quad \text{and} \quad I = I_0\,e^{i\omega t}$$

so that

$$L_0\frac{\partial I}{\partial t} = i\omega L_0 I \quad \text{and} \quad C_0\frac{\partial V}{\partial t} = i\omega C_0 V$$

The voltage and current changes across the line element length dx are now given by

$$\frac{\partial V}{\partial x} = -L_0\frac{\partial I}{\partial t} - R_0 I = -(R_0 + i\omega L_0)I \qquad (6.1a)$$

$$\frac{\partial I}{\partial x} = -C_0\frac{\partial V}{\partial t} - G_o V = -(G_0 + i\omega C_0)V \qquad (6.2a)$$

since $(G_0\,dx)V$ is the current shunted across the condenser. Inserting $\partial/\partial x$ of equation (6.1a) into equation (6.2a) gives

$$\frac{\partial^2 V}{\partial x^2} = -(R_0 + i\omega L_0)\frac{\partial I}{\partial x} = (R_0 + i\omega L_0)(G_0 + i\omega C_0)V = \gamma^2 V$$

where $\gamma^2 = (R_0 + i\omega L_0)(G_0 + i\omega C_0)$, so that γ is a complex quantity which may be written

$$\gamma = \alpha + ik$$

Inserting $\partial/\partial x$ of equation (6.2a) into equation (6.1a) gives

$$\frac{\partial^2 I}{\partial x^2} = -(G_0 + i\omega C_0)\frac{\partial V}{\partial x} = (R_0 + i\omega L_0)(G_0 + i\omega C_0)I = \gamma^2 I$$

an equation similar to that for V.

The equation

$$\frac{\partial^2 V}{\partial x^2} - \gamma^2 V = 0 \qquad (6.5)$$

has solutions for the x-dependence of V of the form

$$V = A\,e^{-\gamma x} \quad \text{or} \quad V = B\,e^{+\gamma x}$$

where A and B are constants.

We know already that the time-dependence of V is of the form $e^{i\omega t}$, so that the complete solution for V may be written

$$V = (A\,e^{-\gamma x} + B\,e^{\gamma x})\,e^{i\omega t}$$

or, since $\gamma = \alpha + ik$,

$$V = (A\,e^{-\alpha x}\,e^{-ikx} + B\,e^{\alpha x}\,e^{+ikx})\,e^{i\omega t}$$

$$= A\,e^{-\alpha x}\,e^{i(\omega t - kx)} + B\,e^{\alpha x}\,e^{i(\omega t + kx)}$$

The behaviour of V is shown in fig. 6.6—a wave travelling to the right with an amplitude decaying exponentially with distance because of the term $e^{-\alpha x}$ and a wave travelling to the left with an amplitude decaying exponentially with distance because of the term $e^{\alpha x}$.

In the expression $\gamma = \alpha + ik$, γ is called the propagation constant, α is called the attenuation or absorption coefficient and k is the wave number.

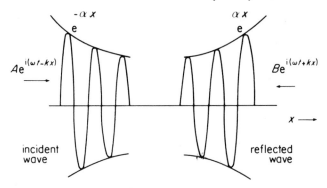

Fig. 6.6. Voltage and current waves in both directions along a transmission line with resistance. The effect of the dissipation term is shown by the exponentially decaying wave in each direction

The behaviour of the current wave I is exactly similar and since power is the product VI, the power loss with distance varies as $(e^{-\alpha x})^2$, that is, as $e^{-2\alpha x}$.

We would expect this behaviour from our discussion of damped simple harmonic oscillations. When the transmission line properties are purely inductive (inertial) and capacitative (elastic), a pure wave equation with a sine or cosine solution will follow. The introduction of a resistive or loss element produces an exponential decay with distance along the transmission line in exactly the same way as an oscillator is damped with time.

Such a loss mechanism, resistive, viscous, frictional or diffusive, will always result in energy loss from the propagating wave. These are all examples of random collision processes which operate in only one direction in the sense that they are thermodynamically irreversible. At the end of this chapter we shall discuss their effects in more detail.

Characteristic Impedance of a Transmission Line with Resistance

In a lossless line we saw that the ratio $V_+/I_+ = Z_0 = \sqrt{L_0/C_0} = Z_0$ ohms, a purely resistive term. In what way does the introduction of the resistance into the line affect the characteristic impedance?

The solution to the equation $\partial^2 I/\partial x^2 = \gamma^2 I$ may be written (for the x-dependence of I) as

$$I = (A' e^{-\gamma x} + B' e^{\gamma x})$$

so that equation (6.2a)

$$\frac{\partial I}{\partial x} = -(G_0 + i\omega C_0) V$$

gives

$$-\gamma(A' e^{-\gamma x} - B' e^{\gamma x}) = -(G_0 + i\omega C_0) V$$

or

$$\frac{\sqrt{(R_0 + i\omega L_0)(G_0 + i\omega C_0)}}{G_o + i\omega C_0}(A' e^{-\gamma x} - B' e^{\gamma x}) = V = V_+ + V_-$$

But, except for the $e^{i\omega t}$ term,

$$A' e^{-\gamma x} = I_+$$

the current wave in the positive x-direction, so that

$$\sqrt{\frac{R_0 + i\omega L_0}{G_0 + i\omega C_0}} I_+ = V_+$$

or

$$\frac{V_+}{I_+} = \sqrt{\frac{R_0 + i\omega L_0}{G_0 + i\omega C_0}} = Z_0'$$

for a transmission line with resistance. Similarly $B' e^{\gamma x} = I_-$ and

$$\frac{V_-}{I_-} = -\sqrt{\frac{R_0 + i\omega L_0}{G_0 + i\omega C_0}} = -Z_0'$$

The presence of the resistance term in the complex characteristic impedance means that power will be lost through Joule dissipation and that energy will be absorbed from the wave system.

We shall discuss this aspect in some detail in the next chapter on electromagnetic waves, but for the moment we shall examine absorption from a different (although equivalent) viewpoint.

(Problems 6.12 and 6.13)

THE DIFFUSION EQUATION AND ENERGY ABSORPTION IN WAVES

On page 24 of Chapter 1 we discussed quite briefly the effect of random processes. We shall now look at this in more detail. The wave equation

$$\frac{\partial^2 \phi}{\partial x^2} = \frac{1}{c^2}\frac{\partial^2 \phi}{\partial t^2}$$

is only one of a family of equations which have a double differential with respect to space on the left hand side.

In three dimensions the left hand side would be of the form

$$\frac{\partial^2 \phi}{\partial x^2} + \frac{\partial^2 \phi}{\partial y^2} + \frac{\partial^2 \phi}{\partial z^2}$$

which, in vector language, is called the divergence of the gradient or div grad and is written $\nabla^2 \phi$.

Five members of this family of equations may be written (in one dimension) as

(1) Laplace's Equation

$$\frac{\partial^2 \phi}{\partial x^2} = 0 \quad \text{(for } \phi(x) \text{ only)}$$

(2) Poisson's Equation

$$\frac{\partial^2 \phi}{\partial x^2} = \text{constant} \quad \text{(for } \phi(x) \text{ only)}$$

(3) Helmholtz Equation

$$\frac{\partial^2 \phi}{\partial x^2} = \text{constant} \times \phi$$

(4) Diffusion Equation

$$\frac{\partial^2 \phi}{\partial x^2} = +\text{ve constant} \times \frac{\partial \phi}{\partial t}$$

(5) Wave Equation

$$\frac{\partial^2 \phi}{\partial x^2} = +\text{ve constant} \times \frac{\partial^2 \phi}{\partial t^2}$$

Laplace's and Poisson's equations occur very often in electrostatic field theory and are used to find the values of the electric field and potential at any point. We have already met the Helmholtz equation in this chapter as equation (6.5), where the constant was positive (written γ^2) and we have seen its behaviour when the constant is negative, for it is then equivalent to the equation for simple harmonic motion except that here the variable is space and not time. The constant in the wave equation is of course $1/c^2$ where c is the wave velocity. Where the wave equation has an 'acceleration' or $\partial^2 \phi / \partial t^2$ term on the right hand side, the diffusion equation has a 'velocity' or $\partial \phi / \partial t$ term.

All equations, however, have the same term $\partial^2 \phi / \partial x^2$ on the left hand side, and we must ask 'what is its physical significance?'

We know that the values of the scalar ϕ will depend upon the point in space at which it is measured. Suppose we choose some point at which ϕ has the value ϕ_0 and surround this point by a small cube of side l, over the volume of which ϕ may take other values. If the average value of ϕ over the small cube is written

$\bar{\phi}$, then the difference between the average $\bar{\phi}$ and the value at the centre of the cube ϕ_0 is given by

$$\bar{\phi} - \phi_0 = \text{constant} \times \left(\frac{\partial^2 \phi}{\partial x^2} + \frac{\partial^2 \phi}{\partial y^2} + \frac{\partial^2 \phi}{\partial z^2} \right)_0$$

This statement is proved in the appendix at the end of this section and is readily understood by those familiar with triple integration. The left hand side of any of these equations therefore measures the value

$$\bar{\phi} - \phi_0$$

In Laplace's equation the difference is zero, so that ϕ has a constant value over the volume considered. Poisson's equation tells us that the difference is constant and Helmholtz equation states that the value of ϕ at any point in the volume is proportional to this difference. The first two equations are 'steady state', i.e. they do not vary with time.

The Helmholtz equation states that if the constant is positive the behaviour of ϕ with space grows or decays exponentially, e.g. γ^2 is positive in equation (6.5), but if the constant is negative, ϕ will vary sinusoidally or cosinusoidally with space as the displacement varies with time in simple harmonic motion and the equation becomes the time independent wave equation for standing waves. This equation says nothing about the time behaviour of ϕ, which will depend only upon the function ϕ itself.

Both the diffusion and wave equations are time-derivative dependent. The diffusion equation states that the 'velocity' or change of ϕ with time at a point in the volume is proportional to the difference $\bar{\phi} - \phi_0$, whereas the wave equation states that the 'acceleration' $\partial^2 \phi / \partial t^2$ depends on this difference.

The wave equation recalls the simple harmonic oscillator, where the difference from the centre ($\bar{x} = 0$) was a measure of the force or acceleration term; both the oscillator and the wave equation have time varying sine and cosine solutions with maximum velocity $\partial \phi / \partial t$ at the zero displacement from equilibrium, that is where the difference $\bar{\phi} - \phi_0 = 0$.

The diffusion equation, however, describes a different kind of behaviour. It describes a non-equilibrium situation which is moving towards equilibrium at a rate governed by its distance from equilibrium, so that it reaches equilibrium in a time which is theoretically infinite. Readers will have already met this situation in Newton's Law of Cooling, where a hot body at temperature T_0 stands in a room of lower temperature $\bar{T}$. The rate at which the body cools, i.e. the value of $\partial T / \partial t$, depends on $\bar{T} - T_0$; a cooling graph of this experiment is given in fig. 6.7. The greatest rate of cooling occurs when the temperature difference is greatest and the process slows down as the system approaches equilibrium. Here, of course, $\bar{T} - T_0$ and $\partial T / \partial t$ are both negative.

All non-equilibrium processes of this kind are unidirectional in the sense that they are thermodynamically irreversible. They involve the transport of mass in diffusion, the transport of momentum in friction or viscosity and the transport

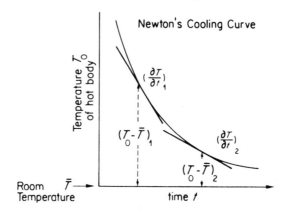

Fig. 6.7. Newton's cooling curve
shows that the rate of cooling of a
hot body $\partial T/\partial t$ depends on the
temperature difference between
the body and its surrounding, this
difference being directly measured by $\partial^2 T/\partial x^2$

of energy in conductivity. All such processes involve the loss of useful energy
and the generation of entropy.

They are all processes which are governed by random collisions, and we
found in the first chapter, where we added vectors of constant length and
random phase, that the average distance travelled by particles involved in these
processes was proportional, not to the time, but to the square root of the time.

Rewriting the diffusion equation as

$$\frac{\partial^2 \phi}{\partial x^2} = \frac{1}{d}\frac{\partial \phi}{\partial t}$$

we see that the dimensions of the constant d, called the diffusivity, are given by

$$\frac{\phi}{\text{length}^2} = \frac{1}{d}\frac{\phi}{\text{time}}$$

so that d has the dimensions of $\text{length}^2/\text{time}$. The interpretation of this as the
square of a characteristic length varying with the square root of time has
already been made in Chapter 1.

In a viscous process d is given by η/ρ, where η is the coefficient of viscosity
and ρ is the density. In thermal conductivity $d = K/\rho C_p$, where K is the
coefficient of thermal conductivity, ρ is the density and C_p is the specific heat at
constant pressure.

A magnetic field which is non-uniformly distributed in a conductor has a
diffusivity $d = (\mu\sigma)^{-1}$, where μ is the permeability and σ is the conductivity.

Brownian motion is one of the best known examples of random collision processes. The distance x travelled in time t by a particle suffering multiple random collisions is given by Einstein's diffusivity relation

$$d = \frac{\bar{x}^2}{t} = \frac{2RT}{6\pi\eta N}$$

The gas law, $pV = RT$, gives RT as the energy of a mole of such particles at temperature T; a mole contains N particles, where N is Avogadro's number and $RT/N = kT$, the average energy of the individual particles, where k is Boltzmann's constant.

The process is governed, therefore, by the ratio of the energy of the particles to the coefficient of viscosity, which measures the frictional force. The higher the temperature, the greater is the energy, the less the effect of the frictional force and the greater the average distance travelled.

Wave Equation with Diffusion Effects

In natural systems we can rarely find pure waves which propagate free from the energy-loss mechanisms we have been discussing, but if these losses are not too serious we can describe the total propagation in space and time by a combination of the wave and diffusion equations.

If we try to solve the combined equation

$$\frac{\partial^2 \phi}{\partial x^2} = \frac{1}{c^2}\frac{\partial^2 \phi}{\partial t^2} + \frac{1}{d}\frac{\partial \phi}{\partial t}$$

we shall not obtain a pure sine or cosine solution.

Let us try the solution

$$\phi = \phi_m \, e^{i(\omega t - \gamma x)}$$

where ϕ_m is the maximum amplitude. This gives

$$i^2 \gamma^2 = i^2 \frac{\omega^2}{c^2} + i\frac{\omega}{d}$$

or

$$\gamma^2 = \frac{\omega^2}{c^2} - i\frac{\omega}{d}$$

giving a complex value for γ. But $\omega^2/c^2 = k^2$, where k is the wave number, and if we put $\gamma = k - i\alpha$ we obtain

$$\gamma^2 = k^2 - 2ik\alpha - \alpha^2 \approx k^2 - ik\alpha \quad \text{if } \alpha \ll k$$

The solution for ϕ then becomes

$$\phi = \phi_m \, e^{i(\omega t - \gamma x)} = \phi_m \, e^{-\alpha x} \, e^{i(\omega t - kx)}$$

i.e. a sine or cosine oscillation of maximum amplitude ϕ_m which decays exponentially with distance. The physical significance of the condition $\alpha \ll k = 2\pi/\lambda$ is that many wavelengths λ are contained in the distance $1/\alpha$ before the amplitude decays to $\phi_m e^{-1}$ at $x = 1/\alpha$. Diffusion mechanisms will cause attenuation or energy loss from the wave; the energy in a wave is proportional to the square of its amplitude and therefore decays as $e^{-2\alpha x}$.

(Problems 6.14, 6.15, and 6.16)

Appendix

Physical interpretation of

$$\frac{\partial^2 \phi}{\partial x^2} + \frac{\partial^2 \phi}{\partial y^2} + \frac{\partial^2 \phi}{\partial z^2} \equiv \nabla^2 \phi$$

At a certain point O of the scalar field, $\phi = \phi_0$. Constructing a cube around the point O having sides of length l gives for the average value over the cube volume

$$\bar{\phi} l^3 = \int \int \int_{-l/2}^{+l/2} \phi \, dx \, dy \, dz$$

Expanding ϕ about the point O by a Taylor series gives

$$\phi = \phi_0 + \left(\frac{\partial \phi}{\partial x}\right)_0 x + \left(\frac{\partial \phi}{\partial y}\right)_0 y + \left(\frac{\partial \phi}{\partial z}\right)_0 z$$

$$+ \frac{1}{2}\left[\left(\frac{\partial^2 \phi}{\partial x^2}\right)_0 x^2 + \left(\frac{\partial^2 \phi}{\partial y^2}\right)_0 y^2 + \left(\frac{\partial^2 \phi}{\partial z^2}\right)_0 z^2\right]$$

$$+ \left(\frac{\partial^2 \phi}{\partial x \partial y}\right)_0 xy + \left(\frac{\partial^2 \phi}{\partial y \partial z}\right)_0 yz + \left(\frac{\partial^2 \phi}{\partial z \partial x}\right)_0 zx + \ldots$$

Integrating from $-l/2$ to $+l/2$ removes all the functions of the form

$$\left(\frac{\partial \phi}{\partial x}\right)_0 x \quad \text{and} \quad \left(\frac{\partial^2 \phi}{\partial x \partial y}\right)_0 xy$$

whose integrals are zero, leaving, since

$$\int \int \int_{-l/2}^{+l/2} x^2 \, dx \, dy \, dz = \frac{l^5}{12}$$

$$\bar{\phi} l^3 = \phi_0 l^3 + \frac{l^5}{24}\left(\frac{\partial^2 \phi}{\partial x^2} + \frac{\partial^2 \phi}{\partial y^2} + \frac{\partial^2 \phi}{\partial z^2}\right)_0$$

i.e.

$$\bar{\phi} - \phi_0 = \frac{l^2}{24}(\nabla^2 \phi)_0$$

where l is a constant.

Problem 6.1

The figure shows the mesh representation of a transmission line of inductance L_0 per

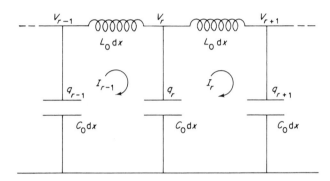

unit length and capacitance C_0 per unit length. Use equations of the form

$$I_{r-1} - I_r = \frac{d}{dt}q_r = C_0 \, dx \frac{d}{dt}V_r$$

and

$$L_0 \, dx \frac{d}{dt}I_r = V_r - V_{r+1}$$

together with the method of the final section of Chapter 3 to show that the voltage and current wave equations are

$$\frac{\partial^2 V}{\partial x^2} = L_0 C_0 \frac{\partial^2 V}{\partial t^2}$$

and

$$\frac{\partial^2 I}{\partial x^2} = L_0 C_0 \frac{\partial^2 I}{\partial t^2}$$

Problem 6.2

Show that the characteristic impedance for a pair of Lecher wires of radius r and separation d in a medium of permeability μ and permittivity ϵ is given by

$$Z_0 = \frac{1}{\pi}\sqrt{\frac{\mu}{\epsilon}}\log_e\frac{d}{r}$$

Problem 6.3

In a short-circuited lossless transmission line integrate the magnetic (inductive) energy $\frac{1}{2}L_0 I^2$ and the electric (potential) energy $\frac{1}{2}C_0 V^2$ over the last quarter wavelength (0 to $-\lambda/4$) to show that they are equal.

Problem 6.4

Show, in problem 6.3, that the sum of the instantaneous values of the two energies over the last quarter wavelength is equal to the maximum value of either.

Problem 6.5

Show that the impedance of a real transmission line seen from a position x on the line is given by

$$Z_x = Z_0 \frac{A \, e^{-\gamma x} - B \, e^{+\gamma x}}{A \, e^{-\gamma x} + B \, e^{+\gamma x}}$$

where γ is the propagation constant and A and B are the current amplitudes at $x = 0$ of the waves travelling in the positive and negative x-directions respectively. If the line has a length l and is terminated by a load Z_L, show that

$$Z_L = Z_0 \frac{A \, e^{-\gamma l} - B \, e^{\gamma l}}{A \, e^{-\gamma l} + B \, e^{\gamma l}}$$

Problem 6.6

Show that the input impedance of the line of problem 6.5, that is, the impedance of the line at $x = 0$, is given by

$$Z_i = Z_0 \left(\frac{Z_0 \sinh \gamma l + Z_L \cosh \gamma l}{Z_0 \cosh \gamma l + Z_L \sinh \gamma l} \right)$$

$$(Note\colon 2 \cosh \gamma l = e^{\gamma l} + e^{-\gamma l}$$

$$2 \sinh \gamma l = e^{\gamma l} - e^{-\gamma l})$$

Problem 6.7

If the transmission line of problem 6.6 is short-circuited, show that its input impedance is given by

$$Z_{sc} = Z_0 \tanh \gamma l$$

and when it is open-circuited the input impedance is

$$Z_{oc} = Z_0 \coth \gamma l$$

By taking the product of these quantities, suggest a method for measuring the characteristic impedance of the line.

Problem 6.8

Show that the input impedance of a short-circuited loss-free line of length l is given by

$$Z_i = i \sqrt{\frac{L_0}{C_0}} \tan \frac{2\pi l}{\lambda}$$

and by sketching the variation of the ratio $Z_i/\sqrt{L_0/C_0}$ with l, show that for l just greater than $(2n+1)\lambda/4$, Z_i is capacitive, and for l just greater than $n\lambda/2$ it is inductive. (This provides a positive or negative reactance to match another line.)

Problem 6.9

Show that a line of characteristic impedance Z_0 may be matched to a load Z_L by a loss-free quarter wavelength line of characteristic impedance Z_m if $Z_m^2 = Z_0 Z_L$. (Hint—calculate the input impedance at the $Z_0 Z_m$ junction.)

Problem 6.10

Show that a short-circuited quarter wavelength loss-free line has an infinite impedance and that if it is bridged across another transmission line it will not affect the fundamental wavelength but will short-circuit any undesirable second harmonic.

Problem 6.11

Show that a loss-free line of characteristic impedance Z_0 and length $n\lambda/2$ may be used to couple two high frequency circuits without affecting other impedances.

Problem 6.12

In a transmission line with losses where $R_0/\omega L_0$ and $G_0/\omega C_0$ are both small quantities expand the expression for the propagation constant

$$\gamma = [(R_0 + i\omega L_0)(G_0 + i\omega C_0)]^{1/2}$$

to show that the attenuation constant

$$\alpha = \frac{R_0}{2}\sqrt{\frac{C_0}{L_0}} + \frac{G_0}{2}\sqrt{\frac{L_0}{C_0}}$$

and the wave number

$$k = \omega\sqrt{L_0 C_0} = \frac{\omega}{v}$$

Show that for $G_0 = 0$ the Q value of such a line is given by $k/2\alpha$.

Problem 6.13

Expand the expression for the characteristic impedance of the transmission line of problem 6.12 in terms of the characteristic impedance of a lossless line to show that if

$$\frac{R_0}{L_0} = \frac{G_0}{C_0}$$

the impedance remains real because the phase effects introduced by the series and shunt losses are equal but opposite.

Problem 6.14

The wave description of an electron ot total energy E in a potential well of depth V over the region $0 < x < l$ is given by Schrödinger's time independent wave equation

$$\frac{\partial^2 \psi}{\partial x^2} + \frac{8\pi^2 m}{h^2}(E - V)\psi = 0$$

where m is the electron mass and h is Plancks constant. (Note that $V = 0$ within the well.)

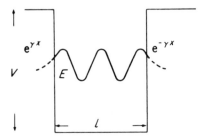

Show that for $E > V$ (inside the potential well) the solution for ψ is a standing wave solution but for $E < V$ (outside the region $0 < x < l$) the x dependence of ψ is $e^{\pm \gamma x}$ where

$$\gamma = \frac{2\pi}{h}\sqrt{2m(V-E)}$$

Problem 6.15

A localized magnetic field H in an electrically conducting medium of permeability μ and conductivity σ will diffuse through the medium in the x-direction at a rate given by

$$\frac{\partial H}{\partial t} = \frac{1}{\mu\sigma}\frac{\partial^2 H}{\partial x^2}$$

Show that the time of decay of the field is given approximately by $L^2\mu\sigma$, where L is the extent of the medium, and show that for a copper sphere of radius 1 metre this time is less than 100 seconds.

$$\mu\ (\text{copper}) = 1\cdot26 \times 10^{-6}\ \text{henries per metre}$$

$$\sigma\ (\text{copper}) = 5\cdot8 \times 10^{7}\ \text{siemens per metre}$$

(If the earth's core were molten iron its field would freely decay in approximately 15×10^3 years. In the sun the local field would take 10^{10} years to decay. When σ is very high the local field will change only by being carried away by the movement of the medium—such a field is said to be 'frozen' into the medium—the field lines are stretched and exert a restoring force against the motion.)

Problem 6.16

A point x_0 at the centre of a large slab of material of thermal conductivity k, specific heat C and density ρ has an infinitely high temperature T at a time t_0. If the heat diffuses through the medium at a rate given by

$$\frac{\partial T}{\partial t} = \frac{k}{\rho C}\frac{\partial^2 T}{\partial x^2} = d\frac{\partial^2 T}{\partial x^2}$$

show that the heat flow along the x-axis is given by

$$f(\alpha, t) = \frac{r}{\sqrt{\pi}}e^{-(r\alpha)^2}$$

where

$$\alpha = (x - x_0) \quad \text{and} \quad r = \frac{1}{2\sqrt{dt}}$$

by inserting this solution in the differential equation. The solution is a Guassian function; its behaviour with x and t in this problem is shown in fig. 9.12. At (x_0, t_0) the function is the Dirac delta function. The Guassian curves decay in height and widen with time as the heat spreads through the medium, the total heat, i.e. the area under the Gaussian curve, remaining constant.

Summary of Important Results

Lossless transmission line

Inductance per unit length $= L_0$ or μ

Capacitance per unit length $= C_0$ or ϵ

Wave equation

$$\frac{\partial^2 V}{\partial x^2} = \frac{1}{v^2}\frac{\partial^2 V}{\partial t^2} \ \text{(voltage)}$$

$$\frac{\partial^2 I}{\partial x^2} = \frac{1}{v^2}\frac{\partial^2 I}{\partial t^2} \ \text{(current)}$$

phase velocity

$$v^2 = \frac{1}{L_0 C_0} \ \text{or} \ \frac{1}{\mu\epsilon}$$

Characteristic impedance

$$Z_0 = \frac{V}{I} = \sqrt{\frac{L_0}{C_0}} \ \text{or} \ \sqrt{\frac{\mu}{\epsilon}} \ \text{(for right going wave)}$$

$(-Z_0 \text{ for left going wave})$

Transmission line with losses

Resistance R_0 per unit length

Shunt conductance G_0 per unit length

Wave equation takes form

$$e^{i\omega t}\left(\frac{\partial^2 V}{\partial x^2} - \gamma^2 V\right) = 0 \quad \text{(same for } I\text{)}$$

where $\gamma = \alpha + ik$ is the propagation constant

$\alpha = $ attenuation coefficient

$k = $ wave number

giving

$$V = A\,e^{-\alpha x}\,e^{i(\omega t - kx)} + B\,e^{\alpha x}\,e^{i(\omega t + kx)}$$

Characteristic impedance

$$Z_0' = \frac{V}{I} = \sqrt{\frac{R_0 + i\omega L_0}{G_0 + i\omega C_0}} \quad \text{(right going wave)}$$

$(-Z_0'$ for left going wave$)$

Wave attenuation

Energy absorption in a medium described by diffusion equation

$$\frac{\partial^2 \phi}{\partial x^2} = \frac{1}{d}\frac{\partial \phi}{\partial t}$$

Add to wave equation to account for attenuation giving

$$\frac{\partial^2 \phi}{\partial x^2} = \frac{1}{c^2}\frac{\partial^2 \phi}{\partial t^2} + \frac{1}{d}\frac{\partial \phi}{\partial t}$$

with exponentially decaying solution

$$\phi = \phi_m\,e^{-\alpha x}\,e^{i(\omega t - kx)}$$

Chapter 7

Electromagnetic Waves

Earlier chapters have shown that the velocity of waves through a medium is determined by the inertia and the elasticity of the medium. These two properties are capable of storing wave energy in the medium, and in the absence of energy dissipation they also determine the impedance presented by the medium to the waves. In addition, when there is no loss mechanism a pure wave equation with a sine or cosine solution will always be obtained, but this equation will be modified by any resistive or loss term to give an oscillatory solution which decays with time or distance.

These physical processes describe exactly the propagation of electromagnetic waves through a medium. The magnetic inertia of the medium, as in the case of the transmission line, is provided by the inductive property of the medium, i.e. the permeability μ, which has the units of henries per metre. The elasticity or capacitive property of the medium is provided by the permittivity ϵ, with units of farads per metre. The storage of magnetic energy arises through the permeability μ; the potential or electric field energy is stored through the permittivity ϵ.

If the material is defined as a dielectric, only μ and ϵ are effective and a pure wave equation for both the magnetic field vector H and the electric field vector E will result. If the medium is a conductor, having conductivity σ (the inverse of resistivity) with dimensions of siemens per metre, in addition to μ and ϵ, then some of the wave energy will be dissipated and absorption will take place.

In this chapter we will consider first the propagation of electromagnetic waves in a medium characterized by μ and ϵ only, and then treat the general case of a medium having μ, ϵ and σ properties.

Maxwell's Equations

Electromagnetic waves arise whenever an electric charge changes its velocity. Electrons moving from a higher to a lower energy level in an atom will radiate a wave of a particular frequency and wavelength. A very hot ionized gas

187

consisting of charged particles will radiate waves over a continuous spectrum as the paths of individual particles are curved in mutual collisions. This radiation is called 'Bremsstrahlung'. The radiation of electromagnetic waves from an aerial is due to the oscillatory motion of charges in an alternating current flowing in the aerial.

Fig. 7.1 shows the frequency spectrum of electromagnetic waves. All of these waves exhibit the same physical characteristics.

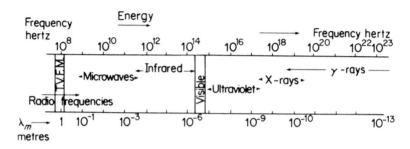

Fig. 7.1. Wavelengths and frequencies in the electro-magnetic spectrum

It is quite remarkable that the whole of electromagnetic theory can be described by the four vector relations in Maxwell's equations. In examining these relations in detail we shall see that two are steady state, that is, independent of time, and that two are time-varying.

The two time-varying equations are mathematically sufficient to produce separate wave equations for the electric and magnetic field vectors, E and H, but the steady state equations help to identify the wave nature as transverse.

The first time-varying equation relates the *time* variation of the magnetic induction, $\mu H = B$, with the *space* variation of E, that is

$$\frac{\partial}{\partial t}(\mu H) \text{ is connected with } \frac{\partial E}{\partial z} \text{(say)}$$

This is nothing but a form of Lenz's or Faraday's Law, as we shall see.

The second time-varying equation states that the *time* variation of ϵE defines the *space* variation of H, that is

$$\frac{\partial}{\partial t}(\epsilon E) \text{ is connected with } \frac{\partial H}{\partial z} \text{(say)}$$

Again we shall see that this is really a statement of Ampere's Law.

These equations show that the variations of E in time and space affect those of H and vice versa. E and H cannot be considered as isolated quantities but are interdependent.

The product ϵE has dimensions

$$\frac{\text{farads}}{\text{metre}} \times \frac{\text{volts}}{\text{metre}} = \frac{\text{charge}}{\text{area}}$$

This charge per unit area is called the displacement charge $D = \epsilon E$.
Physically it appears in a dielectric when an applied electric field polarizes
the constituent atoms or molecules and charge moves across any plane in the
dielectric which is normal to the applied field direction. If the applied field is
varying or alternating with time we see that the dimensions of

$$\frac{\partial \mathbf{D}}{\partial t} = \frac{\partial}{\partial t}(\epsilon \mathbf{E}) = \frac{\text{charge}}{\text{time} \times \text{area}}$$

current per unit area. This current is called the displacement current. It is
comparatively simple to visualize this current in a dielectric where physical
charges may move—it is not easy to associate a displacement current with free
space in the absence of a material.

Consider what happens in the electric circuit of fig. 7.2 when the switch is
closed and the battery begins to charge the condenser C to a potential V. A
current I obeying Ohm's Law ($V = IR$) will flow through the connecting leads
as long as the condenser is charging and a compass needle or other magnetic
field detector placed near the leads will show the presence of the magnetic field
associated with that current. But suppose a magnetic field detector (shielded
from all outside effects) is placed in the region between the condenser plates
where no ohmic or conduction current is flowing. Would it detect a magnetic
field? The answer is yes; all the magnetic field effects from a current exist in this

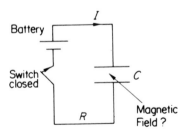

Fig. 7.2. In this circuit, when the switch is
closed the conduction current charges the
condenser. Throughout charging the quan-
tity $\epsilon \mathbf{E}$ in the volume of the condenser is
changing and the displacement current per
unit area $\partial/\partial t$ ($\epsilon \mathbf{E}$) is associated with the
magnetic field present between the con-
denser plates

region as long as the condenser is charging, that is, as long as the potential difference and the electric field between the condenser plates are changing.

It was Maxwell's major contribution to electromagnetic theory to assert that the existence of a time-changing electric field in free space gave rise to a displacement current. The same result follows from considering the conservation of charge. The flow of charge into any small volume in space must equal that flowing out. If the volume includes the top plate of the condenser the ohmic current through the leads produces the flow into the volume, while the displacement current represents the flow out.

In future, therefore, two different kinds of current will have to be considered:
(1) the familar conduction current obeying Ohm's Law ($V = IR$) and
(2) the displacement current of density $\partial \mathbf{D}/\partial t$.

In a medium of permeability μ and permittivity ϵ, but where the conductivity $\sigma = 0$, *the displacement current will be the only current flowing.* In this case a pure wave equation for E and H will follow and there will be no energy loss or attenuation.

When $\sigma \neq 0$ a resistive element allows the conduction current to flow, energy loss will follow, a diffusion term is added to the wave equation and the wave amplitude will attenuate exponentially with distance. We shall see that the relative magnitude of these two currents is frequency-dependent and that their ratio governs whether the medium behaves as a conductor or as a dielectric.

Electromagnetic Waves in a Medium having Finite Permeability μ and Permittivity ϵ but with Conductivity $\sigma=0$

We shall consider a system of plane waves and choose the plane xy as that region over which the wave properties are constant. These properties will not vary with respect to x and y and all derivatives $\partial/\partial x$ and $\partial/\partial y$ will be zero.

The first time-varying equation of Maxwell is written in vector notation as

$$\text{curl } \mathbf{E} = \nabla \times \mathbf{E} = -\frac{\partial \mathbf{B}}{\partial t} = -\mu \frac{\partial \mathbf{H}}{\partial t}$$

This represents three component equations:

$$\left.\begin{aligned}
-\mu \frac{\partial}{\partial t} H_x &= \frac{\partial}{\partial y} E_z - \frac{\partial}{\partial z} E_y \\[2mm]
-\mu \frac{\partial}{\partial t} H_y &= \frac{\partial}{\partial z} E_x - \frac{\partial}{\partial x} E_z \\[2mm]
-\mu \frac{\partial}{\partial t} H_z &= \frac{\partial}{\partial x} E_y - \frac{\partial}{\partial y} E_x
\end{aligned}\right\} \qquad (7.1)$$

where the subscripts represent the component directions. The dimensions of these equations may be written

$$-\frac{\mu H}{\text{time}} = \frac{E}{\text{length}}$$

and multiplying each side by (length)^2 gives

$$-\frac{\mu H}{\text{time}} \times \text{area} = E \times \text{length}$$

i.e.

$$\frac{\text{total magnetic flux}}{\text{time}} = \text{volts}$$

This is dimensionally of the form of Lenz's or Faraday's Law.

The second time-varying equation of Maxwell is written in vector notation as

$$\text{curl } \mathbf{H} = \nabla \times \mathbf{H} = \frac{\partial \mathbf{D}}{\partial t} = \epsilon \frac{\partial \mathbf{E}}{\partial t}$$

This represents three component equations:

$$\left. \begin{aligned} \epsilon \frac{\partial}{\partial t} E_x &= \frac{\partial}{\partial y} H_z - \frac{\partial}{\partial z} H_y \\[1mm] \epsilon \frac{\partial}{\partial t} E_y &= \frac{\partial}{\partial z} H_x - \frac{\partial}{\partial x} H_z \\[1mm] \epsilon \frac{\partial}{\partial t} E_z &= \frac{\partial}{\partial x} H_y - \frac{\partial}{\partial y} H_x \end{aligned} \right\} \qquad (7.2)$$

The dimensions of these equations may be written

$$\frac{\text{current } I}{\text{area}} = \frac{H}{\text{length}}$$

and multiplying both sides by a length gives

$$\frac{\text{current}}{\text{length}} = \frac{I}{\text{length}} = H$$

which is dimensionally of the form of Ampere's Law (i.e., the circular magnetic field at radius r due to the current I flowing in a straight wire is given by $H = I/2\pi r$). Maxwell's first steady state equation may be written

$$\text{div } \mathbf{D} = \nabla . \mathbf{D} = \epsilon \left(\frac{\partial E_x}{\partial x} + \frac{\partial E_y}{\partial y} + \frac{\partial E_z}{\partial z} \right) = \rho \qquad (7.3)$$

where ρ is the charge density. This states that over a small volume element $dx \, dy \, dz$ of charge density ρ the change of displacement depends upon the value of ρ.

When $\rho = 0$ the equation becomes

$$\epsilon \left(\frac{\partial E_x}{\partial x} + \frac{\partial E_y}{\partial y} + \frac{\partial E_z}{\partial z} \right) = 0 \tag{7.3a}$$

so that if the displacement $D = \epsilon E$ is graphically represented by flux lines which must begin and end on electric charges, the number of flux lines entering the volume element $dx\, dy\, dz$ must equal the number leaving it.

The second steady state equation is written

$$\text{div } \mathbf{B} = \nabla . \mathbf{B} = \mu \left(\frac{\partial H_x}{\partial x} + \frac{\partial H_y}{\partial y} + \frac{\partial H_z}{\partial z} \right) = 0 \tag{7.4}$$

Again this states that an equal number of magnetic induction lines enter and leave the volume $dx\, dy\, dz$. This is a physical consequence of the non-existence of isolated magnetic poles, i.e. a single north pole or south pole.

Whereas the charge density ρ in equation (7.3) can be positive, i.e. a source of flux lines (or displacement), or negative, i.e. a sink of flux lines (or displacement), no separate source or sink of magnetic induction can exist in isolation, every source being matched by a sink of equal strength.

The Wave Equation for Electromagnetic Waves

Since, with these plane waves, all derivatives with respect to x and y are zero, equations (7.1) and (7.4) give

$$-\mu \frac{\partial H_z}{\partial t} = 0 \quad \text{and} \quad \frac{\partial H_z}{\partial z} = 0$$

therefore H_z is constant in space and time and because we are considering only the oscillatory nature of H a constant H_z can have no effect on the wave motion. We can therefore put $H_z = 0$. A similar consideration of equations (7.2) and (7.3a) leads to the result that $E_z = 0$.

The absence of variation in H_z and E_z means that the oscillations or variations in H and E occur in directions perpendicular to the z-direction. We shall see that this leads to the conclusion that electromagnetic waves are transverse waves.

In addition to having plane waves we shall simplify our picture by considering only *plane-polarized* waves.

We can choose the electric field vibration to be in either the x or y direction. Let us consider E_x only, with $E_y = 0$. In this case equations (7.1) give

$$-\mu \frac{\partial H_y}{\partial t} = \frac{\partial E_x}{\partial z} \tag{7.1a}$$

and equations (7.2) give

$$\epsilon \frac{\partial E_x}{\partial t} = -\frac{\partial H_y}{\partial z} \tag{7.2a}$$

Using the fact that

$$\frac{\partial^2}{\partial z\, \partial t} = \frac{\partial^2}{\partial t\, \partial z}$$

it follows by taking $\partial/\partial t$ of equation (7.1a) and $\partial/\partial z$ of equation (7.2a) that

$$\frac{\partial^2}{\partial z^2} H_y = \mu\epsilon \frac{\partial^2}{\partial t^2} H_y \quad \text{(the wave equation for } H_y)$$

Similarly, by taking $\partial/\partial t$ of (7.2a) and $\partial/\partial z$ of (7.1a), we obtain

$$\frac{\partial^2}{\partial z^2} E_x = \mu\epsilon \frac{\partial^2}{\partial t^2} E_x \quad \text{(the wave equation for } E_x)$$

Thus the vectors E_x and H_y both obey the same wave equation, propagating in the z-direction with the same velocity $v^2 = 1/\mu\epsilon$. In free space the velocity is that of light, that is, $c^2 = 1/\mu_0\epsilon_0$, where μ_0 is the permeability of free space and ϵ_0 is the permittivity of free space.

The solutions to these wave equations may be written, for plane waves, as

$$E_x = E_0 \sin \frac{2\pi}{\lambda}(vt - z)$$

$$H_y = H_0 \sin \frac{2\pi}{\lambda}(vt - z)$$

where E_0 and H_0 are the maximum amplitude values of E and H. Note that the

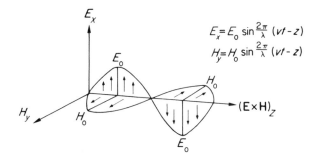

Fig. 7.3. In a plane-polarized electro-magnetic wave the electric field vector E_x and magnetic field vector H_y are perpendicular to each other and vary sinusoidally. In a non-conducting medium they are in phase. The vector product, $\mathbf{E} \times \mathbf{H}$, is called the Poynting vector, and gives both the direction and quantity of energy flow per second across unit area in watts/metre^2

sine (or cosine) solutions means that no attenuation occurs: only displacement currents are involved and there are no conductive or ohmic currents.

We can represent the electromagnetic wave (E_x, H_y) travelling in the z-direction in fig. 7.3, and recall that because E_z and H_z are constant (or zero) the electromagnetic wave is a transverse wave.

The direction of propagation of the waves will always be in the $\mathbf{E} \times \mathbf{H}$ direction; in this case, $\mathbf{E} \times \mathbf{H}$ has magnitude $E_x H_y$ and is in the z-direction.

This product has the dimensions

$$\frac{\text{voltage} \times \text{current}}{\text{length} \times \text{length}} = \frac{\text{electrical power}}{\text{area}}$$

measured in units of watts per metre^2.

The vector product, $\mathbf{E} \times \mathbf{H}$, is called the Poynting vector; this measures the flow of energy per second across unit area.

Illustration of Poynting Vector

We can illustrate the flow of electromagnetic energy in terms of the Poynting vector by considering the simple circuit of fig. 7.4, where the parallel plate condenser of area A and separation d, containing a dielectric of permittivity ϵ, is being charged to a voltage V.

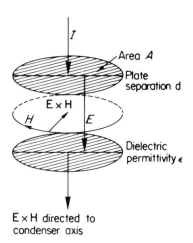

E × H directed to
condenser axis

Fig. 7.4. During charging the Poynting vector $\mathbf{E} \times \mathbf{H}$ is directed into the condenser volume. At the end of the charging the energy is totally electrostatic and equals the product of the condenser volume, Ad, and the electrostatic energy per unit volume, $\frac{1}{2}\epsilon E^2$

Throughout the charging process current flows, and the electric and magnetic field vectors show that the Poynting vector is always directed into the volume Ad occupied by the dielectric.

The capacitance C of the condenser is $\epsilon A/d$ and the total energy of the condenser at potential V is $\frac{1}{2}CV^2$ joules, which is stored as electrostatic energy. But $V = Ed$, where E is the final value of the electric field, so that the total energy

$$\tfrac{1}{2}CV^2 = \tfrac{1}{2}\left(\frac{\epsilon A}{d}\right)E^2 d^2 = \tfrac{1}{2}(\epsilon E^2)Ad$$

where Ad is the volume of the condenser.

The electrostatic energy per unit volume stored in the condenser is therefore $\frac{1}{2}\epsilon E^2$ and results from the flow of electromagnetic energy during charging.

Impedance of a Dielectric to Electromagnetic Waves

If we put the solutions

$$E_x = E_0 \sin\frac{2\pi}{\lambda}(vt - z)$$

and

$$H_y = H_0 \sin\frac{2\pi}{\lambda}(vt - z)$$

in equation (7.1a) where

$$-\mu\frac{\partial H_y}{\partial t} = \frac{\partial E_x}{\partial z}$$

then

$$-\mu v H_y = -E_x, \quad \text{and since} \quad v^2 = \frac{1}{\mu\epsilon}$$

$$\sqrt{\mu}\, H_y = \sqrt{\epsilon}\, E_x$$

that is

$$\frac{E_x}{H_y} = \sqrt{\frac{\mu}{\epsilon}} = \frac{E_0}{H_0}$$

which has the dimensions of ohms.

The value $\sqrt{\mu/\epsilon}$ therefore represents the *characteristic impedance* of the medium to electromagnetic waves (compare this with the equivalent result $V/I = \sqrt{L_0/C_0} = Z_0$ for the transmission line of the last chapter).

In free space

$$\frac{E_x}{H_y} = \sqrt{\frac{\mu_0}{\epsilon_0}} = 376\cdot7 \text{ ohms}$$

so that free space presents an impedance of 376·7 ohms to electromagnetic waves travelling through it.

It follows from

$$\frac{E_x}{E_y} = \sqrt{\frac{\mu}{\epsilon}} \quad \text{that} \quad \frac{E_x^2}{H_y^2} = \frac{\mu}{\epsilon}$$

and therefore

$$\epsilon E_x^2 = \mu H_y^2$$

Both of these quantities have the dimensions of energy per unit volume, for instance ϵE_x^2 has dimensions

$$\frac{\text{farads}}{\text{metre}} \times \frac{\text{volts}^2}{\text{metres}^2} = \frac{\text{joules}}{\text{metres}^3}$$

as we saw in the illustration of the Poynting vector. Thus for a dielectric the electrostatic energy $\frac{1}{2}\epsilon E_x^2$ per unit volume in an electromagnetic wave equals the magnetic energy per unit volume $\frac{1}{2}\mu H_y^2$ and the total energy is the sum $\frac{1}{2}\epsilon E_x^2 + \frac{1}{2}\mu H_y^2$.

This gives the instanteous value of the energy per unit volume and we know that, in the wave,

$$E_x = E_0 \sin (2\pi/\lambda)(vt - z)$$

and

$$H_y = H_0 \sin (2\pi/\lambda)(vt - z)$$

so that the *time averaged value* of the energy per unit volume is

$$\frac{1}{2}\epsilon \bar{E}_x^2 + \frac{1}{2}\mu \bar{H}_y^2 = \frac{1}{4}\epsilon E_0^2 + \frac{1}{4}\mu H_0^2$$

$$= \frac{1}{2}\epsilon E_0^2 \text{ joules m}^{-3}$$

Now the amount of energy in an electromagnetic wave which crosses unit area in unit time is called the intensity, I, of the wave and is evidently $(\frac{1}{2}\epsilon E_0^2)v$ where v is the velocity of the wave.

This gives the time averaged value of the Poynting vector and, for an electromagnetic wave in free space we have

$$I = \frac{1}{2}c\epsilon_0 E_0^2 = \frac{1}{2}c\mu_0 H_0^2 \text{ watts m}^{-2}$$

(Problems 7.1, 7.2, 7.3, 7.4, 7.5, 7.6, 7.7, 7.8, 7.9, 7.10)

Electromagnetic Waves in a Medium of Properties μ, ϵ and σ (where $\sigma \neq 0$)

From a physical point of view the electric vector in electromagnetic waves plays a much more significant role than the magnetic vector, e.g., most optical effects are associated with the electric vector. We shall therefore concentrate our discussion on the electric field behaviour.

In a medium of conductivity $\sigma = 0$ we have obtained the wave equation

$$\frac{\partial^2 E_x}{\partial z^2} = \mu\epsilon \frac{\partial^2 E_x}{\partial t^2}$$

where the right hand term, rewritten

$$\mu\frac{\partial}{\partial t}\left[\frac{\partial}{\partial t}(\epsilon E_x)\right]$$

shows that we are considering a term

$$\mu\frac{\partial}{\partial t}\left[\frac{\text{displacement current}}{\text{area}}\right]$$

When $\sigma \neq 0$ we must also consider the conduction currents which flow. These currents are given by Ohm's Law as $I = V/R$, and we define the current density, that is, the current per unit area, as

$$J = \frac{I}{\text{Area}} = \frac{1}{R \times \text{Length}} \times \frac{V}{\text{Length}} = \sigma E$$

where σ is the conductivity $1/(R \times \text{Length})$ and E is the electric field. $J = \sigma E$ is another form of Ohm's Law.

With both displacement and conduction currents flowing, Maxwell's second time-varying equation reads, in vector form,

$$\nabla \times \mathbf{H} = \frac{\partial}{\partial t}\mathbf{D} + \mathbf{J} \tag{7.5}$$

each term on the right-hand side having dimensions of current per unit area. The presence of the conduction current modifies the wave equation by adding a second term of the same form to its right hand side, namely

$$\mu\frac{\partial}{\partial t}\left(\frac{\text{current}}{\text{area}}\right) \text{ which is } \mu\frac{\partial}{\partial t}(\mathbf{J}) = \mu\frac{\partial}{\partial t}(\sigma \mathbf{E})$$

The final equation is therefore given by

$$\frac{\partial^2}{\partial z^2}E_x = \mu\epsilon\frac{\partial^2}{\partial t^2}E_x + \mu\sigma\frac{\partial}{\partial t}E_x \qquad (7.6)$$

and this equation may be derived formally by writing the component equation of (7.5) as

$$\epsilon\frac{\partial E_x}{\partial t} + \sigma E_x = -\frac{\partial H_y}{\partial z} \qquad (7.5a)$$

together with

$$-\mu\frac{\partial H_y}{\partial t} = \frac{\partial E_x}{\partial z} \qquad (7.1a)$$

and taking $\partial/\partial t$ of (7.5a) and $\partial/\partial z$ of (7.1a). We see immediately that the presence of the resistive or dissipation term, which allows conduction currents to flow, will add a diffusion term of the type discussed in the last chapter to the pure wave equation. The product $(\mu\sigma)^{-1}$ is called the magnetic diffusivity, and has the dimensions $L^2 T^{-1}$, as we expect of all diffusion coefficients.

We are now going to look for the behaviour of E_x in this new equation, with the assumption that its time-variation is simple harmonic, so that $E_x = E_0\,e^{i\omega t}$. Using this value in equation (7.6) gives

$$\frac{\partial^2 E_x}{\partial z^2} - (i\omega\mu\sigma - \omega^2\mu\epsilon)E_x = 0$$

which is in the form of equation (6.5), written

$$\frac{\partial^2 E_x}{\partial z^2} - \gamma^2 E_x = 0$$

where $\gamma^2 = i\omega\mu\sigma - \omega^2\mu\epsilon$.

We saw in Chapter 6 that this produced a solution with the term $e^{-\gamma z}$ or $e^{+\gamma z}$, but we concentrate on the E_x oscillation in the positive z-direction by writing

$$E_x = E_0\,e^{i\omega t}\,e^{-\gamma z}$$

In order to assign a suitable value to γ we must go back to equation (7.6) and consider the relative magnitudes of the two right-hand side terms. If the medium is a dielectric, only displacement currents will flow. When the medium is a conductor, the ohmic currents of the second term on the right hand side will be dominant. The ratio of the magnitudes of the conduction current density to the displacement current density is the ratio of the two right-hand side terms. This ratio is

$$\frac{J}{\partial D/\partial t} = \frac{\sigma E_x}{\partial/\partial t(\epsilon E_x)} = \frac{\sigma E_x}{\partial/\partial t(\epsilon E_o\,e^{i\omega t})} = \frac{\sigma E_x}{i\omega\epsilon E_x} = \frac{\sigma}{i\omega\epsilon}$$

We see immediately from the presence of i that the phase of the displacement current is $90°$ ahead of that of the ohmic or conduction current. It is also $90°$ ahead of the electric field E_x so the displacement current dissipates no power.

For a conductor, where $\mathbf{J} \gg \partial \mathbf{D}/\partial t$, we have $\sigma \gg \omega\epsilon$, and $\gamma^2 = i\sigma(\omega\mu) - \omega\epsilon(\omega\mu)$ becomes

$$\gamma^2 \approx i\sigma\omega\mu$$

to a high order of accuracy.

Now

$$\sqrt{i} = \frac{1+i}{\sqrt{2}}$$

so that

$$\gamma = (1+i)\left(\frac{\omega\mu\sigma}{2}\right)^{1/2}$$

and

$$E_x = E_0\, e^{i\omega t}\, e^{-\gamma z}$$
$$= E_0\, e^{-(\omega\mu\sigma/2)^{1/2}z}\, e^{i[\omega t - (\omega\mu\sigma/2)^{1/2}z]}$$

a progressive wave in the positive z-direction with an amplitude decaying with the factor $e^{-(\omega\mu\sigma/2)^{1/2}z}$.

Note that the product $\omega\mu\sigma$ has dimensions L^{-2}.

(Problem 7.11)

Skin Depth

After travelling a distance

$$\delta = \left(\frac{2}{\omega\mu\sigma}\right)^{1/2}$$

in the conductor the electric field vector has decayed to a value $E_x = E_0\, e^{-1}$; this distance is called the *skin depth* (fig. 7.5).

For copper, with $\mu \approx \mu_0$ and $\sigma = 5\cdot8 \times 10^7$ siemens/metre at a frequency of 60 Hertz, $\delta \approx 9$ mm; at 1 MHertz, $\delta \approx 6\cdot6 \times 10^{-5}$ metres and at 30,000 MHertz (radar wavelength of 1 cm), $\delta \approx 3\cdot8 \times 10^{-7}$ metres.

Thus high frequency electromagnetic waves propagate only a very small distance in a conductor. The electric field is confined to a very small region at the surface; significant currents will flow only at the surface and the resistance of the conductor therefore increases with frequency. We see also why a conductor can act to 'shield' a region from electromagnetic waves.

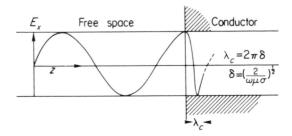

Fig. 7.5. Electromagnetic waves in a dielectric strike the plane surface of a conductor, and the electric field vector E_0 is damped to a value $E_0\, e^{-1}$ in a distance of $(2/\omega\mu\sigma)^{1/2}$, the 'skin depth'. This explains the electrical shielding properties of a conductor. λ_c is the wavelength in the conductor

Electromagnetic Wave Velocity in a Conductor and Anomalous Dispersion

The phase velocity of the wave v is given by

$$v = \frac{\omega}{k} = \frac{\omega}{(\omega\mu\sigma/2)^{1/2}} = \omega\delta = \left(\frac{2\omega}{\mu\sigma}\right)^{1/2} = \nu\lambda_c$$

When δ is small v is small, and the refractive index c/v of a conductor can be very large. We shall see later that this can explain the high optical reflectivities of good conductors. The velocity $v = \omega\delta = 2\pi\nu\delta$, so that λ_c in the conductor is $2\pi\delta$ and can be very small. Since v is a function of the frequency an electrical conductor is a dispersive medium to electromagnetic waves. Moreover, as the table below shows us, $\partial v/\partial\lambda$ is negative, so that the conductor is anomalously dispersive and the group velocity is greater than the wave velocity. Since $c^2/v^2 = \mu\epsilon/\mu_0\epsilon_0 = \mu_r\epsilon_r$, where the subscript r defines non-dimensional relative values and $\mu/\mu_0 = \mu_r$, $\epsilon/\epsilon_o = \epsilon_r$, then for $\mu_r \approx 1$

$$\epsilon_r v^2 = c^2$$

and

$$\frac{\partial}{\partial\lambda}\epsilon_r = -\frac{2}{v}\epsilon_r\frac{\partial v}{\partial\lambda}$$

which confirms our statement in the chapter on group velocity that for $\partial\epsilon_r/\partial\lambda$ positive a medium is anomalously dispersive. We see too that $c^2/v^2 = \epsilon_r = n^2$, where n is the refractive index, so that the curve in fig. 2.9 showing the reactive behaviour of the oscillator impedance at displacement resonance is also showing the behaviour of n. This relative value of the permittivity is, of course, familiarly known as the dielectric constant when the frequency is low. This identity is lost at higher frequencies because the permittivity is frequency-dependent.

frequency Hertz	$\lambda_{\text{free space}}$	δ (m)	$v_{\text{conductor}} = \omega\delta$ (m/sec)	refractive index $(c/v_{\text{conductor}})$
60	5000 km	9×10^{-3}	3·2	$9·5 \times 10^{7}$
10^{6}	300 m	$6·6 \times 10^{-5}$	$4·1 \times 10^{2}$	$7·3 \times 10^{5}$
3×10^{10}	10^{-2} m	$3·9 \times 10^{-7}$	$7·1 \times 10^{4}$	$4·2 \times 10^{3}$

Note that $\lambda_c = 2\pi\delta$ is very small, and that when an electromagnetic wave strikes a conducting surface the electric field vector will drop to about 1% of its surface value in a distance equal to $\frac{3}{4}\lambda_c = 4·6\delta$. Effectively therefore, the electromagnetic wave travels less than one wavelength into the conductor.

(Problems 7.12, 7.13, 7.14)

When is a Medium a Conductor or a Dielectric?

We have already seen that in any medium having $\mu\epsilon$ and σ properties the magnitude of the ratio of the conduction current density to the displacement current density

$$\frac{\mathbf{J}}{\partial\mathbf{D}/\partial t} = \frac{\sigma}{\omega\epsilon}$$

a non-dimensional quantity.

We may therefore represent the medium by the simple circuit in fig. 7.6, where the total current is divided between the two branches, a capacitative branch of reactance $1/\omega e$ (ohms-metres) and a resistive branch of conductance σ (siemens/metre). If σ is large the resistivity is small, and most of the current flows through the σ branch and is conductive. If the capacitative reactance $1/\omega\epsilon$ is so small that it takes most of the current, this current is the displacement current and the medium behaves as a dielectric.

Quite arbitrarily we say that if

$$\frac{\mathbf{J}}{\partial\mathbf{D}/\partial t} = \frac{\sigma}{\omega\epsilon} > 100$$

then conduction currents dominate and the medium is a conductor. If

$$\frac{\partial\mathbf{D}/\partial t}{\mathbf{J}} = \frac{\omega\epsilon}{\sigma} > 100$$

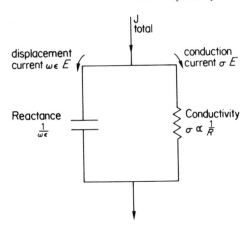

Fig. 7.6. A simple circuit showing the response of a conducting medium to an electromagnetic wave. The total current density J is divided by the parallel circuit in the ratio $\sigma/\omega\epsilon$ (the ratio of the conduction current density to the displacement current density). A large conductance σ (small resistance) gives a large conduction current while a small capacitative reactance $1/\omega\epsilon$ allows a large displacement current to flow. For a conductor $\sigma/\omega\epsilon \geq 100$; for a dielectric $\omega\epsilon/\sigma \geq 100$. Note the frequency dependence of this ratio. At $\omega \approx 10^{20}$ rad/sec copper is a dielectric to X-rays

then displacement currents dominate and the material behaves as a dielectric. Between these values exist a range of quasi-conductors; some of the semi-conductors fall into this category.

The ratio $\sigma/\omega\epsilon$ is, however, frequency dependent, and a conductor at one frequency may be a dielectric at another.

For copper, which has $\sigma = 5 \cdot 8 \times 10^7$ siemens/metre and $\epsilon \approx \epsilon_0 = 9 \times 10^{-12}$ farads/metre,

$$\frac{\sigma}{\omega\epsilon} \approx \frac{10^{18}}{\text{frequency}}$$

so up to a frequency of 10^{16} Hertz (the frequency of ultraviolet light) $\sigma/\omega\epsilon > 100$, and copper is a conductor. At a frequency of 10^{20} Hertz, however, (the frequency of X-rays) $\omega\epsilon/\sigma > 100$, and copper behaves as a dielectric. This explains why X-rays travel distances equivalent to many wavelengths in copper.

Typically an insulator has $\sigma \approx 10^{-15}$ siemens/metre and $\epsilon \approx 10^{-11}$ farads/metre, which gives

$$\frac{\omega\epsilon}{\sigma} \approx 10^4 \omega$$

so the conduction current is negligible at all frequencies.

Why will an Electromagnetic Wave not Propagate into a Conductor?

To answer this question we need only consider the simple circuit where a condenser C discharges through a resistance R. The voltage equation gives

$$\frac{q}{C} + IR = 0$$

and since $I = dq/dt$, we have

$$\frac{dq}{dt} = -\frac{q}{RC} \quad \text{or} \quad q = q_0\, e^{-t/RC}$$

where q_0 is the initial charge.

We see that an electric field will exist between the plates of the condenser only for a time $t \sim RC$ and will disappear when the charge has had time to distribute itself uniformly throughout the circuit. An electric field can only exist in the presence of a non-uniform charge distribution.

If we take a slab of any medium and place a charge of density q at a point within the slab, the medium will behave as an RC circuit and the equation

$$q = q_0\, e^{-t/RC}$$

becomes

$$q = q_0\, e^{-\sigma/\omega\epsilon} \to q_0\, e^{-\sigma t/\epsilon}\left(\frac{\epsilon \equiv C}{\sigma \equiv 1/R}\right)$$

The charge will distribute itself uniformly in a time $t \sim \epsilon/\sigma$, and the electric field will be maintained for that time only. The time ϵ/σ is called the *relaxation time* of the medium (RC time of the electrical circuit) and it is a measure of the maximum time for which an electric field can be maintained before the charge distribution becomes uniform.

Any electric field of a frequency ν, where $1/\nu = t > \epsilon/\sigma$, will not be maintained; only a high frequency field where $1/\nu = t < \epsilon/\sigma$ will establish itself.

Impedance of a Conducting Medium to Electromagnetic Waves

The impedance of a lossless medium is a real quantity. For the transmission line of Chapter 6 the characteristic impedance

$$Z_0 = \frac{V_+}{I_+} = \sqrt{\frac{L_0}{C_0}} \text{ ohms}$$

for an electromagnetic wave in a dielectric

$$Z = \frac{E_x}{H_y} = \sqrt{\frac{\mu}{\epsilon}} \text{ ohms}$$

with E_x and H_y in phase.

We saw in the case of the transmission line that when the loss mechanisms of a series resistance R_0 and a shunt conductance G_0 were introduced the impedance became the complex quantity

$$\mathbf{Z} = \sqrt{\frac{R_0 + i\omega L_0}{G_0 + i\omega C_0}}$$

We now ask what will be the impedance of a conducting medium of properties μ, ϵ and σ to electromagnetic waves? If the ratio of E_x to H_y is a complex quantity, it implies that a phase difference exists between the two field vectors.

We have already seen that in a conductor

$$E_x = E_0 \, e^{i\omega t} \, e^{-\gamma z}$$

where $\gamma = (1+i)(\omega\mu\sigma/2)^{1/2}$, and we shall now write $H_y = H_0 \, e^{i(\omega t - \phi)} \, e^{-\gamma z}$, suggesting that H_y lags E_x by a phase angle ϕ. This gives the impedance of the conductor as

$$\mathbf{Z}_c = \frac{E_x}{H_y} = \frac{E_0}{H_0} e^{i\phi}$$

Equation (7.1a) gives

$$\frac{\partial E_x}{\partial z} = -\mu \frac{\partial H_y}{\partial t}$$

so that

$$-\gamma E_x = -i\omega\mu H_y$$

and

$$\mathbf{Z}_c = \frac{E_x}{H_y} = \frac{\mathrm{i}\omega\mu}{\gamma} = \frac{\mathrm{i}(\omega\mu)}{(1+\mathrm{i})(\omega\mu\sigma/2)^{1/2}} = \frac{\mathrm{i}(1-\mathrm{i})}{(1+\mathrm{i})(1-\mathrm{i})}\left(\frac{2\omega\mu}{\sigma}\right)^{1/2}$$

$$= \frac{(1+\mathrm{i})}{2}\left(\frac{2\omega\mu}{\sigma}\right)^{1/2} = \frac{1+\mathrm{i}}{\sqrt{2}}\left(\frac{\omega\mu}{\sigma}\right)^{1/2}$$

$$= \left(\frac{\omega\mu}{\sigma}\right)^{1/2}\left(\frac{1}{\sqrt{2}}+\mathrm{i}\frac{1}{\sqrt{2}}\right) = \left(\frac{\omega\mu}{\sigma}\right)^{1/2}\mathrm{e}^{\mathrm{i}\phi}$$

a vector of magnitude $(\omega\mu/\sigma)^{1/2}$ and phase angle $\phi = 45°$. Thus the magnitude

$$Z_c = \frac{E_0}{H_0} = \left(\frac{\omega\mu}{\sigma}\right)^{1/2}$$

and H_y lags E_x by 45°.

We can also express $\mathbf{Z}_c$ by

$$\mathbf{Z}_c = R + \mathrm{i}X = \left(\frac{\omega\mu}{2\sigma}\right)^{1/2} + \mathrm{i}\left(\frac{\omega\mu}{2\sigma}\right)^{1/2}$$

and also write it

$$\mathbf{Z}_c = \frac{1+\mathrm{i}}{\sqrt{2}}\left(\frac{\omega\mu}{\sigma}\right)^{1/2}$$

$$= \sqrt{\frac{\mu_0}{\epsilon_0}\frac{\epsilon_0}{\epsilon}\frac{\mu}{\mu_0}\frac{\omega\epsilon}{\sigma}}\,\mathrm{e}^{\mathrm{i}\phi}$$

of magnitude

$$|Z_c| = 376{\cdot}6 \text{ ohms } \sqrt{\frac{\mu_r}{\epsilon_r}}\sqrt{\frac{\omega\epsilon}{\sigma}}$$

At a wavelength $\lambda = 10^{-1}$ m, i.e. at a frequency $\nu = 3000$ MHertz, the value of $\omega\epsilon/\sigma$ for copper is $2{\cdot}9 \times 10^{-9}$ and $\mu_r \approx \epsilon_r \approx 1$. This gives a magnitude $Z_c = 0{\cdot}02$ ohms at this frequency; for $\sigma = \infty$, $Z_c = 0$, and the electric field vector E_x vanishes, so we can say that when Z_c is small or zero the conductor behaves as a short circuit to the electric field. This sets up large conduction currents and the magnetic energy is increased.

In a dielectric, the impedance

$$Z = \frac{E_x}{H_y} = \sqrt{\frac{\mu}{\epsilon}}$$

led to the equivalence of the electric and magnetic field energy densities, that is, $\frac{1}{2}\mu H_y^2 = \frac{1}{2}\epsilon E_x^2$. In a conductor, the magnitude of the impedance

$$Z_c = \left|\frac{E_x}{H_y}\right| = \left(\frac{\omega\mu}{\sigma}\right)^{1/2}$$

so that the ratio of the magnetic to the electric field energy density in the wave is

$$\frac{\frac{1}{2}\mu H_y^2}{\frac{1}{2}\epsilon E_x^2} = \frac{\mu}{\epsilon}\frac{\sigma}{\omega\mu} = \frac{\sigma}{\omega\epsilon}$$

We already know that this ratio is very large for a conductor for it is the ratio of conduction to displacement currents, so that in a conductor the magnetic field energy dominates the electric field energy and increases as the electric field energy decreases.

Reflexion and Transmission of Electromagnetic Waves at a Boundary

(i) Normal Incidence

An infinite plane boundary separates two media of impedances Z_1 and Z_2 (real or complex) in fig. 7.7.

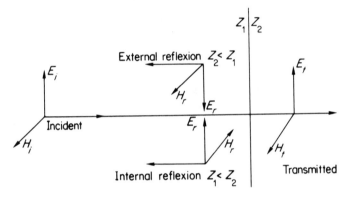

Fig. 7.7. Reflexion and transmission of an electromagnetic wave incident normally on a plane between media of impedances Z_1 and Z_2. The Poynting vector of the reflected wave $(\mathbf{E}\times\mathbf{H})_r$ shows that either $\mathbf{E}$ or $\mathbf{H}$ may be reversed in phase, depending on the relative magnitudes of Z_1 and Z_2

The electromagnetic wave normal to the boundary has the components shown where subscripts i, r and t denote incident, reflected and transmitted respectively. Note that the vector direction $(\mathbf{E}_r \times \mathbf{H}_r)$ must be opposite to that of $(\mathbf{E}_i \times \mathbf{H}_i)$ to satisfy the energy flow condition of the Poynting vector.

The boundary conditions, from electromagnetic theory, are that the components of the field vectors $\mathbf{E}$ and $\mathbf{H}$ tangential or parallel to the boundary are continuous across the boundary.

Thus

$$E_i + E_r = E_t$$

and

$$H_i + H_r = H_t$$

where

$$\frac{E_i}{H_i} = Z_1, \quad \frac{E_r}{H_r} = -Z_1 \quad \text{and} \quad \frac{E_t}{H_t} = Z_2$$

From these relations it is easy to show that the amplitude reflexion coefficient

$$R = \frac{E_r}{E_i} = \frac{Z_2 - Z_1}{Z_2 + Z_1}$$

and the amplitude transmission coefficient

$$T = \frac{E_t}{E_i} = \frac{2Z_2}{Z_2 + Z_1}$$

in agreement with the reflexion and transmission coefficients we have found for other waves. If the wave is travelling in air and strikes a perfect conductor of $Z_2 = 0$ at normal incidence then

$$\frac{E_r}{E_i} = \frac{Z_2 - Z_1}{Z_2 + Z_1} = -1$$

giving complete reflexion and

$$\frac{E_t}{E_i} = \frac{2Z_2}{Z_2 + Z_1} = 0$$

Thus good conductors are very good reflectors of electromagnetic waves, e.g. lightwaves are well reflected from metal surfaces. (See Summary on page 412.)

(ii) Oblique Incidence and Fresnel's Equations for dielectrics

When the incident wave is oblique and not normal to the infinite boundary of fig. 7.7 we may still use the boundary conditions of the preceding section for these apply to the *tangential* components of **E** and **H** at the boundary and remain valid.

In fig. 7.8(a) **H** is perpendicular to the plane of the paper with tangential components H_i, H_r and H_t but the tangential components of **E** become

$$E_i \cos \theta, E_r \cos \theta \quad \text{and } E_t \cos \phi \text{ respectively.}$$

In fig. 7.8(b) **E** is perpendicular to the plane of the paper with tangential components E_i, E_r and E_t but the tangential components of **H** become $H_i \cos \theta$, $H_r \cos \theta$ and $H_t \cos \phi$.

Using these components in the expressions for the reflexion and transmission coefficients we have, for fig. 7.8(a)

$$\frac{E_r \cos \theta}{E_i \cos \theta} = \frac{E_t \cos \phi / H_t - E_i \cos \theta / H_i}{E_t \cos \phi / H_t + E_i \cos \theta / H_i}$$

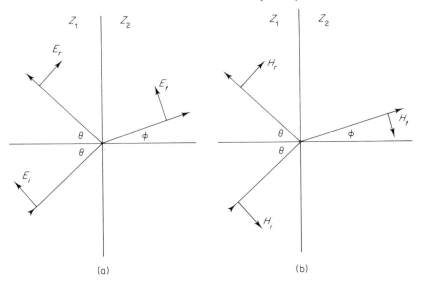

Fig. 7.8. Incident, reflected and transmitted components of a plane polarized electromagnetic wave at oblique incidence to the plane boundary separating media of impedances Z_1 and Z_2. The electric vector lies in the plane of incidence in (a) and is perpendicular to the plane of incidence in (b)

so

$$R_{\parallel} = \frac{E_r}{E_i} = \frac{Z_2 \cos \phi - Z_1 \cos \theta}{Z_2 \cos \phi + Z_1 \cos \theta}$$

where $R_{\parallel}$ is the reflexion coefficient amplitude when **E** lies in the plane of incidence.

For the transmission coefficient in fig. 7.8(a)

$$\frac{E_t \cos \phi}{E_i \cos \theta} = \frac{2E_t \cos \phi / H_t}{E_i \cos \theta / H_i + E_t \cos \phi / H_t}$$

so

$$T_{\parallel} = \frac{E_t}{E_i} = \frac{2Z_2 \cos \theta}{Z_1 \cos \theta + Z_2 \cos \phi}$$

A similar procedure for fig. 7.8(b) where **E** is perpendicular to the plane of incidence yields

$$R_{\perp} = \frac{Z_2 \cos \theta - Z_1 \cos \phi}{Z_2 \cos \theta + Z_1 \cos \phi}$$

and

$$T_\perp = \frac{2Z_2 \cos \theta}{Z_2 \cos \theta + Z_1 \cos \phi}$$

Now the relation between the refractive index n of the dielectric and its impedance Z is given by

$$n = \frac{c}{v} = \sqrt{\frac{\mu \epsilon}{\mu_0 \epsilon_0}} = \sqrt{\epsilon_r} = \frac{Z \,(\text{free space})}{Z \,(\text{dielectric})}$$

where

$$\frac{\mu}{\mu_0} = \mu_r \approx 1.$$

Hence we have

$$\frac{Z_1}{Z_2} = \frac{n_2}{n_1} = \frac{\sin \theta}{\sin \phi}$$

from Snell's Law and we may write the reflexion and transmission amplitude coefficients as

$$R_\| = \frac{\tan (\phi - \theta)}{\tan (\phi + \theta)}, \qquad T_\| = \frac{4 \sin \phi \cos \theta}{\sin 2\phi + \sin 2\theta}$$

$$R_\perp = \frac{\sin (\phi - \theta)}{\sin (\phi + \theta)}, \qquad T_\perp = \frac{2 \sin \phi \cos \theta}{\sin (\phi + \theta)}.$$

In this form the expressions for the coefficients are known as Fresnel's Equations. They are plotted in fig. 7.9 for $n_2/n_1 = 1 \cdot 5$ and they contain several significant features.

When θ is very small and incidence approaches the normal we have $\theta \to 0$ and $\phi \to 0$ so that

$$\sin (\phi - \theta) \sim \tan (\phi - \theta) \sim (\phi - \theta)$$

and

$$R_\| \sim R_\perp \sim \frac{(\phi - \theta)}{(\phi + \theta)} \sim \frac{\dfrac{1}{n_2} - \dfrac{1}{n_1}}{\dfrac{1}{n_2} + \dfrac{1}{n_1}} = \frac{n_1 - n_2}{n_1 + n_2}$$

Thus the reflected intensity

$$R^2_{\theta \to 0} = \frac{I_r}{I_i} = \left(\frac{n_1 - n_2}{n_1 + n_2} \right)^2$$

$$\sim 0 \cdot 4 \quad \text{at an air–glass interface.}$$

We note also that when $\tan (\theta + \phi) = \infty$ and $\theta + \phi = 90°$ then $R_\| = 0$.

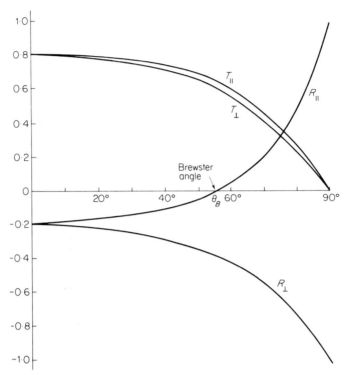

Fig. 7.9. Amplitude coefficients R and T of reflexion and transmission for $n_2/n_1 = 1\cdot5$. $R_\parallel$ and $T_\parallel$ refer to the case when the electric field vector E lies in the plane of incidence. $R_\perp$ and $T_\perp$ apply when E is perpendicular to the plane of incidence. The Brewster angle θ_B defines $\theta + \phi = 90°$ when $R_\parallel = 0$ and the reflected light is polarized with the E vector perpendicular to the plane of incidence. $R_\parallel$ changes sign (phase) at θ_B. When $\theta < \theta_B$, $\tan(\phi - \theta)$ is negative for $n_2/n_1 = 1\cdot5$. When $\theta + \phi \geqslant 90°$, $\tan(\phi + \theta)$ is also negative

In this case only $R_\perp$ is finite and the reflected light is completely plane polarized with the electric vector perpendicular to the plane of incidence. This condition defines the value of the Brewster or polarizing angle θ_B for, when θ and ϕ are complementary $\cos\theta_B = \sin\phi$ so

$$n_1 \sin\theta_B = n_2 \sin\phi = n_2 \cos\theta_B$$

and

$$\tan\theta_B = n_2/n_1$$

which, for air to glass defines $\theta_B = 56°$.

A typical modern laboratory use of the Brewster angle is the production of linearly polarized light from a He–Ne laser. If the window at the end of the laser tube is tilted so that the angle of incidence for the emerging light is θ_B and $R_\parallel = 0$ then the light with its electric vector parallel to the plane of incidence is totally transmitted while some of the light with transverse polarization $(R_\perp)$ is reflected back into the laser off-axis. If the light makes multiple transits along the length of the tube before it emerges the transmitted beam is strongly polarized in one plane.

More general but less precise uses involve the partial polarization of light reflected from wet road and other shiny surfaces where refractive indices are in the range $n = 1\cdot3-1\cdot6$. Polarized windscreens and spectacles are effective in reducing the glare from such reflections.

Reflexion from a Conductor (Normal Incidence)

For Z_2 a conductor and Z_1 free space, the refractive index

$$n = \frac{Z_1}{Z_2} = \frac{\beta}{\alpha + i\alpha}$$

is complex, where

$$\beta = \sqrt{\frac{\mu_0}{\epsilon_0}}$$

and

$$\alpha = \left(\frac{\omega\mu}{2\sigma}\right)^{1/2}$$

A complex refractive index must always be interpreted in terms of absorption because a complex impedance is determined by a complex propagation constant, e.g. here $Z_2 = i\omega\mu/\gamma$, so that

$$n = \frac{Z_1}{Z_2} = \sqrt{\frac{\mu_0}{\epsilon_o}}\,\frac{1}{i\omega\mu}(1+i)\left(\frac{\omega\mu\sigma}{2}\right)^{1/2} = (1-i)\left(\frac{\sigma}{2\omega\epsilon_0}\right)^{1/2}$$

where

$$\frac{(\mu\mu_0)^{1/2}}{\mu} \approx 1$$

The ratio E_r/E_i is therefore complex (there is a phase difference between the incident and reflected vectors) with a value

$$\frac{E_r}{E_i} = \frac{Z_2 - Z_1}{Z_2 + Z_1} = \frac{\alpha + i\alpha - \beta}{\alpha + i\alpha + \beta} = \frac{1 - \beta/\alpha + i}{1 + \beta/\alpha + i}$$

where $\beta/\alpha \gg 1$.

Since E_r/E_i is complex, the value of the reflected intensity $I_r = (E_r/E_i)^2$ is found by taking the ratio of the squares of the moduli of the numerator and the denominator, so that

$$I_r = \frac{|E_r|^2}{|E_i|^2} = \frac{|Z_2 - Z_1|^2}{|Z_2 + Z_1|^2} = \frac{(1 - \beta/\alpha)^2 + 1}{(1 + \beta/\alpha)^2 + 1}$$

$$= 1 - \frac{4\beta/\alpha}{2 + 2\beta/\alpha + (\beta/\alpha)^2} \to 1 - \frac{4\alpha}{\beta} \quad \text{(for } \beta/\alpha \gg 1)$$

so that

$$I_r = 1 - 4\left(\frac{\omega\mu}{2\sigma}\right)^{1/2}\left(\frac{\epsilon_0}{\mu_0}\right)^{1/2}$$

Problems 7.15, 7.16, 7.17, 7.18, 7.19, 7.20, 7.21, 7.22, 7.23

Problem 7.1
Show that the concept of $B^2/2\mu$ (magnetic energy per unit volume) as a magnetic pressure accounts for the fact that two parallel wires carrying currents in the same direction are forced together and that reversing one current will force them apart. (Consider a point midway between the two wires.) Show that it also explains the motion of a conductor carrying a current which is situated in a steady externally applied magnetic field.

Problem 7.2
At a distance r from a charge e on a particle of mass m the electric field value is $E = e/4\pi\epsilon_0 r^2$. Show by integrating the electrostatic energy density over the spherical volume of radius a to infinity and equating it to the value mc^2 that the 'classical' radius of the electron is given by

$$a = 2 \cdot 82 \times 10^{-15} \, \text{m}$$

Problem 7.3
The rate of generation of heat in a long cylindrical wire carrying a current I is $I^2 R$, where R is the resistance of the wire. Show that this Joule heating can be described in terms of the flow of energy into the wire from surrounding space and is equal to the product of the Poynting vector and the surface area of the wire.

Problem 7.4
Show that when a current is increasing in a long uniformly wound solenoid of coil radius r the total inward energy flow rate over a length l (the Poynting vector times the surface area $2\pi rl$) gives the time rate of change of the magnetic energy stored in that length of the solenoid.

Problem 7.5
The plane polarized electromagnetic wave (E_x, H_y) of this chapter travels in free space. Show that its Poynting vector (energy flow in watts per m^2) is given by

$$S = E_x H_y = c(\tfrac{1}{2}\epsilon_0 E_x^2 + \tfrac{1}{2}\mu_0 H_y^2) = c\epsilon_0 E_x^2$$

where c is the velocity of light. The intensity in such a wave is given by

$$I = \bar{S}_{av} = c\epsilon_0 \overline{E^2} = \tfrac{1}{2}c\epsilon_0 E_{max}^2$$

Show that

$$\bar{S} = 1 \cdot 327 \times 10^{-3} \, E_{max}^2$$

$$E_{max} = 27 \cdot 45 \, \bar{S}^{1/2} \text{ volts metre}^{-1}$$

$$H_{max} = 7 \cdot 3 \times 10^{-2} \, \bar{S}^{1/2} \text{ amps metre}^{-1}$$

Problem 7.6

A light pulse from a ruby laser consists of a linearly polarized wave train of constant amplitude lasting for 10^{-4} seconds and carrying energy of $0 \cdot 3$ joules. The diameter of the circular cross section of the beam is 5×10^{-3} metres. Use the results of problem 7.5 to calculate the energy density in the beam to show that the root mean square value of the electric field in the wave is

$$2 \cdot 4 \times 10^5 \text{ volts metre}^{-1}$$

Problem 7.7

One square metre of the earth's surface is illuminated by the sun at normal incidence by an energy flux of $1 \cdot 35$ kilowatts. Show that the amplitude of the electric field at the earth's surface is 1010 volts/metre and that the associated magnetic field in the wave has an amplitude of $2 \cdot 7$ amps/metre (See Problem 7.5). The electric field energy density $\tfrac{1}{2}\epsilon E^2$ has the dimensions of a pressure. Calculate the **radiation** pressure of sunlight upon the earth.

Problem 7.8

If the total power lost by the sun is equal to the power received per unit area of the earth's surface multiplied by the surface area of a sphere of radius equal to the earth sun distance (15×10^7 km), show that the mass per second converted to radiant energy and lost by the sun is $4 \cdot 2 \times 10^9$ kilograms. (See problem 7.5.)

Problem 7.9

A radio station radiates an average power of 10^5 watts uniformly over a hemisphere concentric with the station. Find the magnitude of the Poynting vector and the amplitude of the electric and magnetic fields of the plane electromagnetic wave at a point 10 kilometres from the station. (See problem 7.5.)

Problem 7.10

A plane polarized electromagnetic wave propagates along a transmission line consisting of two parallel strips of a perfect conductor containing a medium of permeability μ and permittivity ϵ. A section of one cubic metre in the figure shows the appropriate field vectors. The electric field E_x generates equal but opposite surface charges on the conductors of magnitude ϵE_x coulombs per square metre. The motion of these surface charges in the direction of wave propagation gives rise to a surface current (as in the discussion associated with fig. 6.1). Show that the magnitude of this current is H_y and that the characteristic impedance of the transmission line is

$$\frac{E_x}{H_y} = \sqrt{\frac{\mu}{\epsilon}}$$

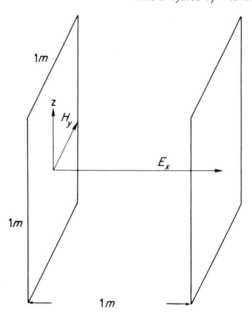

Problem 7.11

Show that equation (7.6) is dimensionally of the form

$$V = L\frac{dI}{dt}$$

where V is a voltage, L is an inductance and I is a current.

Problem 7.12

Show that when a group of electromagnetic waves of nearly equal frequencies propagates in a conducting medium the group velocity is twice the wave velocity.

Problem 7.13

A medium has a conductivity $\sigma = 10^{-1}$ siemens/metre and a relative permittivity $\epsilon_r = 50$, which is constant with frequency. If the relative permeability $\mu_r = 1$, is the medium a conductor or a dielectric at a frequency of (a) 50 kiloHertz, and (b) 10^4 MegaHertz?

$$[\epsilon_0 = (36\pi \times 10^9)^{-1} \text{ farads/metre}; \mu_0 = 4\pi \times 10^{-7} \text{ henries/metre.}]$$

Answer: (a) $\sigma/\omega\epsilon = 720$ (conductor)
(b) $\sigma/\omega\epsilon = 3.6 \times 10^{-3}$ (dielectric).

Problem 7.14

The electrical properties of the Atlantic Ocean are given by

$$\epsilon_r = 81, \qquad \mu_r = 1, \qquad \sigma = 4 \cdot 3 \text{ siemens/metre}$$

Show that it is a conductor up to a frequency of about 10 MegaHertz. What is the longest electromagnetic wavelength you would expect to propagate under water?

Problem 7.15

Show that when a plane electromagnetic wave travelling in air is reflected normally from a plane conducting surface the transmitted magnetic field value $H_t \approx 2H_i$, and that a magnetic standing wave exists in air with a very large standing wave ratio. If the wave is travelling in a conductor and is reflected normally from a plane conductor-air interface, show that $E_t \approx 2E_i$. Show that these two cases are respectively analogous to a short-circuited and an open-circuited transmission line.

Problem 7.16

Show that in a conductor the average value of the Poynting vector is given by

$$S_{av} = \tfrac{1}{2}E_0 H_0 \cos 45°$$

$$= \tfrac{1}{2}H_0^2 \times (\text{real part of } Z_c) \text{ watts/metre}^2$$

where E_0 and H_0 are the peak field values. A plane 1000 MHertz wave travelling in air with $E_0 = 1$ volt/metre is incident normally on a large copper sheet. Show firstly that the real part of the conductor impedance is $8 \cdot 2 \times 10^{-3}$ ohms and then (remembering from problem 7.15 that H_0 doubles in the conductor) show that the average power absorbed by the copper per square metre is $1 \cdot 6 \times 10^{-7}$ watts.

Problem 7.17

For a good conductor $\epsilon_r = \mu_r = 1$. Show that when an electromagnetic wave is reflected normally from such a conducting surface its fractional loss of energy $(1 - \text{reflection coefficient } I_r)$ is exactly $\sqrt{8\omega\epsilon/\sigma}$. Note that the ratio of the displacement current density to the conduction current density is therefore a direct measure of the reflectivity of the surface.

Problem 7.18

Using the value of the Poynting vector in the conductor from problem 7.16, show that the ratio of this value to the value of the Poynting vector in air is exactly $\sqrt{8\omega\epsilon/\sigma}$, as expected from problem 7.17.

Problem 7.19

The electromagnetic wave of problems 7.17 and 7.18 has electric and magnetic field magnitudes in the conductor given by

$$E_x = A\, e^{-kz}\, e^{i(\omega t - kz)}$$

and

$$H_y = A\left(\frac{\sigma}{\omega\mu}\right)^{1/2} e^{-kz}\, e^{i(\omega t - kz)}\, e^{-i\pi/4}$$

where $k = (\omega\mu\sigma/2)^{1/2}$.

Show that the average value of the Poynting vector in the conductor is given by

$$S_{av} = \tfrac{1}{2}A^2\left(\frac{\sigma}{2\omega\mu}\right)^{1/2} e^{-2kz} \quad (\text{watts/metre}^2)$$

This is the power absorbed per unit area by the conductor. We know, however, that the wave propagates only a distance of the order of the skin depth, so that this power is rapidly transformed. The rate at which it changes with distance is given by $\partial S_{av}/\partial z$, which gives the energy transformed per unit volume in unit time. Show that this quantity is equal to the conductivity σ times the square of the mean value of the electric field vector **E**, that is, the Joule heating from currents flowing in the surface of the conductor down to a depth of the order of the skin depth.

Problem 7.20

Show that when light travelling in free space is normally incident on the surface of a dielectric of refractive index n the reflected intensity

$$I_r = \left(\frac{E_r}{E_i}\right)^2 = \left(\frac{1-n}{1+n}\right)^2$$

and the transmitted intensity

$$I_t = \left(\frac{E_t}{E_i}\right)^2 = \frac{4n}{(1+n)^2}$$

(Note $I_r + I_t = 1$.)

Problem 7.21

Show that if the medium of problem 7.20 is galss ($n = 1\cdot5$) then $I_r = 4\%$ and $I_t = 96\%$. If an electromagnetic wave of 100 MegaHertz is normally incident on water ($\epsilon_r = 81$) show that $I_r = 65\%$ and $I_t = 35\%$.

Problem 7.22

Light passes normally through a glass plate suffering only one air to glass and one glass to air reflection. What is the loss of intensity?

Problem 7.23

A radiating antenna in simplified form is just a length x_0 of wire in which an oscillating current is maintained. The expression for the radiating power is that used on page 39 for an oscillating electron

$$P = \frac{dE}{dt} = \frac{q^2\omega^4 x_0^2}{12\pi\epsilon_0 c^3}$$

where q is the electron charge and ω is the oscillation frequency. The current I in the antenna may be written $I_0 = \omega q$. If $P = \frac{1}{2}RI_0^2$ show that the radiation resistance of the antenna is given by

$$R = \frac{2\pi}{3}\sqrt{\frac{\mu_0}{\epsilon_0}}\left(\frac{x_0}{\lambda}\right)^2 = 787\left(\frac{x_0}{\lambda}\right)^2 \text{ ohms}$$

where λ is the radiated wavelength (an expression valid for $\lambda \gg x_0$).

If the antenna is 30 metres long and transmits at a frequency of 5×10^5 Hertz with a root mean square current of 20 amps, show that its radiation resistance is $1\cdot97$ ohms and that the power radiated is 400 watts. (Verify that $\lambda \gg x_0$.)

Summary of Important Results

Dielectric; μ *and* ϵ ($\sigma = 0$)

Wave equation

$$\frac{\partial^2 E_x}{\partial z^2} = \mu\epsilon\frac{\partial^2 E_x}{\partial t^2} \qquad \left(v^2 = \frac{1}{\mu\epsilon}\right)$$

$$\frac{\partial^2 H_y}{\partial z^2} = \mu\epsilon\frac{\partial^2 H_y}{\partial t^2}$$

Impedance

$$\frac{E_x}{H_y} = \sqrt{\frac{\mu}{\epsilon}} \quad (376.7 \text{ ohms for free space})$$

Energy density $\frac{1}{2}\epsilon E_x^2 + \frac{1}{2}\mu H_y^2$

Mean energy flow $= \text{Intensity} = \bar{S} = v(\text{mean energy density})$

$$= v\left(\frac{1}{2}\epsilon E_x^2 + \frac{1}{2}\mu H_y^2\right)_{\text{average}}$$

$$= v\epsilon \overline{E_x^2} = \frac{1}{2}v\epsilon E_{x(\text{max})}^2$$

Conductor; μ ϵ *and* σ

Add diffusion equation to wave equation for loss effects from σ

$$\frac{\partial^2 E_x^2}{\partial z^2} = \mu\epsilon\frac{\partial^2 E_x^2}{\partial t^2} + \mu\sigma\frac{\partial E_x}{\partial t}$$

giving

$$E_x = E_0\,e^{-kz}\,e^{i(\omega t - kz)}$$

where

$$k^2 = \omega\mu\sigma/2$$

Skin depth

$$\delta = \frac{1}{k} \quad \text{giving } E_x = E_0\,e^{-1}$$

Criterion for conductor/dielectric behaviour is ratio

$$\frac{\text{conduction current}}{\text{displacement current}} = \frac{\sigma}{\omega\epsilon} \quad (\text{note frequency dependence})$$

Impedance Z_c (conductor)

$$\mathbf{Z}_c = \frac{1+i}{\sqrt{2}}\left(\frac{\omega\mu}{\sigma}\right)^{1/2}$$

with magnitude $Z_c = 376\cdot6\,\sqrt{\mu_r/\epsilon_r}\,\sqrt{\omega\epsilon/\sigma}$ ohms

Reflexion and transmission coefficients (normal incidence),

$$R = \frac{E_r}{E_i} = \frac{Z_2 - Z_1}{Z_2 + Z_1} \quad (E\text{'s and } Z\text{'s may be complex})$$

$$T = \frac{E_t}{E_i} = \frac{2Z_2}{Z_2 + Z_1}$$

Fresnel's Equations (dielectrics)

$$R_{\parallel} = \frac{\tan(\phi - \theta)}{\tan(\phi + \theta)}, \qquad T_{\parallel} = \frac{4 \sin \phi \cos \theta}{\sin 2\phi + \sin 2\theta}$$

$$R_{\perp} = \frac{\sin(\phi - \theta)}{\sin(\phi + \theta)}, \qquad T_{\perp} = \frac{2 \sin \phi \cos \theta}{\sin(\phi + \theta)}$$

Refractive index

$$n = \frac{c}{v} = \frac{Z \text{ (free space)}}{Z \text{ (dielectric)}}$$

Chapter 8

Waves in More than
One Dimension

Plane Wave Representation in Two and Three Dimensions

Fig. 8.1 shows that in two dimensions waves of velocity c may be represented by lines of constant phase propagating in a direction $\mathbf{k}$ which is normal to each line, where the magnitude of $\mathbf{k}$ is the wave number $k = 2\pi/\lambda$. The direction cosines of $\mathbf{k}$ are given by

$$l = \frac{k_1}{k}, \quad m = \frac{k_2}{k} \quad \text{where } k^2 = k_1^2 + k_2^2$$

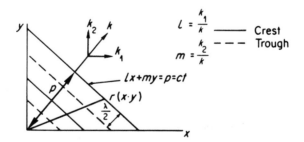

Fig. 8.1. Crests and troughs of a two-dimensional plane wave propagating in a general direction $\mathbf{k}$ (direction cosines l and m). The wave is specified by $lx + my = p = ct$, where p is its perpendicular distance from the origin, travelled in a time t at a velocity c

219

and any point $\mathbf{r}(x, y)$ on the line of constant phase satisfies the equation

$$lx + my = p = ct$$

where p is the perpendicular distance from the line to the origin. The displacements at all points $\mathbf{r}(x, y)$ on a given line are in phase and the phase difference ϕ between the origin and a given line is

$$\phi = \frac{2\pi}{\lambda}\,(\text{path difference}) = \frac{2\pi}{\lambda}p = \mathbf{k} \cdot \mathbf{r} = k_1 x + k_2 y$$

$$= kp$$

Hence, the bracket $(\omega t - \phi) = (\omega t - kx)$ used in a one dimensional wave is replaced by $(\omega t - \mathbf{k} \cdot \mathbf{r})$ in waves of more than one dimension, e.g. we shall use the exponential expression

$$e^{i(\omega t - \mathbf{k} \cdot \mathbf{r})}$$

In three dimensions all points $\mathbf{r}(x, y, z)$ in a given wavefront will lie on *planes* of constant phase satisfying the equation

$$lx + my + nz = p = ct$$

where the vector $\mathbf{k}$ which is normal to the plane and in the direction of propagation has direction cosines

$$l = \frac{k_1}{k}, \qquad m = \frac{k_2}{k}, \qquad n = \frac{k_3}{k}$$

(so that $k^2 = k_1^2 + k_2^2 + k_3^2$) and the perpendicular distance p is given by

$$kp = \mathbf{k} \cdot \mathbf{r} = k_1 x + k_2 y + k_3 z$$

Wave Equation in Two Dimensions

We shall consider waves propagating on a stretched plane membrane of negligible thickness having a mass ρ per unit area and stretched under a uniform tension T. This means that if a line of unit length is drawn in the surface of the membrane then the material on one side of this line exerts a force T on the material on the other side in a direction perpendicular to that of the line.

If the equilibrium position of the membrane is the xy plane the vibration displacements perpendicular to this plane will be given by z where z depends on the position x, y. In fig. 8.2a where the small rectangular element $ABCD$ of sides δx and δy is vibrating, forces $T\,\delta x$ and $T\,\delta y$ are shown acting on the sides in directions which tend to restore the element to its equilibrium position.

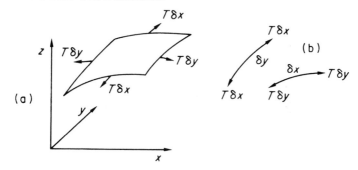

Fig. 8.2. Rectangular element of a uniform membrane vibrating in the z-direction subject to one restoring force, $T\delta x$, along its sides of length δy and another, $T\delta y$, along its sides of length δx

In deriving the equation for waves on a string we saw that the tension T along a curved element of string of length $\mathrm{d}x$ produced a force perpendicular to x of

$$T\frac{\partial^2 y}{\partial x^2}\mathrm{d}x$$

where y was the perpendicular displacement. Here in fig. 8.2b by exactly similar arguments we see that a force $T\delta y$ acting on a membrane element of length ∂x produces a force

$$T\delta y\frac{\partial^2 z}{\partial x^2}\delta x$$

where z is the perpendicular displacement, whilst another force $T\delta x$ acting on a membrane element of length δy produces a force

$$T\delta x\frac{\partial^2 z}{\partial y^2}\delta y.$$

The sum of these restoring forces which act in the z-direction is equal to the mass of the element $\rho\,\delta x\,\delta y$ times its perpendicular acceleration in the z-direction, so that

$$T\frac{\partial^2 z}{\partial x^2}\delta x\,\delta y + T\frac{\partial^2 z}{\partial y^2}\delta x\,\delta y = \rho\,\delta x\,\delta y\frac{\partial^2 z}{\partial t^2}$$

giving the wave equation in two dimensions as

$$\frac{\partial^2 z}{\partial x^2}+\frac{\partial^2 z}{\partial y^2}=\frac{\rho}{T}\frac{\partial^2 z}{\partial t^2}=\frac{1}{c^2}\frac{\partial^2 z}{\partial t^2}$$

where

$$c^2 = \frac{T}{\rho}$$

The displacement of waves propagating on this membrane will be given by

$$z = A\, e^{i(\omega t - \mathbf{k}\cdot\mathbf{r})} = A\, e^{i[\omega t - (k_1 x + k_2 y)]}$$

where

$$k^2 = k_1^2 + k_2^2$$

The reader should verify that this expression for z is indeed a solution to the two-dimensional wave equation.

(Problem 8.1)

Wave Guides

Reflection of a 2D Wave at Rigid Boundaries

Let us first consider a 2D wave propagating in a vector direction $\mathbf{k}(k_1, k_2)$ in the xy plane along a membrane of width b stretched under a tension T between two long rigid rods which present an infinite impedance to the wave.

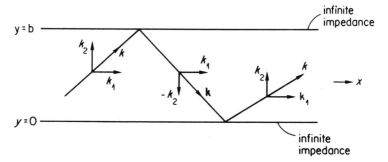

Fig. 8.3. Propagation of a two-dimensional wave along a stretched membrane with infinite impedances at $y = 0$ and $y = b$ giving reversal of k_2 at each reflection

We see from fig. 8.3 that upon reflexion from the line $y = b$ the component k_1 remains unaffected whilst k_2 is reversed to $-k_2$. Reflexion at $y = 0$ leaves k_1 unaffected whilst $-k_2$ is reversed to its original value k_2. The wave system on the membrane will therefore be given by the superposition of the incident and reflected waves, that is, by

$$z = A_1 \sin\left[\omega t - (k_1 x + k_2 y)\right] + A_2 \sin\left[\omega t - (k_1 x - k_2 y)\right]$$

subject to the boundary conditions that

$$z = 0 \quad \text{at} \quad y = 0 \quad \text{and} \quad y = b$$

the positions of the frame of infinite impedance.

The condition $z = 0$ at $y = 0$ requires

$$A_2 = -A_1$$

and $z = 0$ at $y = b$ gives

$$\sin k_2 b = 0$$

or

$$k_2 = \frac{n\pi}{b}$$

(Problem 8.2)

With these values of A_2 and k_2 the displacement of the wave system is given by

$$z = -2A_1 \sin k_2 y \cos (\omega t - k_1 x)$$

which represents a wave travelling along the x direction with a phase velocity

$$v_p = \frac{\omega}{k_1} = \left(\frac{k}{k_1}\right)v$$

where the velocity on an infinitely wide membrane is given by

$$v = \frac{\omega}{k} \quad \text{which is} \quad < v_p$$

because

$$k^2 = k_1^2 + k_2^2$$

Now

$$k^2 = k_1^2 + \frac{n^2\pi^2}{b^2}$$

so

$$k_1 = \left(k^2 - \frac{n^2\pi^2}{b^2}\right)^{1/2} = \left(\frac{\omega^2}{v^2} - \frac{n^2\pi^2}{b^2}\right)^{1/2}$$

and the group velocity for the wave in the x direction

$$v_g = \frac{\partial \omega}{\partial k_1} = \frac{k_1}{\omega}v^2 = \left(\frac{k_1}{k}\right)v$$

giving the product

$$v_p v_g = v^2$$

Since k_1 must be real for the wave to propagate we have, from

$$k_1^2 = k^2 - \frac{n^2\pi^2}{b^2}$$

the condition that

$$k^2 = \frac{\omega^2}{v^2} \geqslant \frac{n^2\pi^2}{b^2}$$

that is

$$\omega \geqslant \frac{n\pi v}{b}$$

or

$$\nu \geqslant \frac{nv}{2b}$$

where n defines the mode number in the y direction. Thus only waves of certain frequencies ν are allowed to propagate along the membrane which acts as a *wave guide*.

There is a cut off frequency $n\pi v/b$ for each mode of number n and the wave guide acts as a frequency filter (recall the discussion on similar behaviour in wave propagation on the loaded string in Chapter 3). The presence of the $\sin k_2 y$ term in the expression for the displacement z shows that the amplitude varies across the transverse y direction as shown in fig. 8.4 for the mode values

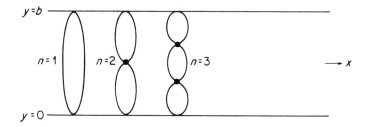

Fig. 8.4. Variation of amplitude with y-direction for two-dimensional wave propagating along the membrane of fig. 8.3. Normal modes ($n = 1, 2$ and 3 shown) are set up along any axis bounded by infinite impedances

$n = 1, 2, 3$. Thus, along any direction in which the waves meet rigid boundaries a standing wave system will be set up analogous to that on a string of fixed length and we shall discuss the implication of this in the section on normal modes and the method of separation of variables.

Wave guides are used for all wave systems, particularly in those with acoustical and electromagnetic applications. Fibre optics is based on wave guide principles, but the major use of wave guides has been with electromagnetic waves in telecommunications. Here the reflecting surfaces are the sides of a copper tube of circular or rectangular cross section. Note that in this case the free space velocity becomes the velocity of light

$$c = \frac{\omega}{k} < v_p$$

the phase velocity, but the relation $v_p v_g = c^2$ ensures that energy in the wave always travels with a group velocity $v_g < c$.

(Problems 8:3, 8.4, 8.5, 8.6, 8.7, 8.8, 8.9, 8.10, 8.11)

Normal Modes and the Method of Separation of Variables

We have just seen that when waves propagate in more than one dimension a standing wave system will be set up along any axis which is bounded by infinite impedances.

In Chapter 4 we found that standing waves could exist on a string of fixed length l where the displacement was of the form

$$y = A \left. \begin{matrix} \sin \\ \cos \end{matrix} \right\} kx \left. \begin{matrix} \sin \\ \cos \end{matrix} \right\} \omega_n t$$

where A is constant and where $\begin{smallmatrix} \sin \\ \cos \end{smallmatrix}\}$ means that either solution may be used to fit the boundary conditions in space and time. When the string is fixed at both ends, the condition $y = 0$ at $x = 0$ removes the $\cos kx$ solution, and $y = 0$ at $x = l$ requires $k_n l = n\pi$ or $k_n = n\pi/l = 2\pi/\lambda_n$, giving $l = n\lambda_n/2$. Since the wave velocity $c = \nu_n \lambda_n$, this permits frequencies $\omega_n = 2\pi\nu_n = \pi nc/l$, defined as *normal modes of vibration* or *eigenfrequencies*.

We can obtain this solution in a way which allows us to extend the method to waves in more than one dimension. We have seen that the wave equation

$$\frac{\partial^2 \phi}{\partial x^2} = \frac{1}{c^2} \frac{\partial^2 \phi}{\partial t^2}$$

has a solution which is the product of two terms, one a function of x only and the other a function of t only.

Let us write $\phi = X(x)T(t)$ and apply the method known as separation of variables.

The wave equation then becomes

$$\frac{\partial^2 X}{\partial x^2} \cdot T = \frac{1}{c^2} X \frac{\partial^2 T}{\partial t^2}$$

or

$$X_{xx}T = \frac{1}{c^2}XT_{tt}$$

where the double subscript refers to double differentiation with respect to the variables. Dividing by $\phi = X(x)T(t)$ we have

$$\frac{X_{xx}}{X} = \frac{1}{c^2}\frac{T_{tt}}{T}$$

where the left-hand side depends on x only and the right-hand side depends on t only. However, both x and t are independent variables and the equality between both sides can only be true when both sides are independent of x and t and are equal to a constant, which we shall take, for convenience, as $-k^2$. Thus

$$\frac{X_{xx}}{X} = -k^2, \quad \text{giving} \quad X_{xx} + k^2 X = 0$$

and

$$\frac{1}{c^2}\frac{T_{tt}}{T} = -k^2 \quad \text{giving} \quad T_{tt} + c^2 k^2 T = 0$$

$X(x)$ is therefore of the form $e^{\pm ikx}$ and $T(t)$ is of the form $e^{\pm ickt}$, so that $\phi = A\,e^{\pm ikx}\,e^{\pm ickt}$, where A is constant, and we choose a particular solution in a form already familiar to us by writing

$$\phi = A\,e^{i(ckt-kx)}$$
$$= A\,e^{i(\omega t-kx)},$$

where $\omega = ck$, or we can write

$$\phi = A \left.\begin{matrix} \sin \\ \cos \end{matrix}\right\} kx \left.\begin{matrix} \sin \\ \cos \end{matrix}\right\} ckt$$

as above.

Two-Dimensional Case

In extending this method to waves in 2 dimensions we consider the wave equation in the form

$$\frac{\partial^2 \phi}{\partial x^2} + \frac{\partial^2 \phi}{\partial y^2} = \frac{1}{c^2}\frac{\partial^2 \phi}{\partial t^2}$$

and we write $\phi = X(x)Y(y)T(t)$, where $Y(y)$ is a function of y only.
    Differentiating twice and dividing by $\phi = XYT$ gives

$$\frac{X_{xx}}{X} + \frac{Y_{yy}}{Y} = \frac{1}{c^2}\frac{T_{tt}}{T}$$

where the left-hand side depends on x and y only and the right-hand side depends on t only. Since x, y and t are independent variables each side must be equal to a constant, $-k^2$ say. This means that the left-hand side terms in x and y differ by only a constant for all x and y, so that each term is itself equal to a constant. Thus we can write

$$\frac{X_{xx}}{X} = -k_1^2, \qquad \frac{Y_{yy}}{Y} = -k_2^2$$

and

$$\frac{1}{c^2}\frac{T_{tt}}{T} = -(k_1^2 + k_2^2) = -k^2$$

giving

$$X_{xx} + k_1^2 X = 0$$
$$Y_{yy} + k_2^2 Y = 0$$
$$T_{tt} + c^2 k^2 T = 0$$

or

$$\phi = A\, e^{\pm ik_1 x}\, e^{\pm ik_2 y}\, e^{\pm ickt}$$

where $k^2 = k_1^2 + k_2^2$. Typically we may write

$$\phi = A\, {\sin \brace \cos} k_1 x\, {\sin \brace \cos} k_2 y\, {\sin \brace \cos} ckt.$$

Three-Dimensional Case

The three-dimensional treatment is merely a further extension. The wave equation is

$$\frac{\partial^2 \phi}{\partial x^2} + \frac{\partial^2 \phi}{\partial y^2} + \frac{\partial^2 \phi}{\partial z^2} = \frac{1}{c^2}\frac{\partial^2 \phi}{\partial t^2}$$

with a solution

$$\phi = X(x)\,Y(y)\,Z(z)\,T(t)$$

yielding

$$\phi = A\, {\sin \brace \cos} k_1 x\, {\sin \brace \cos} k_2 y\, {\sin \brace \cos} k_3 z\, {\sin \brace \cos} ckt,$$

where $k_1^2 + k_2^2 + k_3^2 = k^2$.

Using vector notation we may write

$$\phi = A\, e^{i(\omega t - \mathbf{k}.\mathbf{r})}, \quad \text{where} \quad \mathbf{k}.\mathbf{r} = k_1 x + k_2 y + k_3 z$$

Normal Modes in Two Dimensions on a Rectangular Membrane

Suppose waves proceed in a general direction **k** on the rectangular membrane of sides a and b shown in fig. 8.5. Each dotted wave line is separated by a

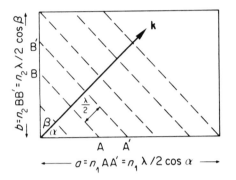

Fig. 8.5. Normal modes on a rectangular membrane in a direction **k** satisfying boundary conditions of zero displacement at the edges of length $a = n_1\lambda/2 \cos \alpha$ and $b = n_2\lambda/2 \cos \beta$

distance $\lambda/2$ and a standing wave system will exist whenever $a = n_1\text{AA}'$ and $b = n_2\text{BB}'$, where n_1 and n_2 are integers.

But

$$\text{AA}' = \frac{\lambda}{2 \cos \alpha} = \frac{\lambda}{2} \frac{k}{k_1} = \frac{\lambda}{2} \frac{2\pi}{\lambda} \frac{1}{k_1} = \frac{\pi}{k_1}$$

so that

$$a = \frac{n_1 \pi}{k_1} \quad \text{and} \quad k_1 = \frac{n_1 \pi}{a}.$$

Similarly

$$k_2 = \frac{n_2 \pi}{b}$$

Hence

$$k^2 = k_1^2 + k_2^2 = \frac{4\pi^2}{\lambda^2} = \pi^2\left(\frac{n_1^2}{a^2} + \frac{n_2^2}{b^2}\right)$$

or

$$\frac{2}{\lambda} = \sqrt{\frac{n_1^2}{a^2} + \frac{n_2^2}{b^2}}$$

defining the frequency of the n_1th mode on the x-axis and the n_2th mode on the y-axis, that is, the $(n_1 n_2)$ normal mode, as

$$\nu = \frac{c}{2}\sqrt{\frac{n_1^2}{a^2}+\frac{n_2^2}{b^2}}, \quad \text{where } c^2 = \frac{T}{\rho}$$

If $\mathbf{k}$ is not normal to the direction of either a or b we can write the general solution for the waves as

$$z = A \left.\begin{matrix}\sin\\\cos\end{matrix}\right\}k_1 x \left.\begin{matrix}\sin\\\cos\end{matrix}\right\}k_2 y \left.\begin{matrix}\sin\\\cos\end{matrix}\right\}ckt$$

with the boundary conditions $z = 0$ at $x = 0$ and a; $z = 0$ at $y = 0$ and b.

The condition $z = 0$ at $x = y = 0$ requires a $\sin k_1 x \sin k_2 y$ term, and the condition $z = 0$ at $x = a$ defines $k_1 = n_1 \pi / a$. The condition $z = 0$ at $y = b$ gives $k_2 = n_2 \pi / b$, so that

$$z = A \sin\frac{n_1 \pi x}{a} \sin\frac{n_2 \pi y}{b} \sin ckt$$

The fundamental vibration is given by $n_1 = 1$, $n_2 = 1$, so that

$$\nu = \sqrt{\left(\frac{1}{a^2}+\frac{1}{b^2}\right)\frac{T}{4\rho}}$$

In the general mode $(n_1 n_2)$ zero displacement or nodal lines occur at

$$x = 0, \quad \frac{a}{n_1}, \quad \frac{2a}{n_1}, \quad \ldots a$$

and

$$y = 0, \quad \frac{b}{n_2}, \quad \frac{2b}{n_2}, \quad \ldots b$$

Some of these normal modes are shown in fig. 8.6, where the shaded and plain areas have opposite displacements as shown.

The complete solution for a general displacement would be the sum of individual normal modes, as with the simpler case of waves on a string (see the chapter on Fourier Series) where boundary conditions of space and time would have to be met. Several modes of different values $(n_1 n_2)$ may have the same frequency, e.g., in a square membrane the modes (4,7) (7,4) (1,8) and (8,1) all have equal frequencies. If the membrane is rectangular and $a = 3b$, modes (3,3) and (9,1) have equal frequencies.

These modes are then said to be *degenerate*, a term used in describing equal energy levels for electrons in an atom which are described by different quantum numbers.

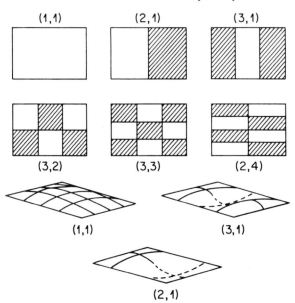

Fig. 8.6. Some normal modes on a rectangular membrane with shaded and clear sections having opposite sinusoidal displacements as indicated

Normal Modes in Three Dimensions

In three dimensions a normal mode is described by the numbers n_1, n_2, n_3, with a frequency

$$\nu = \frac{c}{2}\sqrt{\frac{n_1^2}{l_1^2} + \frac{n_2^2}{l_2^2} + \frac{n_3^2}{l_3^2}}, \tag{8.1}$$

where l_1, l_2 and l_3 are the lengths of the sides of the rectangular enclosure. If we now form a rectangular lattice with the x-, y- and z-axes marked off in units of

$$\frac{c}{2l_1}, \quad \frac{c}{2l_2} \quad \text{and} \quad \frac{c}{2l_3}$$

respectively (fig. 8.7), we can consider a vector of components n_1 units in the x-direction, n_2 units in the y-direction and n_3 units in the z-direction to have a length

$$\nu = \frac{c}{2}\sqrt{\frac{n_1^2}{l_1^2} + \frac{n_2^2}{l_2^2} + \frac{n_3^2}{l_3^2}}$$

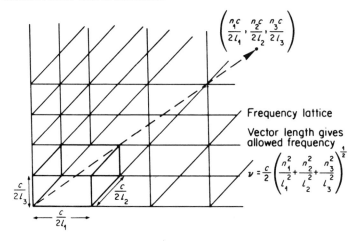

$$\left(\frac{n_1 c}{2l_1}, \frac{n_2 c}{2l_2}, \frac{n_3 c}{2l_3}\right)$$

Frequency lattice

Vector length gives allowed frequency

$$\nu = \frac{c}{2}\left(\frac{n_1^2}{l_1^2} + \frac{n_2^2}{l_2^2} + \frac{n_3^2}{l_3^2}\right)^{\frac{1}{2}}$$

Fig. 8.7. Lattice of rectangular cells in frequency space. The length of the vector joining the origin to any cell corner is the value of the frequency of an allowed normal mode. The vector direction gives the propagation direction of that particular mode

Each frequency may thus be represented by a line joining the origin to a point $cn_1/2l_1$, $cn_2/2l_2$, $cn_3/2l_3$ in the rectangular lattice.

The length of the line gives the magnitude of the frequency, and the vector direction gives the direction of the standing waves.

Each point will be at the corner of a rectangular unit cell of sides $c/2l_1$, $c/2l_2$ and $c/2l_3$ with a volume $c^3/8l_1l_2l_3$. There are as many cells as points (i.e., as frequencies) since each cell has eight points at its corners and each point serves as a corner to 8 cells.

A very important question now arises: how many normal modes (stationary states in quantum mechanics) can exist in the frequency range ν to $\nu + d\nu$?

The answer to this question is the total number of all those positive integers n_1, n_2, n_3 for which, from equation (8.1),

$$\nu^2 < \frac{c^2}{4}\left(\frac{n_1^2}{l_1^2} + \frac{n_2^2}{l_2^2} + \frac{n_3^2}{l_3^2}\right) < (\nu + d\nu)^2$$

This total is the number of possible points (n_1, n_2, n_3) lying in the positive octant between two concentric spheres of radii ν and $\nu + d\nu$. The other octants will merely repeat the positive octant values because the n's appear as squared quantities.

Hence the total number of possible points or cells will be

$$\frac{1}{8}\frac{\text{(volume of spherical shell)}}{\text{volume of cell}}$$

$$= \frac{4\pi\nu^2 \, d\nu}{8} \cdot \frac{8 l_1 l_2 l_3}{c^3}$$

$$= 4\pi l_1 l_2 l_3 \cdot \frac{\nu^2 \, d\nu}{c^3}$$

so that the number of possible normal modes in the frequency range ν to $\nu + d\nu$ *per unit volume* of the enclosure

$$= \frac{4\pi\nu^2 \, d\nu}{c^3}$$

Note that this result, *per unit volume of the enclosure,* is independent of any particular system; we shall consider two very important applications.

(1) Frequency Distribution of Energy Radiated from a Hot Body. Planck's Law

The electromagnetic energy radiated from a hot body at temperature T in the small frequency interval ν to $\nu + d\nu$ may be written $E_\nu \, d\nu$. If this quantity is measured experimentally over a wide range of ν a curve T_1 in fig. 8.8 will result. The general shape of the curve is independent of the temperature, but as T is increased the maximum of the curve increases and shifts towards a higher frequency.

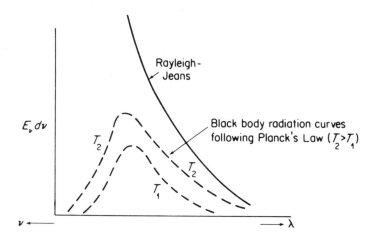

Fig. 8.8. Planck's black body radiation curve plotted for two different temperatures $T_2 > T_1$, together with the curve of the classical Rayleigh–Jeans explanation leading to the 'ultra-violet catastrophe'

The early attempts to describe the shape of this curve were based on two results we have already used.

In the chapter on coupled oscillations we associated normal modes with 'degrees of freedom', the number of ways in which a system could take up energy. In kinetic theory, assigning an energy $\frac{1}{2}kT$ to each degree of freedom of a monatomic gas at temperature T leads to the gas law $pV = RT = NkT$ where N is Avogadro's number, k is Boltzmann's constant and R is the gas constant.

If we assume that each frequency radiated from a hot body is associated with the normal mode of an oscillator with two degrees of freedom and two transverse planes of polarization, the energy radiated per frequency interval $d\nu$ may be considered as the product of the number of normal modes or oscillators in the interval $d\nu$ and an energy contribution of kT from each oscillator for each plane of polarization. This gives

$$E_\nu \, d\nu = \frac{4\pi\nu^2 \, d\nu 2kT}{c^3} = \frac{8\pi\nu^2 kT \, d\nu}{c^3}$$

a result known as the Rayleigh–Jeans Law.

This, however, gives the energy density proportional to ν^2 which, as the solid curve in fig. 8.8 shows, becomes infinite at very high frequencies, a physically absurd result known as the *ultraviolet catastrophe*.

The correct solution to the problem was a major advance in physics. Planck had introduced the quantum theory, which predicted that the average energy value kT should be replaced by the factor $h\nu/(e^{h\nu/kT} - 1)$, where h is Planck's constant (the unit of action). The experimental curve is thus accurately described by Planck's Radiation Law

$$E_\nu \, d\nu = \frac{8\pi\nu^2}{c^3} \frac{h\nu}{e^{h\nu/kT} - 1} \, d\nu$$

(2) Debye Theory of Specific Heats

The success of the modern theory of the specific heats of solids owes much to the work of Debye, who considered the thermal vibrations of atoms in a solid lattice in terms of a vast complex of standing waves over a great range of frequencies. This picture corresponds in three dimensions to the problem of atoms spaced along a one dimensional line (Chapter 4). In the specific heat theory each atom was allowed two transverse vibrations (perpendicular planes of polarization) and one longitudinal vibration.

The number of possible modes or oscillations per unit volume in the frequency interval ν to $\nu + d\nu$ is then given by

$$dn = 4\pi\nu^2 \, d\nu \left(\frac{2}{c_T^3} + \frac{1}{c_L^3} \right) \tag{8.2}$$

where c_T and c_L are respectively the transverse and longitudinal wave velocities.

Each mode has an average energy (from Planck's Law) of $\bar{\epsilon} = h\nu/(e^{h\nu/kT} - 1)$ and the total energy in the frequency range ν to $\nu + d\nu$ for a gram atom of the solid of volume V_A is then

$$V_A \bar{\epsilon} \, dn = 4\pi V_A \left(\frac{2}{c_T^3} + \frac{1}{c_L^3} \right) \frac{h\nu^3}{e^{h\nu/kT} - 1} \, d\nu.$$

The total energy per gram atom over all permitted frequencies is then

$$E_A = \int V_A \bar{\epsilon} \, dn = 4\pi V_A \left(\frac{2}{c_T^3} + \frac{1}{c_L^3} \right) \int_0^{\nu_m} \frac{h\nu^3}{e^{h\nu/kT} - 1} \, d\nu,$$

where ν_m is the maximum frequency of the oscillations.

There are N atoms per gram atom of the solid (N is Avogadro's number) and each atom has three allowed oscillation modes, so an approximation to ν_m is found by writing the integral of equation 8.2 for a gram atom as

$$\int dn = 3N = 4\pi V_A \left(\frac{2}{c_T^3} + \frac{1}{c_L^3} \right) \int_0^{\nu_m} \nu^2 \, d\nu = \frac{4\pi V_A}{3} \left(\frac{2}{c_T^3} + \frac{1}{c_L^3} \right) \nu_m^3$$

The values of c_T and c_L can be calculated from the elastic constants of the solid (see Chapter 5 on longitudinal waves) and ν_m can then be found.

The value of E_A thus becomes

$$E_A = \frac{9N}{\nu_m^3} \int_0^{\nu_m} \frac{h\nu}{e^{h\nu/kT} - 1} \nu^2 \, d\nu$$

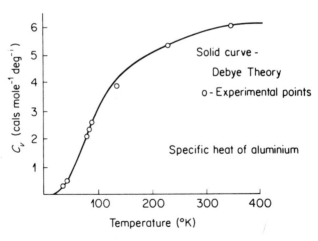

Fig. 8.9. Debye theory of specific heat of solids. Experimental values versus theoretical curve for aluminium

and the variation of E_A with the temperature T is the molar specific heat of the substance at constant volume. The specific heat of aluminium calculated by this method is compared with experimental results in fig. 8.9.

(Problems 8.12, 8.13, 8.14, 8.15, 8.16, 8.17, 8.18, 8.19)

Reflexion and Transmission of a Three-Dimensional Wave at a Plane Boundary

To illustrate such an event we choose a physical system of great significance, the passage of a light wave from air to glass. More generally, Figure 8.10 shows a plane polarized electromagnetic wave E_i incident at an angle θ to the normal of the plane boundary $z = 0$ separating two dielectrics of impedance Z_1 and Z_2, giving reflected and transmitted rays E_r and E_t respectively. The boundary condition requires that the tangential electric field E_x is continuous at $z = 0$. The propagation direction k_i of E_i lies wholly in the plane of the paper ($y = 0$)

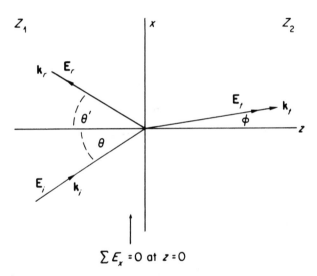

Fig. 8.10. Plane-polarized electromagnetic wave propagating in the plane of the paper is represented by vector E_i and is reflected as vector E_r and transmitted as vector E_t at a plane interface between media of impedances Z_1 and Z_2. No assumptions are made about the planes of propagation of E_r and E_t. From the boundary condition that the electric field component E_x is continuous at the plane $z = 0$ it follows that (1) vectors E_i E_r and E_t propagate in the same plane; (2) $\theta = \theta'$ (angle of incidence = angle of reflection); (3) Snell's law $(\sin \theta / \sin \phi) = n_2 / n_1$ where n is the refractive index

but no assumptions are made about the directions of the reflected and transmitted waves (nor about the planes of oscillation of their electric field vectors).

We write

$$\mathbf{E}_i = A_i\, e^{i[\omega t - k_i(l_i x + n_i z)]} = A_i\, e^{i[\omega t - k_i(x\,\sin\theta + z\,\cos\theta)]}$$

$$\mathbf{E}_r = A_r\, e^{i[\omega t - k_r(l_r x + m_r y + n_r z)]}$$

and

$$\mathbf{E}_t = A_t\, e^{i[\omega t - k_t(l_t x + m_t y + n_t z)]}$$

where $\mathbf{k}_r$ and $\mathbf{k}_t$ are respectively the reflected and transmitted propagation vectors and l, m, n are the appropriate direction cosines.

Since E_x is continuous at $z = 0$ for all x, y, t we have

$$A_i\, e^{i[\omega t - k_i(x\,\sin\theta)]} + A_r\, e^{i[\omega t - k_r(l_r x + m_r y)]}$$
$$= A_t\, e^{i[\omega t - k_t(l_t x + m_t y)]}$$

an identity which is only possible if the indices of all three terms are identical that is

$$\omega t - k_i x\,\sin\theta \equiv \omega t - k_r(l_r x + m_r y)$$
$$\equiv \omega t - k_t(l_t x + m_t y)$$

Equating the coefficients of x in this identity gives

$$k_i\,\sin\theta = k_r l_r = k_t l_t$$

whilst equal coefficients of y give

$$0 = k_r m_r = k_t m_t$$

This second relation gives

$$m_r = m_t = 0$$

so that the reflected and transmitted rays have no component in the y direction and lie wholly in the xz plane of incidence, that is, incident reflected and transmitted (refracted) rays are coplanar.

Now the magnitude

$$k_i = k_r = \frac{2\pi}{\lambda_1}$$

since both incident and reflected waves are travelling in medium Z_1. Hence

$$k_1\,\sin\theta = k_r l_r$$

gives

$$k_i\,\sin\theta = k_r\,\sin\theta'$$

that is

$$\theta = \theta'$$

so the angle of incidence equals the angle of reflection.

The magnitude

$$k_t = \frac{2\pi}{\lambda_2}$$

so that

$$k_i \sin \theta = k_t l_t$$

gives

$$\frac{2\pi}{\lambda_1} \sin \theta = \frac{2\pi}{\lambda_2} \sin \phi$$

or

$$\frac{\sin \theta}{\sin \phi} = \frac{\lambda_1}{\lambda_2} = \frac{n_2}{n_1} \left[\frac{\text{Refractive Index (medium 2)}}{\text{Refractive Index (medium 1)}} \right]$$

a relationship between the angles of incidence and refraction which is well known as Snell's Law.

Total Internal Reflexion and Evanescent Waves

On page 235 we discussed the propagation of an electromagnetic wave across the boundary between air and a dielectric (glass, say). We now consider the reverse process where a wave in the dielectric crosses the interface into air. Snell's Law still holds so we have, in fig. 8.11,

$$n_1 \sin \theta = n_2 \sin \phi$$

where

$$n_1 > n_2 \quad \text{and} \quad n_2/n_1 = n_r < 1$$

Thus

$$\sin \theta = (n_2/n_1) \sin \phi = n_r \sin \phi$$

with $\phi > \theta$. Eventually a critical angle of incidence θ_c is reached where $\phi = 90°$ and $\sin \theta = n_r$; for $\theta > \theta_c$, $\sin \theta > n_r$. For glass to air $n_r = \frac{1}{1 \cdot 5}$ and $\theta_c = 42°$.

It is evident that for $\theta \geqslant \theta_c$ no electromagnetic energy is transmitted into the rarer medium and the incident wave is said to suffer *total internal reflexion*.

In the reflexion coefficients $R_{\parallel}$ and $R_{\perp}$ on page 208 we may replace $\cos \phi$ by

$$(1 - \sin^2 \phi)^{1/2} = [1 - (\sin \theta/n_r)^2]^{1/2}$$

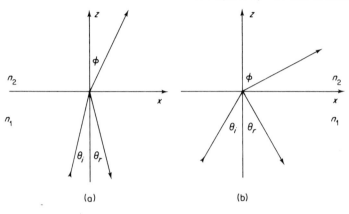

(a)   (b)

$n_1 > n_2$

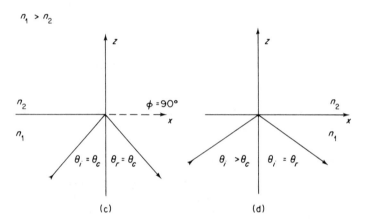

(c)   (d)

Fig. 8.11. When light propagates from a dense to a rare medium $(n_1 > n_2)$ Snell's Law defines $\theta = \theta_c$ as that angle of incidence for which $\phi = 90°$ and the refracted ray is tangential to the plane boundary. Total internal reflection can take place but the boundary conditions still require a transmitted wave known as the evanescent or surface wave. It propagates in the x direction but its amplitude decays exponentially with z

and rewrite

$$R_{\parallel} = \frac{(n_r^2 - \sin^2 \theta)^{1/2} - n_r^2 \cos \theta}{(n_r^2 - \sin^2 \theta)^{1/2} + n_r^2 \cos \theta}$$

and

$$R_{\perp} = \frac{\cos \theta - (n_r^2 - \sin^2 \theta)^{1/2}}{\cos \theta + (n_r^2 - \sin^2 \theta)^{1/2}}$$

Now for $\theta > \theta_c$, $\sin \theta > n_r$ and the bracketed quantities in $R_\parallel$ and $R_\perp$ are negative so that $R_\parallel$ and $R_\perp$ are complex quantities, that is $(E_r)_\parallel$ and $(E_r)_\perp$ have a phase relation which depends on θ.

It is easily checked that the product of R and R^* is unity so we have $R_\parallel R_\parallel^* = R_\perp R_\perp^* = 1$. This means, for both of the examples of fig. 7.8, that the incident and reflected intensities I_i and $I_r = 1$. The transmitted intensity $I_t = 0$ so that no energy is carried across the boundary.

However, if there is no transmitted wave we cannot satisfy the boundary condition $E_i + E_r = E_t$ on page 236, using only incident and reflected waves. We must therefore assert that a transmitted wave does exist but that it cannot on the average carry energy across the boundary.

We now examine the implications of this assertion, using fig. 8.10 above, and we keep the notation of page 236. This gives a transmitted electric field vector

$$E_t = A_t\, e^{i[\omega t - k_t(l_t + m_t y + n_t z)]}$$
$$= A_t\, e^{i[\omega t - k_t(x \sin \phi + z \cos \phi)]}$$

because $y = 0$ in the xz plane, $l_t = \sin \phi$ and $n_t = \cos \phi$. Now

$$\cos \phi = 1 - \sin^2 \phi = 1 - \sin^2 \theta / n_r^2$$
$$\therefore\ k_t \cos \phi = \pm k_t (1 - \sin^2 \theta / n_r^2)^{1/2}$$

which for $\theta > \theta_c$ gives $\sin \theta > n_r$ so that

$$k_t \cos \phi = \mp i k_t \left(\frac{\sin^2 \theta}{n_r^2} - 1\right)^{1/2} = \mp i\beta$$

We also have

$$k_t \sin \phi = k_t \sin \theta / n_r$$

so

$$E_t = A_t\, e^{\mp \beta z}\, e^{i(\omega t - k_t x \sin \theta / n_r)}$$

The alternative factor $e^{+\beta z}$ defines an exponential growth of A_t which is physically untenable and we are left with a wave whose amplitude decays exponentially as it penetrates the less dense medium. The disturbance travels in the x direction along the interface and is known as a *surface* or *evanescent wave*.

It is possible to show from the expressions for $R_\parallel$ and $R_\perp$ on page 238 that except at $\theta = 90°$ the incident and the reflected electric field components for $(E)_\parallel$ in one case and $(E)_\perp$ in the other, do not differ in phase by π radians and cannot therefore cancel each other out. The continuity of the tangential component of **E** at the boundary therefore leaves a component parallel to the interface which propagates as the surface wave. This effect has been observed at optical frequencies.

Moreover, if only a very thin air gap exists between two glass blocks it is possible for energy to flow across the gap allowing the wave to propagate in the second glass block. This process is called *frustrated total internal reflexion* and has its quantum mechanical analogue in the tunelling effect discussed on page 396.

Problem 8.1
Show that

$$z = A\, e^{i\{\omega t - (k_1 x + k_2 y)\}}$$

where $k^2 = \omega^2/c^2 = k_1^2 + k_2^2$ is a solution of the two-dimensional wave equation

$$\frac{\partial^2 z}{\partial x^2} + \frac{\partial^2 z}{\partial y^2} = \frac{1}{c^2}\frac{\partial^2 z}{\partial t^2}$$

Problem 8.2
Show that if the displacement of the waves on the membrane of width b of fig. 8.3 is given by the superposition

$$z = A_1 \sin[\omega t - (k_1 x + k_2 y)] + A_2 \sin[\omega t - (k_1 x - k_2 y)]$$

with the boundary conditions

$$z = 0 \quad \text{at} \quad y = 0 \quad \text{and} \quad y = b$$

then

$$z = -2A_1 \sin k_2 y \cos(\omega t - k_1 x)$$

where

$$k_2 = \frac{n\pi}{b}$$

Problem 8.3
An electromagnetic wave loses negligible energy when reflected from a highly conducting surface. With repeated reflexions it may travel along a transmission line or wave guide consisting of two parallel, infinitely conducting planes (separation a). If the wave

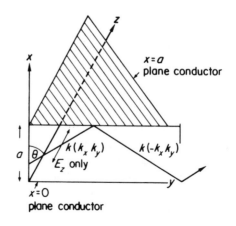

is plane polarized, so that only E_z exists, then the propagating direction **k** lies wholly in the xy plane. The boundary conditions require that the total tangential electric field E_z is zero at the conducting surfaces $x = 0$ and $x = a$. Show that the first boundary condition allows E_z to be written $E_z = E_0(e^{ik_x x} - e^{-ik_x x})\, e^{i(k_y y - \omega t)}$, where $k_x = k \cos\theta$ and $k_y = k \sin\theta$ and that the second boundary condition requires $k_x = n\pi/a$.

If $\lambda_0 = 2\pi c/\omega$, $\lambda_c = 2\pi/k_x$ and $\lambda_g = 2\pi/k_y$ are the wavelengths propagating in the x and y directions respectively show that

$$\frac{1}{\lambda_c^2} + \frac{1}{\lambda_g^2} = \frac{1}{\lambda_0^2}$$

We see that for $n = 1$, $k_x = \pi/a$ and $\lambda_c = 2a$, and that as ω decreases and λ_0 increases, $k_y = k \sin\theta$ becomes imaginary and the wave is damped. Thus $n = 2$ $(k_x = 2\pi/a)$ gives $\lambda_c = a$, the 'critical wavelength', i.e. the longest wavelength propagated by a waveguide of separation a. Such cut-off wavelengths and frequencies are a feature of wave propagation in periodic structures, transmission lines and waveguides.

Problem 8.4

Show, from equations (7.1) and (7.2), that the magnetic field in the plane-polarized electromagnetic wave of problem 8.3 has components in both x- and y-directions. [When an electromagnetic wave propagating in a waveguide has only transverse electric field vectors and no electric field in the direction of propagation it is called a transverse electric (TE) wave. Similarly a transverse magnetic (TM) wave may exist. The plane-polarized wave of problem 8.3 is a transverse electric wave; the corresponding transverse magnetic wave would have H_z, E_x and E_y components. The values of n in problem 8.3 satisfying the boundary conditions are written as subscripts to define the exact mode of propagation, e.g. TE_{10}.]

Problem 8.5

Use the value of the inductance and capacitance of a pair of plane parallel conductors of separation a and width b to show that the characteristic impedance of such a waveguide is given by

$$Z_0 = \frac{a}{b}\sqrt{\frac{\mu}{\epsilon}} \text{ ohms}$$

where μ and ϵ are respectively the permeability and permittivity of the medium between the conductors.

Problem 8.6

Consider either the Poynting vector or the energy per unit volume of an electromagnetic wave to show that the power transmitted by a single positive travelling wave in the waveguide of problem 8.5 is $\frac{1}{2}abE_0^2\sqrt{\epsilon/\mu}$.

Problem 8.7

An electromagnetic wave $(\mathbf{E}, \mathbf{H})$ propagates in the x-direction down a perfectly conducting hollow tube of arbitrary cross section. The tangential component of $\mathbf{E}$ at the conducting walls must be zero at all times.

Show that the solution $\mathbf{E} = E(y, z) \cos(\omega t - k_x x)$ substituted in the wave equation yields

$$\frac{\partial^2 \mathbf{E}}{\partial y^2} + \frac{\partial^2 \mathbf{E}}{\partial z^2} = -k^2 \mathbf{E}$$

where $k^2 = \omega^2/c^2 - k_x^2$ and k_x is the wave number appropriate to the x-direction.

Problem 8.8

If the waveguide of problem 8.7 is of rectangular cross-section of width a in the y-direction and height b in the z-direction, show that the boundary conditions $E_x = 0$ at $y = 0$ and a and at $z = 0$ and b in the wave equation of problem 8.7 gives

$$E_x = A \sin \frac{m\pi y}{a} \sin \frac{n\pi z}{b} \cos (\omega t - k_x x)$$

where

$$k^2 = \pi^2 \left(\frac{m^2}{a^2} + \frac{m^2}{b^2} \right)$$

Problem 8.9

Show, from problems 8.7 and 8.8, that the lowest possible value of ω (the cut-off frequency) for k_x to be real is given by $m = n = 1$.

Problem 8.10

Prove that the product of the phase and group velocity ω/k_x, $\partial\omega/\partial k_x$ of the wave of problems 8.7–8.9 is c^2, where c is the velocity of light.

Problem 8.11

Consider now the extension of problem 8.2 where the waves are reflected at the rigid edges of the rectangular membrane of sides length a and b as shown in the diagram. The final displacement is the result of the superposition

$$z = A_1 \sin [\omega t - (k_1 x + k_2 y)]$$
$$+ A_2 \sin [\omega t - (k_1 x - k_2 y)]$$
$$+ A_3 \sin [\omega t - (-k_1 x - k_2 y)]$$
$$+ A_4 \sin [\omega t - (-k_1 x + k_2 y)]$$

with the boundary conditions

$$z = 0 \quad \text{at} \quad x = 0 \quad \text{and} \quad x = a$$

and

$$z = 0 \quad \text{at} \quad y = 0 \quad \text{and} \quad y = b$$

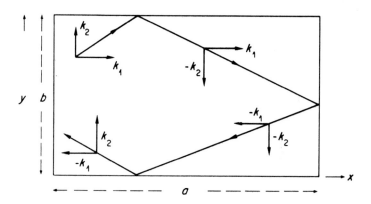

Show that this leads to a displacement

$$z = -4A_1 \sin k_1 x \sin k_2 y \sin \omega t$$

where

$$k_1 = \frac{n_1 \pi}{a} \quad \text{and} \quad k_2 = \frac{n_2 \pi}{b}$$

Problem 8.12

In deriving the result that the average energy of an oscillator at frequency ν and temperature T is given by

$$\bar{\epsilon} = \frac{h\nu}{e^{(h\nu/kT)} - 1}$$

Planck assumed that a large number N of oscillators had energies 0, $h\nu$, $2h\nu$ ... $nh\nu$ distributed according to Boltzmann's Law

$$N_n = N_0\, e^{-nh\nu/kT}$$

where the number of oscillators N_n with energy $nh\nu$ decreases exponentially with increasing n.

Use the geometric progression series

$$N = \sum_n N_n = N_0(1 + e^{-h\nu/kT} + e^{-2h\nu/kT} \ldots)$$

to show that

$$N = \frac{N_0}{1 - e^{-h\nu/kT}}$$

If the total energy of the oscillators in the nth energy state is given by

$$E_n = N_n nh\nu$$

prove that the total energy over all the n energy states is given by

$$E = \sum_n E_n = N_0 \frac{h\nu\, e^{-h\nu/kT}}{(1 - e^{-h\nu/kT})^2}$$

Hence show that the average energy per oscillator

$$\bar{\epsilon} = \frac{E}{N} = \frac{h\nu}{e^{h\nu/kT} - 1}$$

and expand the denominator to show that for $h\nu \ll kT$, that is low frequencies and long wavelengths, Planck's Law becomes the classical expression of Rayleigh–Jeans.

Problem 8.13

The wave representation of a particle, e.g. an electron, in a rectangular potential well throughout which $V = 0$ is given by Schrödinger's time-independent equation

$$\frac{\partial^2 \Psi}{\partial x^2} + \frac{\partial^2 \Psi}{\partial y^2} + \frac{\partial^2 \Psi}{\partial z^2} = -\frac{8\pi^2 m}{h^2} E\Psi$$

where E is the particle energy, m is the mass and h is Planck's constant. The boundary conditions to be satisfied are $\psi = 0$ at $x = y = z = 0$ and at $x = L_x$, $y = L_y$, $z = L_z$, where L_x, L_y and L_z are the dimensions of the well.

Show that

$$\Psi = A \sin \frac{l\pi x}{L_x} \sin \frac{r\pi y}{L_y} \sin \frac{n\pi z}{L_z}$$

is a solution of Schrödinger's equation, giving

$$E = \frac{h^2}{8m}\left(\frac{l^2}{L_x^2} + \frac{r^2}{L_y^2} + \frac{n^2}{L_z^2}\right)$$

When the potential well is cubical of side L,

$$E = \frac{h^2}{8mL^2}(l^2 + r^2 + n^2)$$

and the lowest value of the quantized energy is given by

$$E = E_0 \quad \text{for} \quad l = 1, \qquad r = n = 0$$

Show that the next energy levels are $3E_0$, $6E_0$ (3-fold degenerate), $9E_0$ (three-fold degenerate), $11E_0$ (three-fold degenerate), $12E_0$ and $14E_0$ (six-fold degenerate).

Problem 8.14

Show that at low energy levels (long wavelengths) $h\nu \ll kT$, Planck's radiation law is equivalent to the Rayleigh–Jeans expression.

Problem 8.15

Planck's radiation law, expressed in terms of energy per unit range of wavelength instead of frequency, becomes

$$E_\lambda = \frac{8\pi ch}{\lambda^5(e^{ch/\lambda kT} - 1)}$$

Use the variable $x = ch/\lambda kT$ to show that the total energy per unit volume at temperature $T°$ absolute is given by

$$\int_0^\infty E_\lambda \, d\lambda = aT^4 \text{ joules m}^{-3}$$

where

$$a = \frac{8\pi^5 k^4}{15c^3 h^3}$$

(The constant $ca/4 = \sigma$, Stefan's Constant in the Stefan–Boltzmann Law.) Note that

$$\int_0^\infty \frac{x^3 \, dx}{e^x - 1} = \frac{\pi^4}{15}$$

Problem 8.16

Show that the wavelength λ_m at which E_λ in problem 8.16 is a maximum is given by the solution of

$$\left(1 - \frac{x}{5}\right)e^x = 1, \quad \text{where } x = \frac{ch}{\lambda kT}$$

The solution is $ch/\lambda_m kT = 4.965$.

Problem 8.17

Given that $ch/\lambda_m = 5\,kT$ in problem 8.16, show that if the sun's temperature is about 6000°K, then $\lambda_m \approx 4\cdot 7 \times 10^{-7}$ metres, the green region of the visible spectrum where the human eye is most sensitive (evolution ?).

Problem 8.18

The tungsten filament of an electric light bulb has a temperature of $\approx 2000°K$. Show that in this case $\lambda_m \approx 14 \times 10^{-7}$ metres, well into the infrared. Such a lamp is therefore a good heat source but an inefficient light source.

Problem 8.19

A free electron (travelling in a region of zero potential) has an energy

$$E = \frac{p^2}{2m} = \left(\frac{\hbar^2}{2m}\right)k^2 = E(k)$$

where the wavelength

$$\lambda = h/p = 2\pi/k$$

In a weak periodic potential, for example in a solid which is a good electrical conductor, $E = (\hbar^2/2m^*)k^2$, where m^* is called the effective mass. (For valence electrons $m^* \approx m$.)

Represented as waves, the electrons in a cubic potential well ($V = 0$) of side L have allowed wave numbers k, where

$$k^2 = k_x^2 + k_y^2 + k_z^2 \quad \text{and} \quad k_i = \frac{n_i \pi}{L}$$

(see problem 8.13). For each value of k there are two allowed states (each defining the spin state of the single occupying electron—Pauli's principle). Use the arguments in Chapter 8 to show that the total number of states in k space between the values k and $k + dk$ is given by

$$P(k) = 2\left(\frac{L}{\pi}\right)^3 \frac{4\pi k^2 dk}{8}$$

Use the expression $E = (\hbar^2/2m^*)\,k^2$ to convert this into the number of states $S(E)\,dE$ in the energy interval dE to give

$$S(E) = \frac{A}{2\pi^2}\left(\frac{2m}{\hbar^2}\right)^{3/2}\sqrt{E}$$

where $A = L^3$.

If there are N electrons in the N lowest energy states consistent with Pauli's principle, show that the integral

$$\int_0^{E_f} S(E)\,dE = N$$

gives the Fermi energy level

$$E_f = \frac{\hbar^2}{2m^*}\left(\frac{3\pi^2 N}{A}\right)^{2/3}$$

where E_f is the kinetic energy of the most energetic electron when the solid is in its ground state.

Summary of Important Results

Wave equation in two dimensions

$$\frac{\partial^2 z}{\partial x^2} + \frac{\partial^2 z}{\partial y^2} = \frac{1}{c^2}\frac{\partial^2 z}{\partial t^2}$$

Lines of constant phase $lx + my = ct$ propagate in direction $\mathbf{k}(k_1, k_2)$ where $l = k_1/k$, $m = k_2/k$, $k^2 = k_1^2 + k_2^2$ and $c^2 = \omega^2/k^2$. Solution is

$$z = A\,e^{i(\omega t - \mathbf{k}\cdot\mathbf{r})} \qquad \text{for } \mathbf{r}(x, y)$$

Wave equation in three dimensions

$$\frac{\partial^2 \phi}{\partial x^2} + \frac{\partial^2 \phi}{\partial y^2} + \frac{\partial^2 \phi}{\partial z^2} = \frac{1}{c^2}\frac{\partial^2 \phi}{\partial t^2}$$

Planes of constant phase $lx + my + nz = ct$ propagate in a direction

$$\mathbf{k}(k_1, k_2, k_3) \quad \text{where } l = k_1/k, \qquad m = k_2/k, \qquad n = k_3/k$$
$$k^2 = k_1^2 + k_2^2 + k_3^2 \quad \text{and} \quad c^2 = \omega^2/k^2.$$

Solution is

$$\phi = A\,e^{i(\omega t - \mathbf{k}.\mathbf{r})} \quad \text{for } \mathbf{r}(x, y, z).$$

Wave guides

Reflection from walls $y = 0$, $y = b$ in a two-dimensional wave guide for a wave of frequency ω and vector direction $\mathbf{k}(k_1, k_2)$ gives normal modes in the y direction with $k_2 = n\pi/b$ and propagation in the x direction with phase velocity

$$v_p = \frac{\omega}{k_1} > \frac{\omega}{k} = v$$

and group velocity

$$v_g = \frac{\partial \omega}{\partial k_1} \quad \text{such that} \quad v_p v_g = v^2$$

Cut off frequency

Only frequencies $\omega \geqslant n\pi v/b$ will propagate where n is mode number.

Normal modes in three dimensions

Wave equation separates into three equations (one for each variable x, y, z) to give solution

$$= A \frac{\sin}{\cos} k_1 x \frac{\sin}{\cos} k_2 y \frac{\sin}{\cos} k_3 z \frac{\sin}{\cos} \omega t$$

(Boundary conditions determine final form of solution.)

For waves of velocity c, the number of normal modes per unit volume of an enclosure in the frequency range ν to $\nu + d\nu$

$$= \frac{4\pi \nu^2 \, d\nu}{c^3}$$

Directly applicable to
(a) Planck's Radiation Law
(b) Debye's theory of specific heats of solids
(c) Fermi energy level (problem 8.19)

Chapter 9

Fourier Methods

Fourier Series

In this chapter we are going to look in more detail at the implications of the principles of superposition which we met at the beginning of the book when we added the two separate solutions of the simple harmonic motion equation. Our discussion of monochromatic waves has led to the idea of repetitive behaviour in a simple form. Now we consider more complicated forms of repetition which arise from superposition.

Any function which repeats itself regularly over a given interval of space or time is called a periodic function. This may be expressed by writing it as $f(x) = f(x \pm \alpha)$ where α is the interval or period.

The simplest examples of a periodic function are sines and cosines of fixed frequency and wavelength, where α represents the period τ, the wavelength λ or the phase angle 2π radians, according to the form of x. Most periodic functions, for example the square wave system of fig. 9.1, although quite simple

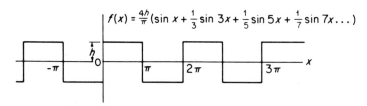

$$f(x) = \frac{4h}{\pi}\left(\sin x + \frac{1}{3}\sin 3x + \frac{1}{5}\sin 5x + \frac{1}{7}\sin 7x \ldots\right)$$

Fig. 9.1. Square wave of height h and its Fourier sine series representation (odd function)

to visualize are more complicated to represent mathematically. Fortunately this can be done for almost all periodic functions of interest in physics using the method of Fourier Series, which states that any periodic function may be represented by the series

$$f(x) = \tfrac{1}{2}a_0 + a_1 \cos x + a_2 \cos 2x \ldots + a_n \cos nx$$

$$+ b_1 \sin x + b_2 \sin 2x \ldots + b_n \sin nx, \qquad (9.1)$$

that is, a constant $\frac{1}{2}a_0$ plus sine and cosine terms of different amplitudes, having frequencies which increase in discrete steps. Such a series must of course, satisfy certain conditions, chiefly those of convergence. These convergence criteria are met for a function with discontinuities which are not too severe and with first and second differential coefficients which are well behaved. At such discontinuities, for instance in the square wave where $f(x) = \pm h$ at $x = 0, \pm 2\pi$, etc., the series represents the mean of the values of the function just to the left and just to the right of the discontinuity.

We may write the series in several equivalent forms:

$$f(x) = \tfrac{1}{2}a_0 + \sum_{n=1}^{\infty} (a_n \cos nx + b_n \sin nx)$$

$$= \tfrac{1}{2}a_0 + \sum_{n=1}^{\infty} c_n \cos (nx - \theta_n)$$

where

$$c_n^2 = a_n^2 + b_n^2$$

and

$$\tan \theta_n = b_n / a_n$$

or

$$f(x) = \sum_{n=-\infty}^{\infty} d_n \, e^{inx}$$

where

$$2d_n = a_n - ib_n \;\; (n \geqslant 0)$$

and

$$2d_n = a_n + ib_n \;\; (n < 0)$$

To find the values of the coefficients a_n and b_n let us multiply both sides of equation (9.1) by $\cos nx$ and integrate with respect to x over the period 0 to 2π (say).

Every term

$$\int_0^{2\pi} \cos mx \cos nx \, \mathrm{d}x = \begin{cases} 0 \text{ if } m \neq n \\ \pi \text{ if } m = n \end{cases}$$

whilst every term

$$\int_0^{2\pi} \sin mx \cos nx \, \mathrm{d}x = 0 \text{ for all } m \text{ and } n.$$

Thus for $m = n$,

$$a_n \int_0^{2\pi} \cos^2 nx = \pi a_n$$

so that

$$a_n = \frac{1}{\pi} \int_0^{2\pi} f(x) \cos nx \, dx$$

Similarly, by multiplying both sides of equation (9.1) by $\sin nx$ and integrating from 0 to 2π we have, since

$$\int_0^{2\pi} \sin mx \sin nx = \begin{cases} 0 \text{ if } m \neq n \\ \pi \text{ if } m = n \end{cases}$$

that

$$b_n = \frac{1}{\pi} \int_0^{2\pi} f(x) \sin nx \, dx$$

Immediately we see that the constant $(n = 0)$, given by $\frac{1}{2}a_0 = 1/2\pi \int_0^{2\pi} f(x) \, dx$, is just the average of the function over the interval 2π. It is, therefore, the steady or 'd.c.' level on which the alternating sine and cosine components of the series are superimposed, and the constant can be varied by moving the function with respect to the x-axis. When a periodic function is symmetric about the x-axis its average value, that is, its steady or d.c. base level, $\frac{1}{2}a_0$, is zero, as in the square wave system of fig. 9.1. If we raise the square waves so that they stand as pulses of height $2h$ on the x-axis, the value of $\frac{1}{2}a_0$ is $h\pi$ (average value over 2π). The values of a_n represent twice the average value of the product $f(x) \cos nx$ over the interval 2π; b_n can be interpreted in a similar way.

We see also that the series representation of the function is the sum of cosine terms which are even functions [$\cos x = \cos (-x)$] and of sine terms which are odd functions [$\sin x = -\sin (-x)$]. Now every function $f(x) = \frac{1}{2}[f(x) + f(-x)] + \frac{1}{2}[f(x) - f(-x)]$, in which the first bracket is even and the second bracket is odd. Thus the cosine part of a Fourier series represents the

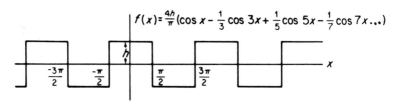

$$f(x) = \frac{4h}{\pi}\left(\cos x - \frac{1}{3}\cos 3x + \frac{1}{5}\cos 5x - \frac{1}{7}\cos 7x \dots\right)$$

Fig. 9.2. The wave of fig 9.1 is now symmetric about the y axis and becomes a cosine series (even function)

even part of the function and the sine terms represent the odd part of the function. Taking the argument one stage further, a function $f(x)$ which is an even function is represented by a Fourier series having only cosine terms; if $f(x)$ is odd it will have only sine terms in its Fourier representation. Whether a function is completely even or completely odd can often be determined by the position of the y-axis. Our square wave of fig. 9.1 is an odd function $[f(x) = -f(-x)]$; it has no constant and is represented by $f(x) = 4h/\pi$ ($\sin x + 1/3 \sin 3x + 1/5 \sin 5x$, etc., but if we now move the y-axis a half period to the right as in fig. 9.2, then $f(x) = f(-x)$, an even function, and the square wave is represented by

$$f(x) = \frac{4h}{\pi}(\cos x - \tfrac{1}{3} \cos 3x + \tfrac{1}{5} \cos 5x - \tfrac{1}{7} \cos 7x + \ldots)$$

If we take the first three or four terms of the series representing the square wave of fig. 9.1 and add them together, the result is fig. 9.3. The fundamental,

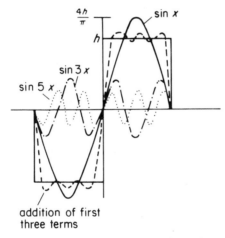

addition of first
three terms

Fig. 9.3. Addition of the first three terms of the Fourier series for the square wave of fig. 9.1 shows that the higher frequencies are responsible for sharpening the edges of the pulse

or first harmonic, has the frequency of the square wave and the higher frequencies build up the squareness of the wave. The highest frequencies are responsible for the sharpness of the vertical sides of the waves; this type of square wave is commonly used to test the frequency response of amplifiers. An amplifier with a square wave input effectively 'Fourier analyses' the input and responds to the individual frequency components. It then puts them together again at its output, and if a perfect square wave emerges from the amplifier it proves that the amplifier can handle the whole range of the frequency components equally well. Loss of sharpness at the edges of the waves shows that the amplifier response is limited at the higher frequency range.

Example of Fourier Series

Consider the square wave of height h in fig. 9.1. The value of the function is given by

$$f(x) = h \quad \text{for} \quad 0 < x < \pi$$

and

$$f(x) = -h \quad \text{for} \quad \pi < x < 2\pi$$

The coefficients of the series representation are given by

$$a_n = \frac{1}{\pi}\left[h\int_0^{\pi} \cos nx \, dx - h\int_{\pi}^{2\pi} \cos nx \, dx\right] = 0$$

because

$$\int_0^{\pi} \cos nx \, dx = \int_{\pi}^{2\pi} \cos nx \, dx = 0$$

and

$$b_n = \frac{1}{\pi}\left[h\int_0^{\pi} \sin nx \, dx - h\int_{\pi}^{2\pi} \sin nx \, dx\right]$$

$$= \frac{h}{n\pi}[[\cos nx]_{\pi}^0 + [\cos nx]_{\pi}^{2\pi}]$$

$$= \frac{h}{n\pi}[(1 - \cos n\pi) + (1 - \cos n\pi)]$$

giving $b_n = 0$ for n even and $b_n = 4h/n\pi$ for n odd. Thus the Fourier series representation of the square wave is given by

$$f(x) = \frac{4h}{\pi}\left(\sin x + \frac{\sin 3x}{3} + \frac{\sin 5x}{5} + \frac{\sin 7x}{7} + \ldots\right)$$

Fourier Series for any Interval

Although we have discussed the Fourier representation in terms of a periodic function its application is much more fundamental, for any section or interval of a well behaved function may be chosen and expressed in terms of a Fourier Series. This series will accurately represent the function *only within the chosen interval*. If applied outside that interval it will not follow the function but will periodically repeat the value of the function within the chosen interval. If we represent this interval by a Fourier cosine series the repetition will be that of an even function, if the representation is a Fourier sine series an odd function repetition will follow.

Suppose now that we are interested in the behaviour of a function over only one-half of its full interval and have no interest in its representation outside this restricted region. In fig. 9.4a the function $f(x)$ is shown over its full space interval $-l/2$ to $+l/2$, but $f(x)$ can be represented completely in the interval 0 to $+l/2$ by either a cosine function (which will repeat itself each half-interval as an even function) or it can be represented completely by a sine function, in which case it will repeat itself each half-interval as an odd function. Neither representation will match $f(x)$ outside the region 0 to $+l/2$, but in the half-interval 0 to $+l/2$ we can write

$$f(x) = f_e(x) = f_0(x)$$

where the subscripts e and o are the even (cosine) or odd (sine) Fourier representations respectively.

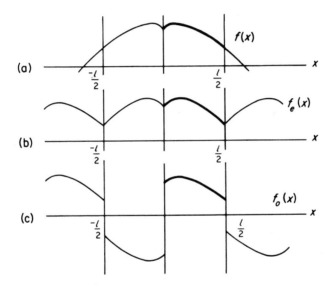

Fig. 9.4. A Fourier series may represent a function over a selected half-interval. The general function in (a) is represented in the half-interval $0 < x < l/2$ by f_e, an even function cosine series in (b), and by f_0, an odd function sine series in (c). These representations are valid only in the specified half-interval. Their behaviour outside that half-interval is purely repetitive and departs from the original function

The arguments of sines and cosines must, of course, be phase angles, and so far the variable x has been measured in radians. Now, however, the interval is specified as a distance and the variable becomes $2\pi x/l$, so that each time x changes by l the phase angle changes by 2π.

Thus

$$f_e(x) = \frac{a_0}{2} + \sum_{n=1}^{\infty} a_n \cos \frac{2\pi nx}{l}$$

where

$$a_n = \frac{1}{\frac{1}{2}\,\text{interval}} \int_{-l/2}^{l/2} f(x) \cos \frac{2\pi nx}{l}\, dx$$

$$= \frac{2}{l}\left[\int_{-l/2}^{0} f_e(x) \cos \frac{2\pi nx}{l}\, dx + \int_{0}^{l/2} f_e(x) \cos \frac{2\pi nx}{l}\, dx \right]$$

$$= \frac{4}{l} \int_{0}^{l/2} f(x) \cos \frac{2\pi nx}{l}\, dx$$

because

$$f(x) = f_e(x) \quad \text{from } x = 0 \text{ to } l/2$$

and

$$f(x) = f(-x) = f_e(x) \quad \text{from } x = 0 \text{ to } -l/2$$

Similarly we can represent $f(x)$ by the sine series

$$f(x) = f_0(x) = \sum_{n=1}^{\infty} b_n \sin \frac{2\pi nx}{l}$$

in the range $x = 0$ to $l/2$ with

$$b_n = \frac{1}{\frac{1}{2}\,\text{interval}} \int_{-l/2}^{l/2} f(x) \sin \frac{2\pi nx}{l}\, dx$$

$$= \frac{2}{l}\left[\int_{-l/2}^{0} f_0(x) \sin \frac{2\pi nx}{l}\, dx + \int_{0}^{l/2} f_0(x) \sin \frac{2\pi nx}{l}\, dx \right]$$

In the second integral $f_0(x) = f(x)$ in the interval 0 to $l/2$ whilst

$$\int_{l/2}^{0} f_0(x) \sin \frac{2\pi nx}{l}\, dx = \int_{l/2}^{0} f_0(-x) \sin \frac{2\pi nx}{l}\, dx = -\int_{l/2}^{0} f_0(x) \sin \frac{2\pi nx}{l}\, dx$$

$$= \int_{0}^{l/2} f_0(x) \sin \frac{2\pi nx}{l}\, dx = \int_{0}^{l/2} f(x) \sin \frac{2\pi nx}{l}\, dx$$

Hence

$$b_n = \frac{4}{l} \int_{0}^{l/2} f(x) \sin \frac{2\pi nx}{l}\, dx$$

If we follow the behaviour of $f_e(x)$ and $f_0(x)$ outside the half-interval 0 to $l/2$ (fig. 9.4a, b) we see that they no longer represent $f(x)$.

Application of Fourier Sine Series to a Triangular Function

Fig. 9.5 shows a function which we are going to describe by a sine series in the

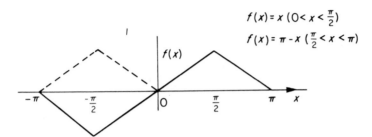

$$f(x) = x \quad (0 < x < \tfrac{\pi}{2})$$
$$f(x) = \pi - x \quad (\tfrac{\pi}{2} < x < \pi)$$

Fig. 9.5. Function representing a plucked string and defined over a limited interval. When the string vibrates all the permitted harmonics contribute to the initial configuration

half-interval 0 to π. The function is

$$f(x) = x \quad \left(0 < x < \frac{\pi}{2}\right)$$

and

$$f(x) = \pi - x \quad \left(\frac{\pi}{2} < x < \pi\right)$$

Writing $f(x) = \sum b_n \sin nx$ gives

$$b_n = \frac{2}{\pi}\int_0^{\pi/2} x \sin nx \, dx + \frac{2}{\pi}\int_{\pi/2}^{\pi} (\pi - x) \sin nx \, dx$$

$$= \frac{4}{n^2 \pi}\sin\frac{n\pi}{2}$$

When n is even $\sin n\pi/2 = 0$, so that only terms with odd values of n are present and

$$f(x) = \frac{4}{\pi}\left(\frac{\sin x}{1^2} - \frac{\sin 3x}{3^2} + \frac{\sin 5x}{5^2} - \frac{\sin 7x}{7^2} + \dots\right)$$

Note that at $x = \pi/2$, $f(x) = \pi/2$, giving

$$\frac{\pi^2}{8} = \frac{1}{1^2} + \frac{1}{3^2} + \frac{1}{5^2} + = \sum_{n=0}^{\infty}\frac{1}{(2n+1)^2}$$

We shall use this result a little later.

Note that the solid line in the interval 0 to $-\pi$ in fig. 9.5 is the Fourier sine representation for $f(x)$ repeated outside the interval 0 to π whilst the dotted line would result if we had represented $f(x)$ in the interval 0 to π by an even cosine series.

(Problems 9.1, 9.2, 9.3, 9.4, 9.5, 9.6, 9.7, 9.8, 9.9)

Application to the Energy in the Normal Modes of a Vibrating String

If we take a string of length l with fixed ends and pluck its centre a distance d we have the configuration of the half interval 0 to π of fig. 9.5 which we represented as a Fourier sine series. Releasing the string will set up its normal mode or standing wave vibrations, each of which we have shown on page 119 to have the displacement

$$y_n = (A_n \cos \omega_n t + B_n \sin \omega_n t) \sin \frac{\omega_n x}{c}$$

where $\omega_n = n\pi c/l$ is the normal mode frequency.

The total displacement, which represents the shape of the plucked string at $t = 0$ is given by the Fourier series

$$y = \sum y_n = \sum (A_n \cos \omega_n t + B_n \sin \omega_n t) \sin \frac{\omega_n x}{c}$$

where the fixed ends of the string allow only the sine terms in x in the series expansion. If the string remains plucked *at rest* only the terms in x with appropriate coefficients are required to describe it, but its vibrational motion after release has a time dependence which is expressed in each harmonic coefficient as

$$A_n \cos \omega_n t + B_n \sin \omega_n t$$

The significance of these coefficients emerges when we consider the initial or boundary conditions in time.

Let us write the total displacement of the string at time $t = 0$ as

$$y_0(x) = \sum y_n(x) = \sum (A_n \cos \omega_n t + B_n \sin \omega_n t) \sin \frac{\omega_n x}{c}$$

$$= \sum A_n \sin \frac{\omega_n x}{c} \quad \text{at } t = 0$$

Similarly we write the velocity of the string at time $t = 0$ as

$$v_0(x) = \frac{\partial}{\partial t} y_0(x) = \sum \dot{y}_n(x)$$

$$= \sum (-\omega_n A_n \sin \omega_n t + \omega n B_n \cos \omega_n t) \sin \frac{\omega_n x}{c}$$

$$= \sum \omega_n B_n \sin \frac{\omega_n x}{c} \quad \text{at } t = 0$$

Both $y_0(x)$ and $v_0(x)$ are thus expressed as Fourier sine series, but if the string is at rest at $t = 0$, then $v_0(x) = 0$ and all the B_n coefficients are zero, leaving only the A_n's. If the displacement of the string $y_0(x) = 0$ at time $t = 0$ whilst the string is moving, then all the A_n's are zero and the Fourier coefficients are the $\omega_n B_n$'s.

We can solve for both A_n and $\omega_n B_n$ in the usual way for if

$$y_0(x) = \sum A_n \sin \frac{\omega_n x}{c}$$

and

$$v_0(x) = \sum \omega_n B_n \sin \frac{\omega_n x}{c}$$

for a string of length l then

$$A_n = \frac{2}{l} \int_0^l y_0(x) \sin \frac{\omega_n x}{c}$$

and

$$\omega_n B_n = \frac{2}{l} \int_0^l v_0(x) \sin \frac{\omega_n x}{e}$$

If the plucked string of mass m (linear density ρ) is released from rest at $t = 0$ $(v_0(x) = 0)$ the energy in each of its normal modes of vibration, given on page 121 as

$$E_n = \tfrac{1}{4} m \omega_n^2 (A_n^2 + B_n^2)$$

is simply

$$E_n = \tfrac{1}{4} m \omega_n^2 A_n^2$$

because all B_n's are zero.

The total vibrational energy of the released string will be the sum $\sum E_n$ over all the modes present in the vibration.

Let us now solve the problem of the plucked string released from rest. The configuration of fig. 9.5 (string length l, centre plucked a distance d) is given by

$$y_0(x) = \frac{2dx}{l} \qquad 0 \leqslant x \leqslant \frac{l}{2}$$

$$= \frac{2d(l-x)}{l} \qquad \frac{l}{2} \leqslant x \leqslant l$$

so

$$A_n = \frac{2}{l}\left[\int_0^{l/2} \frac{2dx}{l} \sin \frac{\omega_n x}{c}\, dx + \int_{l/2}^l \frac{2d(l-x)}{l} \sin \frac{\omega_n x}{c}\, dx \right]$$

$$= \frac{8d}{n^2 \pi^2} \sin \frac{n\pi}{2} \quad \left(\text{for } \omega_n = \frac{n\pi c}{l} \right)$$

We see at once that $A_n = 0$ for n even (when the sine term is zero) so that all even harmonic modes are missing. The physical explanation for this is that the even harmonics would require a node at the centre of the string which is always moving after release.

The displacement of our plucked string is therefore given by the addition of all the permitted (odd) modes as

$$y_0(x) = \sum_{n \text{ odd}} y_n(x) = \sum_{n \text{ odd}} A_n \sin \frac{\omega_n x}{c}$$

where

$$A_n = \frac{8d}{n^2 \pi^2} \sin \frac{n\pi}{2}$$

The energy of the nth mode of oscillation is

$$E_n = \tfrac{1}{4} m \omega_n^2 A_n^2 = \frac{64 d^2 m \omega_n^2}{4(n^2 \pi^2)^2}$$

and the total vibrational energy of the string is given by

$$E = \sum_{n \text{ odd}} E_n = \frac{16 d^2 m}{\pi^4} \sum_{n \text{ odd}} \frac{\omega_n^2}{n^4} = \frac{16 d^2 c^2 m}{\pi^2 l^2} \sum_{n \text{ odd}} \frac{1}{n^2}$$

for

$$\omega_n = \frac{n\pi c}{l}$$

But we saw in the last section that

$$\sum_{n \text{ odd}} \frac{1}{n^2} = \frac{\pi^2}{8}$$

so

$$E = \sum E_n = \frac{2 m c^2 d^2}{l^2} = \frac{2 T d^2}{l}$$

where $T = \rho c^2$ is the constant tension in the string.

This vibrational energy, in the absence of dissipation, must be equal to the potential energy of the plucked string before release and the reader should prove this.

To summarize, our plucked string can be represented as a sine series of Fourier components, each giving an allowed normal mode of vibration when it is released. The concept of normal modes allows the energies of each mode to be added to give the total energy of vibration which must equal the potential energy of the plucked string before release. The energy of the nth mode is proportional to n^{-2} and therefore decreases with increasing frequency. Even modes are forbidden by the initial boundary conditions.

The boundary conditions determine which modes are allowed. If the string were struck by a hammer those harmonics having a node at the point of impact would be absent, as in the case of the plucked string. Pianos are commonly designed with the hammer striking a point one seventh of the way along the string, thus eliminating the seventh harmonic which combines to produce discordant effects.

Fourier Series Analysis of a Rectangular Velocity Pulse on a String

Let us now consider a problem similar to that of the last section except that now the displacement $y_0(x)$ of the string is zero at time $t = 0$ whilst the velocity $v_0(x)$ is non zero. A string of length l, fixed at both ends, is struck by a mallet of width a about its centre point. At the moment of impact the displacement

$$y_0(x) = 0$$

but the velocity

$$v_0(x) = \frac{\partial y_0(x)}{\partial t} = 0 \quad \text{for} \quad \left| x - \frac{l}{2} \right| \geqslant \frac{a}{2}$$

$$= v \quad \text{for} \quad \left| x - \frac{l}{2} \right| < \frac{a}{2}$$

This configuration is shown in fig. 9.6.

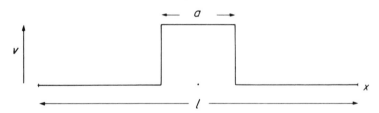

Fig. 9.6. Velocity distribution at time $t = 0$ of a string length l, fixed at both ends and struck about its centre point by a mallet of width a. Displacement $y_0(x) = 0$; velocity $v_0(x) = v$ for $|x - l/2| < a/2$ and zero outside this region

The Fourier series is given by

$$v_0(x) = \sum_n \dot{y}_n = \sum_n \omega_n B_n \sin \frac{\omega_n x}{c}$$

where

$$\omega_n B_n = \frac{2}{l} \int_{+l/2-a/2}^{l/2+a/2} v \sin \frac{\omega_n x}{c} \, dx$$

$$= \frac{4v}{n\pi} \sin \frac{n\pi}{2} \sin \frac{n\pi a}{2l}$$

Again we see that $\omega_n B_n = 0$ for n even ($\sin n\pi/2 = 0$) because the centre point of the string is never stationary, as is required in an even harmonic.

Thus

$$v_0(x) = \sum_{n \text{ odd}} \frac{4v}{n\pi} \sin \frac{n\pi a}{2l} \sin \frac{\omega_n x}{c}$$

The energy per mode of oscillation

$$E_n = \tfrac{1}{4} m\omega_n^2 (A_n^2 + B_n^2)$$

$$= \tfrac{1}{4} m\omega_n^2 B_n^2 \qquad (\text{All } A_n\text{'s} = 0)$$

$$= \tfrac{1}{4} m \frac{16 v^2}{n^2 \pi^2} \sin^2 \frac{n\pi a}{2l}$$

$$= \frac{4 m v^2}{n^2 \pi^2} \sin^2 \frac{n\pi a}{2l}$$

Now

$$n = \frac{\omega_n}{\omega_1} = \frac{\omega_n l}{\pi c}$$

for the fundamental frequency

$$\omega_1 = \frac{\pi c}{l}$$

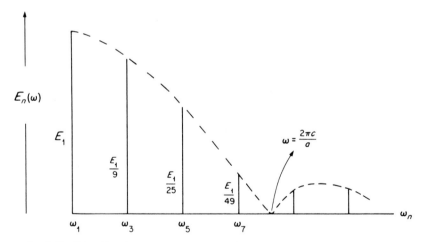

Fig. 9.7. Distribution of the energy in the harmonics ω_n of the string of fig. 9.6. The spectrum $E_n(\omega) \propto \sin^2 \alpha/\alpha^2$ where $\alpha = \omega_n a/2c$. Most of the energy in the string is contained in the frequency range $\Delta\omega \approx 2\pi c/a$, and for $a = \Delta x$ (the spatial width of the pulse), $\Delta x/c = \Delta t$ and $\Delta\omega \Delta t \approx 2\pi$ (bandwidth theorem)

So

$$E_n = \frac{4mv^2c^2}{l^2\omega_n^2}\sin^2\frac{\omega_n a}{2c}$$

Again we see, since $\omega_n \propto n$ that the energy of the nth mode $\propto n^{-2}$ and decreases with increasing harmonic frequency. We may show this by rewriting

$$E_n(\omega) = \frac{mv^2a^2}{l^2}\frac{\sin^2(\omega_n a/2c)}{(\omega_n a/2c)^2}$$

$$= \frac{mv^2a^2}{l^2}\frac{\sin^2\alpha}{\alpha^2}$$

where

$$\alpha = \omega_n a/2c$$

and plotting this expression as an energy-frequency spectrum in fig. 9.7.

The familiar curve of $\sin^2\alpha/\alpha^2$ again appears as the envelope of the energy values for each ω_n.

If the energy at ω_1 is E_1 then $E_3 = E_1/9$ and $E_5 = E_1/25$ so the major portion of the energy in the velocity pulse is to be found in the low frequencies. The first zero of the envelope $\sin^2\alpha/\alpha^2$ occurs when

$$\alpha = \frac{\omega a}{2c} = \pi$$

so the width of the central frequency pulse containing most of the energy is given by

$$\omega \approx \frac{2\pi c}{a}$$

This range of energy bearing harmonics is known as the 'spectral width' of the pulse written

$$\Delta\omega \approx \frac{2\pi c}{a}$$

The 'spatial width' a of the pulse may be written as Δx so we have

$$\Delta x\, \Delta\omega \approx 2\pi c$$

Reducing the width Δx of the mallet will increase the range of frequencies $\Delta\omega$ required to take up the energy in the rectangular velocity pulse. Now c is the velocity of waves on the string so

$$\Delta x/c$$

defines the duration Δt of the pulse giving

$$\Delta\omega\, \Delta t \approx 2\pi$$

or

$$\Delta \nu \, \Delta t \approx 1$$

the Bandwidth Theorem we first met on page 128.
Note that the harmonics have frequencies

$$\omega_n = \frac{n \pi c}{l}$$

so $\pi c/l$ is the harmonic interval. When the length l of the string becomes very long and $l \to \infty$ so that the pulse is isolated and non-periodic, the harmonic interval becomes so small that it becomes differential and the Fourier series summation becomes the Fourier Integral.

The Spectrum of a Fourier Series

The Fourier series can always be represented as a frequency spectrum. In fig. 9.8a the relative amplitudes of the frequency components of the square wave of fig. 9.1 are plotted, each sine term giving a single spectral line. In a similar manner, the distribution of energy with frequency may be displayed for the plucked string of the earlier section. The frequency of the rth mode of vibration is given by $\omega_r = r\pi c/l$, and the energy in each mode varies inversely with r^2, where r is odd. The spectrum of energy distribution is therefore given by fig. 9.8b.

Suppose now that the length of this string is halved but that the total energy remains constant. The frequency of the fundamental is now increased to $\omega_r' = 2r\pi c/l$ and the frequency interval between consecutive spectral lines is doubled (fig. 9.8c). Again, the smaller the region in which a given amount of energy is concentrated the wider the frequency spectrum required to represent it.

Frequently, as in the next section, a Fourier series is expressed in its complex or exponential form

$$f(t) = \sum_{n=-\infty}^{\infty} d_n \, e^{in\omega t}$$

where $2d_n = a_n - ib_n$ $(n \geq 0)$ and $2d_n = a_n + ib_n$ $(n < 0)$.
Because

$$\cos n\omega t = \tfrac{1}{2}(e^{in\omega t} + e^{-in\omega t})$$

and

$$\sin n\omega t = \frac{1}{2i}(e^{in\omega t} - e^{-in\omega t})$$

a frequency spectrum in the complex plane produces two spectral lines for each frequency component $n\omega$, one at $+n\omega$ and the other at $-n\omega$. Fig. 9.8d shows the cosine representation, which lies wholly in the real plane, and fig. 9.8e shows the sine representation, which is wholly imaginary. The amplitudes of the lines

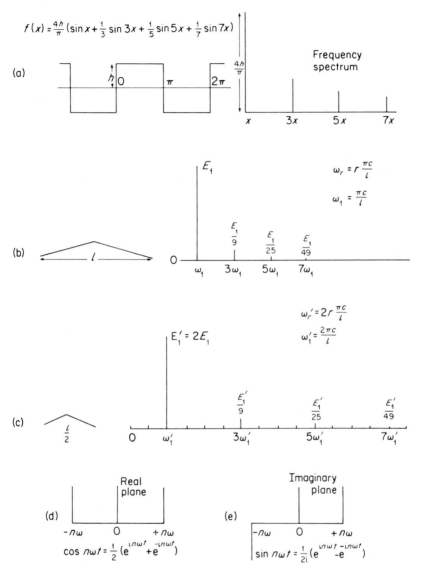

Fig. 9.8. (a) Fourier sine series of a square wave represented as a frequency spectrum; (b) energy spectrum of a plucked string of length l; and (c) the energy spectrum of a plucked string of length $l/2$ with the same total energy as (b), demonstrating the bandwidth theorem that the greater the concentration of the energy in space or time the wider its frequency spectrum. Complex exponential frequency spectrum of (d) $\cos \omega t$ and (e) $\sin \omega t$

in the positive and negative frequency ranges are, of course, complex conjugates, and the modulus of their product gives the square of the true amplitude. The concept of a negative frequency is seen to arise because the $e^{-in\omega t}$ term increases its phase in the opposite sense to that of the positive term $e^{in\omega t}$. The negative amplitude of the negative frequency in the sine representation indicates that it is in antiphase with respect to that of the positive term.

Fourier Integral

At the beginning of this chapter we saw that one Fourier representation of the function could be written

$$f(x) = \sum_{n=-\infty}^{\infty} d_n \, e^{inx}$$

where $2d_n = a_n - ib_n$ $(n \geqslant 0)$ and $2d_n = a_n + ib_n$ $(n < 0)$.

If we use the time as a variable we may rewrite this as

$$f(t) = \sum_{n=-\infty}^{\infty} d_n \, e^{in\omega t}$$

where, if T is the period,

$$d_n = \frac{1}{T} \int_{-T/2}^{T/2} f(t) \, e^{-in\omega t} \, dt$$

(for $n = -2, 1, 0, 1, 2$, etc.).

If we write $\omega = 2\pi\nu_1$, where ν_1 is the fundamental frequency, we can write

$$f(t) = \sum_{n=-\infty}^{\infty} \left[\int_{-T/2}^{T/2} f(t) \, e^{-i2\pi n\nu_1 t} \, dt \right] e^{i2\pi n\nu_1 t} \cdot \frac{1}{T}$$

If we now let the period T approach infinity we are isolating a single pulse by saying that it will not be repeated for an infinite period; the frequency $\nu_1 = 1/T \to 0$, and $1/T$ becomes infinitesimal and may be written $d\nu$.

Furthermore, n times ν_1, when n becomes as large as we please and $1/T = \nu_1 \to 0$, may be written as $n\nu_1 = \nu$, and the sum over n now becomes an integral, since unit change in n produces an infinitesimal change in $n/T = n\nu_1$.

Hence for an infinite period, that is for a single non-periodic pulse, we may write

$$f(t) = \int_{-\infty}^{\infty} \left[\int_{-\infty}^{\infty} f(t) \, e^{-i2\pi\nu t} \, dt \right] e^{i2\pi\nu t} \, d\nu$$

which is called the *Fourier Integral.*

We may express this as

$$f(t) = \int_{-\infty}^{\infty} F(\nu) \, e^{i2\pi\nu t} \, d\nu$$

where

$$F(\nu) = \int_{-\infty}^{\infty} f(t)\, e^{-i2\pi\nu t}\, dt$$

is called the *Fourier Transform* of $f(t)$. We shall discuss the transform in more detail in a later section of this chapter.

We see that when the period is finite and $f(t)$ is periodic, the expression

$$f(t) = \sum_{n=-\infty}^{\infty} d_n\, e^{in\omega t}$$

tells us that the representation is in terms of an infinite number of different frequencies, each frequency separated by a finite amount from its nearest neighbour, but when $f(t)$ is not periodic and has an infinite period then

$$f(t) = \int_{-\infty}^{\infty} F(\nu)\, e^{i2\pi\nu t}\, d\nu$$

and this expression is the integral (not the sum) of an infinite number of frequency components of amplitude $F(\nu)\, d\nu$ infinitely close together, since ν varies continuously instead of in discrete steps.

For a periodic function the amplitude of the Fourier series coefficient

$$d_n = \frac{1}{T} \int_{-T/2}^{T/2} f(t)\, e^{-in\omega t}\, dt$$

whereas the corresponding amplitude in the Fourier integral is

$$F(\nu)\, d\nu = d\left(\frac{1}{T}\right) \int_{-\infty}^{\infty} f(t)\, e^{-in\omega t}\, dt$$

This corroborates the statement we made when discussing the frequency spectrum that the narrower or less extended the pulse the wider the range of frequency components required to represent it. A truly monochromatic wave of one frequency and wavelength (or wave number) requires a wave train of infinite length before it is properly defined.

No wave train of finite length can be defined in terms of *one* unique wavelength.

Since a monochromatic wave, infinitely long, of single frequency and constant amplitude transmits no information, its amplitude must be modified by adding other frequencies (as we have seen in Chapter 4) before the variation in amplitude can convey information. These ideas are expressed in terms of the bandwidth theorem.

Fourier Transforms

We have just seen that the Fourier integral representing a non-periodic wave group can be written

$$f(t) = \int_{-\infty}^{\infty} F(\nu)\, e^{i2\pi\nu t}\, d\nu$$

where its Fourier transform

$$F(\nu) = \int_{-\infty}^{\infty} f(t)\, e^{-i2\pi\nu t}\, dt$$

so that integration with respect to one variable produces a function of the other. Both variables appear as a product in the index of an exponential, and this product must be non-dimensional. Any pair of variables which satisfy this criterion forms a Fourier pair of transforms, since from the symmetry of the expressions we see immediately that if

$$F(\nu) \text{ is the Fourier transform of } f(t)$$

then

$$f(-\nu) \text{ is the Fourier transform of } F(t)$$

If we are given the distribution in time of a function we can immediately express it as a spectrum of frequency, and vice versa. In the same way, a given distribution in space can be expressed as a function of wave numbers (this merely involves a factor, $1/2\pi$, in front of the transform because $k = 2\pi/\lambda$).

A similar factor appears if ω is used instead of ν. If the function of $f(t)$ is even only the cosine of the exponential is operative, and we have a Fourier cosine transform

$$f(t) = \int_{0}^{\infty} F(\nu) \cos 2\pi\nu t\, d\nu$$

and

$$F(\nu) = \int_{0}^{\infty} f(t) \cos 2\pi\nu t\, dt$$

If $f(t)$ is odd only the sine terms operate, and sine terms replace the cosines above. Note that only positive frequencies appear. The Fourier transform of an even function is real and even, whilst that of an odd function is imaginary and odd.

Examples of Fourier Transforms

The two examples of Fourier transforms chosen to illustrate the method are of great physical significance. They are

(1) the 'slit' function of fig. 9.9a, and
(2) the Gaussian function of fig. 9.11.

As shown, they are both even functions and their transforms are therefore real; the physical significance of this is that all the frequency components have the same phase at zero time.

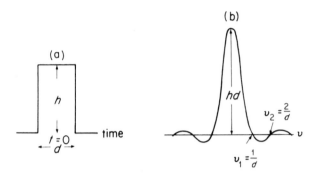

Fig. 9.9. (a) Narrow slit function of extent d in time and of height h, and (b) its Fourier transform

The Slit Function

This is a function having height h over the time range $\pm d/2$. Thus $f(t) = h$ for $|t| < d/2$ and zero for $|t| > d/2$, so that

$$F(\nu) = \int_{-\infty}^{\infty} f(t)\, e^{-i2\pi\nu t}\, dt = \int_{-d/2}^{d/2} h\, e^{-i2\pi\nu t}\, dt$$

$$= \frac{-h}{i2\pi\nu}[e^{-i2\pi\nu d/2} - e^{+i2\pi\nu d/2}] = hd\frac{\sin\alpha}{\alpha}$$

where

$$\alpha = \frac{2\pi\nu d}{2}$$

Again we see the Fourier transformation of a rectangular pulse in time to a $\sin\alpha/\alpha$ pattern in frequency. The Fourier transform of the same pulse in space will give the same distribution as a function of wavelength. Fig. 9.9b shows that as the pulse width decreases in time the separation between the zeros of the transform is increased. The negative values in the spectrum of the transform indicate a phase reversal for the amplitude of the corresponding frequency component.

The Fourier Transform Applied to Optical Diffraction from a Single Slit

This topic belongs more properly to the next chapter where it will be treated by another method, but here we derive the fundamental result as an example of the Fourier Transform. The elegance of this method is seen in problems more complicated than the one-dimensional example considered here. We shall see its extension to two dimensions in Chapter 10 when we consider the diffraction patterns produced by rectangular and circular apertures.

The amplitude of light passing through a single slit may be represented in space by the rectangular pulse of fig. 9.9a where d is now the width of the slit. A plane wave of monochromatic light, wavelength λ, falling normally on a screen which contains the narrow slit of width $d \sim \lambda$, forms a secondary system of plane waves diffracted in all directions with respect to the screen. When these diffracted waves are focused on to a second screen the intensity distribution (square of the amplitude) may be determined in terms of the aperture dimension d, the wavelength λ and the angle of diffraction θ.

In fig. 9.10 the light diffracted through an angle θ is brought to focus at a point P on the screen PP_0. Finding the amplitude of the light at P is the simple problem of adding all the small contributions in the diffracted wavefront taking account of all the phase differences which arise with variation of path length from P to the points in the slit aperture from which the contributions originate. The diffraction amplitude in k or wave number space is the Fourier transform of the pulse, width d, in x space in fig. 9.9b. The conjugate parameters ν and t are exactly reciprocal but the product of x and k involves the term 2π which requires either a constant factor $1/2\pi$ in front of one of the transform integrals

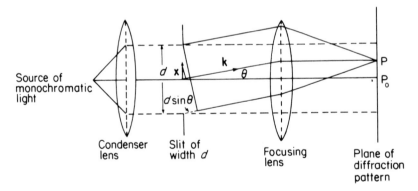

Fig. 9.10. A monochromatic plane wave normally incident on a narrow slit of width d is diffracted an angle θ, and the light in this direction is focused at a point P. The amplitude at P is the superposition of all contributions with their appropriate phases with respect to the central point in the slit. The contribution from a point x in the slit has phase $\phi = 2\pi x \sin \theta/\lambda$ with respect to the central contribution. The phase difference from contributing points on opposite edges of the slit is $\phi = 2\pi d \sin \theta/\lambda = 2\alpha$

or a common factor $1/\sqrt{2\pi}$ in front of each. This factor is however absorbed into the constant value of the maximum intensity and all other intensities are measured relative to it.

The constant pulse height now measures the amplitude h of the small wave sources across the slit width d and the Fourier transform method is the addition by integration of their contributions.

In fig. 9.10 we see that the path difference between the contribution at the centre of the slit and that at a point x in the slit is given by $x \sin \theta$, so that the phase difference is

$$\phi = \frac{2\pi}{\lambda} x \sin \theta = kx \sin \theta$$

The product $kx \sin \theta$ can, however, be expressed in a form more suitable for extension to two- and three-dimensional examples by writing it as $\mathbf{k} \cdot \mathbf{x} = klx$, the scalar product of the vector $\mathbf{k}$, giving the wave propagation direction, and the vector $\mathbf{x}$, l being the direction cosine

$$l = \cos (\pi/2 - \theta)$$

$$= \sin \theta$$

of $\mathbf{k}$ with respect to the x-axis.

Adding all the small contributions across the slit to obtain the amplitude at P by the Fourier transform method gives

$$F(k) = \frac{1}{2\pi} \int f(x) \, e^{-i\phi} \, dx$$

$$= \frac{1}{2\pi} \int_{-d/2}^{+d/2} h \, e^{-iklx} \, dx$$

$$= \frac{h}{-ikl} \frac{1}{2\pi} (e^{-ikld/2} - e^{+ikld/2})$$

$$= \frac{-2i}{-ikl2\pi} \sin \frac{kld}{2}$$

$$= \frac{dh}{2\pi} \frac{\sin \alpha}{\alpha}$$

where

$$\alpha = \frac{kld}{2} = \frac{\pi}{\lambda} d \sin \theta$$

The intensity I at P is given by the square of the amplitude, that is, by the product of $F(k)$ and its complex conjugate $F^*(k)$, so that

$$I = \frac{d^2 h^2}{4\pi^2} \frac{\sin^2 \alpha}{\alpha^2}$$

where I_0, the principal maximum intensity at $\alpha = 0$, (P_0 in fig. 9.10) is now

$$I_0 = \frac{d^2 h^2}{4\pi^2}$$

The Gaussian or Normal Error Distribution

This curve often appears as the wave group description of a particle in wave mechanics. The Fourier transform of a Gaussian distribution is another Gaussian distribution.

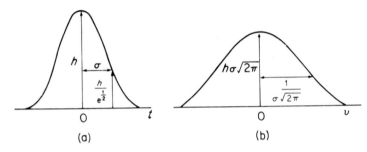

(a)   (b)

Fig. 9.11. (a) A Gaussian function Fourier transforms
(b) into another Gaussian function

In fig. 9.11a the Gaussian function of height h is symmetrically centred at time $t = 0$, and is given by $f(t) = h\,e^{-t^2/2\sigma^2}$, where the width parameter or standard deviation σ is that value of t at which the height of the curve has a value equal to $e^{-1/2}$ of its maximum.

Its transform is

$$F(\nu) = \int_{-\infty}^{\infty} h\,e^{-t^2/2\sigma^2}\,e^{-i2\pi\nu t}\,\mathrm{d}t$$

$$= \int_{-\infty}^{\infty} h\,e^{(-t/2\sigma^2 - i2\pi\nu t + 2\pi^2\nu^2\sigma^2)}\,e^{-2\pi^2\nu^2\sigma^2}\,\mathrm{d}t$$

$$= h\,e^{(-2\pi^2\nu^2\sigma^2)} \int_{-\infty}^{\infty} e^{-(t/\sqrt{2}\sigma + i\sqrt{2}\pi\nu\sigma)^2}\,\mathrm{d}t$$

The integral

$$\int_{-\infty}^{\infty} e^{-x^2}\,\mathrm{d}x = \sqrt{\pi}$$

and substituting, with $x = (t/\sqrt{2}\sigma + i\sqrt{2}\pi\nu\sigma)$ and $\mathrm{d}t = \sqrt{2}\sigma\,\mathrm{d}x$, gives

$$F(\nu) = h\sqrt{2\pi}\sigma\,e^{-2\pi\nu^2\sigma^2}$$

another Gaussian distribution in frequency space (fig. 9.11b) with a new height $h\sigma\sqrt{2\pi}$ and a new width parameter $(\sqrt{2\pi}\sigma)^{-1}$.

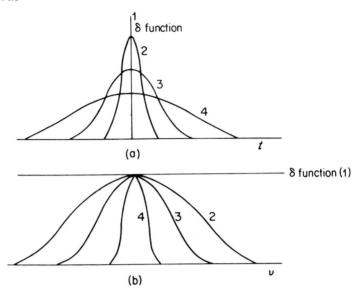

Fig. 9.12. (a) A family of normalized Gaussian functions narrowed in the limit to Dirac's delta function; (b) the family of their Fourier transforms

As in the case of the slit and the diffraction pattern, we see again that a narrow pulse in time (width σ) leads to a wide frequency distribution [width $(\sqrt{2\pi}\sigma)^{-1}$].

When the curve is normalized so that the area under it is unity, h takes the value $(\sqrt{2\pi}\sigma)^{-1}$ because

$$\frac{1}{\sqrt{2\pi}\sigma}\int_{-\infty}^{\infty} e^{-t^2/2\sigma^2}\, dt = 1$$

Thus the height of a normalized curve transforms into a pulse of unit height whereas a pulse of unit height transforms to a pulse of width $(\sqrt{2\pi}\sigma)^{-1}$.

If we consider a family of functions with progressively increasing h values and decreasing σ values, each satisfying the condition of unit area under their curves, we are led in the limit as the height $h \to \infty$ and the width $\sigma \to 0$ to an infinitely narrow pulse of finite area unity which defines the Dirac delta (δ) function. The transform of such a function is the constant unity, and figs. 9.12a and b show the family of normalized Gaussian distributions and their transforms. Fig. 9.13 shows a number of common Fourier transform pairs.

(Problems 9.10, 9.11, 9.12, 9.13, 9.14, 9.15, 9.16, 9.17)

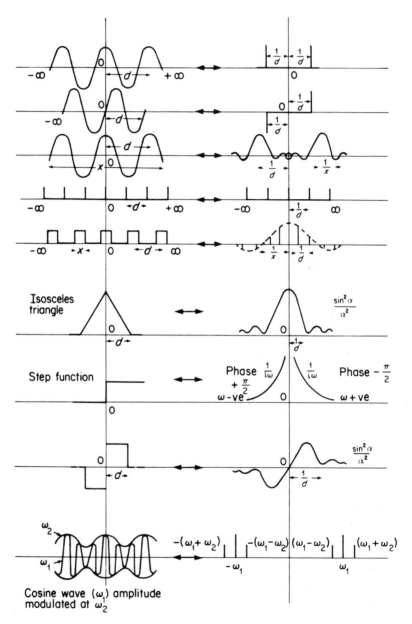

Fig. 9.13. Some common Fourier transform
pairs

Problem 9.1
After inspection of the two wave forms in the diagram what can you say about the values of the constant, absence or presence of sine terms, cosine terms, odd or even harmonics, and range of harmonics required in their Fourier series representation? (Do not use any mathematics.)

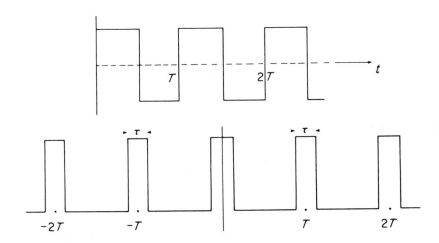

Problem 9.2
Show that if a periodic waveform is such that each half-cycle is identical except in sign with the previous one, its Fourier spectrum contains no even order frequency components. Examine the result physically.

Problem 9.3
A half-wave rectifier removes the negative half-cycles of a pure sinusoidal wave $y = h \sin x$. Show that the Fourier series is given by

$$y = \frac{h}{\pi}\left(1 + \frac{\pi}{1.2}\sin x - \frac{2}{1.3}\cos 2x - \frac{2}{3.5}\cos 4x - \frac{2}{5.7}\cos 6x \ldots\right)$$

Problem 9.4
A full-wave rectifier merely inverts the negative half-cycle in problem 9.2. Show that this doubles the output and removes the undesirable modulating ripple of the first harmonic.

Problem 9.5
Show that $f(x) = x^2$ may be represented in the interval $\pm\pi$ by

$$f(x) = \tfrac{2}{3}\pi^2 + \sum (-1)^n \frac{4}{n^2}\cos nx$$

Problem 9.6
Use the square wave sine series of unit height $f(x) = 4/\pi \ (\sin x + \tfrac{1}{3}\sin 3x + \tfrac{1}{5}\sin 5x)$ to show that

$$1 - \tfrac{1}{3} + \tfrac{1}{5} - \tfrac{1}{7} = \pi/4$$

Problem 9.7

An infinite train of pulses of unit height, with pulse duration 2τ and a period between pulses of T, is expressed as

$$f(t) = 0 \quad \text{for } -\tfrac{1}{2}T < t < -\tau$$
$$= 1 \quad \text{for } -\tau < t < \tau$$
$$= 0 \quad \text{for } \tau < t < \tfrac{1}{2}T$$

and

$$f(t+T) = f(t)$$

Show that this is an even function with the cosine coefficients given by

$$a_n = \frac{2}{n\pi} \sin \frac{2\pi}{T} n\tau$$

Problem 9.8

Show, in problem 9.7, that as τ becomes very small the values of $a_n \to 4\tau/T$ and are independent of n, so that the spectrum consists of an infinite set of lines of constant height and spacing. The representation now has the same form in both time and frequency; such a function is called 'self reciprocal'. What is the physical significance of the fact that as $\tau \to 0$, $a_n \to 0$?

Problem 9.9

The pulses of problems 9.7 and 9.8 now have amplitude $1/2\tau$ with unit area under each pulse. Show that as $\tau \to 0$ the infinite series of pulses is given by

$$f(t) = \frac{1}{T} + \frac{2}{T} \sum_{n=1}^{\infty} \cos 2\pi nt/T$$

Under these conditions the amplitude of the original pulses becomes infinite, the energy per pulse remains finite and for an infinity of pulses in the train the total energy in the waveform is also infinite. The amplitude of the individual components in the frequency representation is finite, representing finite energy, but again, an infinity of components gives an infinite energy.

Problem 9.10

The unit step function is defined by the relation

$$f(t) = 1 \ (t > 0)$$
$$= 0 \ (t < 0)$$

This is a very important function in physics and engineering, but it does not satisfy the criteria for Fourier representation because its integral is not finite. A similar function of finite period will satisfy the criteria. If this function is defined

$$f(t) = 1 \ (0 < t < T)$$
$$= 0 \text{ elsewhere}$$

show that if the transform

$$F(\omega) = \int_{-\infty}^{\infty} f(t)\, e^{-i\omega t}\, dt = \int_{0}^{T} e^{-i\omega t}\, dt$$

$$= \frac{1}{i\omega}[1 - e^{i\omega T}]$$

then for large T

$$f(t) = \tfrac{1}{2} + \frac{1}{2\pi}\int_{-\infty}^{\infty} \frac{1}{i\omega}\, e^{i\omega t}\, d\omega$$

(Note that the integral of the second term gives $-\pi$ for $t < 0$ and $+\pi$ for $t > 0$. This spectral representation is shown in fig. 9.13).

Problem 9.11

Optical wave trains emitted by radiating atoms are of finite length and only an infinite wave train may be defined in terms of one frequency. The radiation from atoms therefore has a frequency bandwidth which contributes to the spectral linewidth. The random phase relationships between these wave trains create incoherence and produce the difficulties in obtaining interference effects from separate sources.

Let a finite length monochromatic wave train of wavelength λ_0 be represented by

$$f(t) = f_0\, e^{i2\pi\nu_0 t}$$

and be a cosine of constant amplitude f_0 extending in time between $\pm \tau/2$. The distance $l = c\tau$ is called the coherence length. This finite train is the superposition of frequency components of amplitude $F(\nu)$ where the transform gives

$$f(t) = \int_{-\infty}^{\infty} F(\nu)\, e^{i2\pi\nu t}\, dt$$

so that

$$F(\nu) = \int_{-\infty}^{\infty} f(t)\, e^{-i2\pi\nu t}\, dt$$

$$= \int_{-\tau/2}^{+\tau/2} f_0\, e^{-i2\pi(\nu - \nu_0)t}\, dt$$

Show that

$$F(\nu) = f_0\tau\frac{\sin[\pi(\nu - \nu_0)\tau]}{\pi(\nu - \nu_0)\tau}$$

and that the relative energy distribution in the spectrum follows the intensity distribution curve in a single slit diffraction pattern.

Problem 9.12

Show that the total width of the first maximum of the energy spectrum of problem 9.11 has a frequency range $2\Delta\nu$ which defines the coherence length l of problem 9.11 as $\lambda_0^2/\Delta\lambda$.

Problem 9.13

For a ruby laser beam the value of $\Delta\nu$ in problem 9.12 is found to be 10^4 Hertz and $\lambda_0 = 6\cdot936 \times 10^{-7}$ metres. Show that $\Delta\lambda = 1\cdot6 \times 10^{-17}$ metres and that the coherence length l of the beam is 3×10^4 metres.

Problem 9.14

The energy of the finite wave train of the damped simple harmonic vibrations of the radiating atom in Chapter 1 was described by $E = E_0\, e^{-\omega_0 t/Q}$. Show from physical arguments that this defines a frequency bandwidth in this train of $\Delta\omega$ about the frequency ω_0, where the quality factor $Q = \omega_0/\Delta\omega$. (Suggested line of argument—at the maximum amplitude all frequency components are in phase. After a time τ the frequency component ω_0 has changed phase by $\omega_0\tau$. Other components have a phase change which interfere destructively. What bandwidth and phase change is acceptable?)

Problem 9.15

Consider problem 9.14 more formally. Let the damped wave be represented as a function of time by

$$f(t) = f_0 \, e^{i2\pi\nu_0 t} \, e^{-t/\tau}$$

where f_0 is constant and τ is the decay constant.

Use the Fourier transform to show that the amplitudes in the frequency spectrum are given by

$$F(\nu) = \frac{f_0}{1/\tau + i2\pi(\nu - \nu_0)}$$

Write the denominator of $F(\nu)$ as $r\,e^{i\theta}$ to show that the energy distribution of frequencies in the region of $\nu - \nu_0$ is given by

$$|F(\nu)|^2 = \frac{f_0^2}{r^2} = \frac{f_0^2}{(1/\tau)^2 + (\omega - \omega_0)^2}$$

Problem 9.16

Show that the expression $|F(\nu)|^2$ of problem 9.15 is the resonance power curve of Chapter 2; show that it has a width at half the maximum value $(f_0\tau)^2$ which gives $\Delta\nu = 1/\pi\tau$, and show that a spectral line which has a value of $\Delta\lambda$ in problem 9.12 equal to 3×10^{-9} metres has a finite wave train of coherence length equal to 32×10^{-6} metres (32 microns) if $\lambda_0 = 5\cdot46 \times 10^{-7}$ metres.

Problem 9.17

The Fourier transform of a rectangular pulse of height h and width d is $F(\nu) = hd \sin\alpha/\alpha$ where $\alpha = 2\pi\nu d/2$. Show that the Fourier transform of an isosceles triangular function height h length $2d$ is given by

$$F(\nu) = 4h^2 d^2 \frac{\sin^2\alpha}{\alpha^2}$$

where $\alpha = 2\pi\nu d/2$.

Observe that the triangular pulse is produced by sliding one rectangular pulse of height h width d along its base over a similar pulse, any coordinate on the triangular pulse being a measure of the overlaps between the pulses (the peak measuring their complete overlap). Such a process is called convolution.

Summary of Important Results

Fourier series

Any function may be represented in the interval $\pm\pi$ by

$$f(x) = \tfrac{1}{2}a_0 + \sum_1^n a_n \cos nx + \sum_1^n b_n \sin nx$$

where

$$a_n = \frac{1}{\pi} \int_0^{2\pi} f(x) \cos nx \, dx$$

and

$$b_n = \frac{1}{\pi} \int_0^{2\pi} f(x) \sin nx \, dx$$

Fourier integral

A single non-periodic pulse may be represented as

$$f(t) = \int_{-\infty}^{+\infty} \left[\int_{-\infty}^{+\infty} f(t) \, e^{-i2\pi\nu t} \, dt \right] e^{i2\pi\nu t} \, d\nu$$

or as

$$f(t) = \int_{-\infty}^{+\infty} F(\nu) e^{i2\pi\nu t} \, d\nu$$

where

$$F(\nu) = \int_{-\infty}^{+\infty} f(t) \, e^{-i2\pi\nu t} \, dt$$

$f(t)$ and $F(\nu)$ are *Fourier Transforms* of each other. When t is replaced by x and ν by k the right hand side of each transform has a factor $1/\sqrt{2\pi}$. The Fourier Transform of a rectangular pulse has the shape of $\sin \alpha / \alpha$. (Important in optical diffraction.)

Chapter 10

Waves in Optical Systems

Light. Waves or Rays?

Light exhibits a dual nature. In practice, its passage through optical instruments such as telescopes and microscopes is most easily shown by geometrical ray diagrams but the fine detail of the images formed by these instruments is governed by diffraction which, together with interference, requires light to propagate as waves. The earlier parts of this chapter will correlate the geometrical optics of these instruments with wavefront propagation. Later we shall consider the effects of interference and diffraction.

The electromagnetic wave nature of light was convincingly settled by Clerk–Maxwell in 1864 but as early as 1690 Huygens was trying to reconcile waves and rays. He proposed that light be represented as a wavefront, each point on this front acting as a source of secondary wavelets whose envelope became the new position of the wavefront, fig. 10.1(a). Light propagation was seen as the progressive development of such a process. In this way reflexion and refraction at a plane boundary separating two optical media could be explained as shown in fig. 10.1(b) and (c).

Huygens' theory was explicit only on those contributions to the new wavefront directly ahead of each point source of secondary waves. No statement was made about propagation in the backward direction nor about contributions in the oblique forward direction. Each of these difficulties is resolved in the more rigorous development of the theory by Kirchhoff which uses the fact that light waves are oscillatory.

The way in which rays may represent the propagation of wavefronts is shown in fig. 10.2 where spherically diverging, plane and spherically converging wavefronts are moving from left to right. All parts of the wavefront (a surface of constant phase) take the same time to travel from the source and all points on the wavefront are the same *optical distance* from the source. This optical distance must take account of the changes of refractive index met by the wavefront as it propagates. If the physical path length is measured

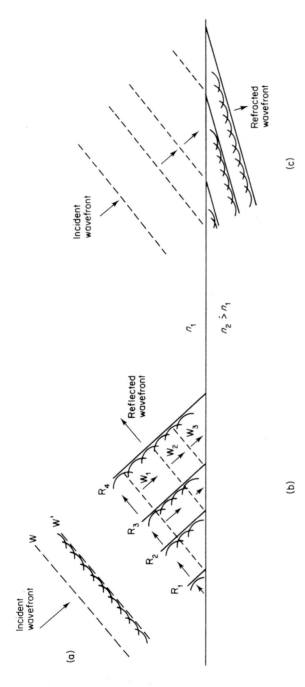

Fig. 10.1. (a) Incident plane wavefront W propagates via Huygens wavelets to W'. (b) At the plane boundary between the media (refractive index $n_2 > n_1$) the incident wavefront W_1 has a reflected section R_1. Increasing sections R_2 and R_3 are reflected until the whole wavefront is reflected as R_4. (c) An increasing section of the incident wavefront is refracted. Incident wavefronts are shown dashed, and reflected and refracted wavefronts as a continuous line

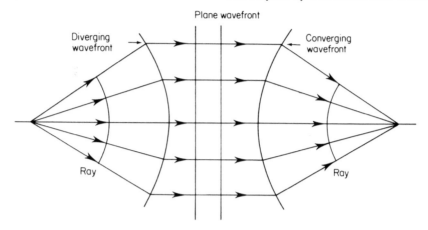

Fig. 10.2. Ray representation of spherically diverging, plane and spherically converging wavefronts

as x in a medium of refractive index n then the *optical path length* in the medium is the product nx. In travelling from one point to another light chooses a unique optical path which may always be defined in terms of Fermat's Principle.

Fermat's Principle

Fermat's Principle states that the optical path length has a stationary value; its first order variation or first derivative in a Taylor series expansion is zero. This means that when an optical path lies wholly within a medium of constant refractive index the path is a straight line, the shortest distance between its end points, and the light travels between these points in the minimum possible time. When the medium has a varying refractive index or the path crosses the boundary between media of different refractive indices the direction of the path always adjusts itself so that the time taken between its end points is a minimum. Fermat's Principle is therefore sometimes known as the Principle of Least Time. Fig. 10.3 shows examples of light paths in a medium of varying refractive index. As examples of light meeting a boundary between two media we use Fermat's Principle to derive the laws of reflexion and refraction.

The Laws of Reflexion

In Fig. 10.4(a) Fermat's Principle requires that the optical path length OSI should be a minimum where O is the object, S lies on the plane reflecting surface and I is the point on the reflected ray at which the image of O is viewed. The plane OSI must be perpendicular to the reflecting surface for, if reflexion takes place at any other point S' on the reflecting surface where

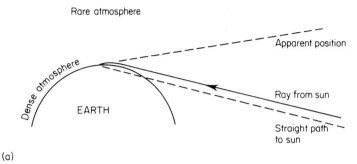

(a)

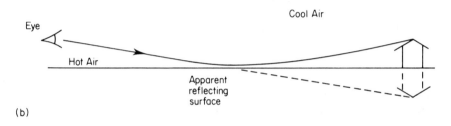

(b)

Fig. 10.3. Light takes the shortest optical path in a medium of varying refractive index. (a) A light ray from the sun bends towards the earth in order to shorten its path in the denser atmosphere. The sun remains visible after it has passed below the horizon. (b) A light ray avoids the denser atmosphere and the road immediately below warm air produces an apparent reflexion

OSS' and ISS' are right angles then evidently OS' > OS and IS' > IS, giving OS'I > OSI.

The laws of reflexion also require, in fig. 10.4(a) that the angle of incidence i equals the angle of reflexion r. If the co-ordinates of O, S and I are those shown and the velocity of light propagation is c then the time taken to traverse OS is

$$t = (x^2 + y^2)^{1/2}/c$$

and the time taken to traverse SI is

$$t' = [(X - x)^2 + y^2]^{1/2}/c$$

so that the total time taken to travel the path OSI is

$$T = t + t'$$

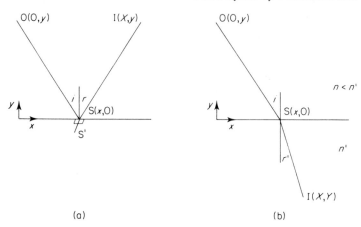

(a)   (b)

Fig. 10.4. The time for light to follow the path OSI is a minimum (a) in reflexion, when OSI forms a plane perpendicular to the reflecting surface and $\hat{i} = \hat{r}$; and (b) in refraction, when $n \sin i = n' \sin r'$ (Snell's Law)

The position of S is now varied along the x axis and we seek, via Fermat's Principle of Least Time, that value of x which minimizes T, so that

$$\frac{\mathrm{d}T}{\mathrm{d}x} = \frac{x}{c(x^2 + y^2)^{1/2}} - \frac{X - x}{c[(X - x)^2 + y^2]^{1/2}} = 0$$

But

$$\frac{x}{(x^2 + y^2)^{1/2}} = \sin i$$

and

$$\frac{X - x}{[(X - x)^2 + y^2]^{1/2}} = \sin r$$

Hence

$$\sin i = \sin r$$

and

$$\hat{i} = \hat{r}$$

The Law of Refraction

Exactly similar arguments lead to Snell's Law, already derived on page 237. Here we express it as

$$n \sin i = n' \sin r'$$

where i is the angle of incidence in the medium of refractive index n and r' is the angle of refraction in the medium of refractive index n' $(n' > n)$. In fig. 10.4(b) a plane boundary separates the media and light from O $(0, y)$ is refracted at S $(x, 0)$ and viewed at I (X, Y) on the refracted ray. If v and v' are respectively the velocities of light propagation in the media n and n' then OS is traversed in the time

$$t = (x^2 + y^2)^{1/2}/v$$

and SI is traversed in the time

$$t' = [(X - x)^2 + Y^2]^{1/2}/v'$$

The total time to travel from O to I is $T = t + t'$ and we vary the position of S along the x axis which lies on the plane boundary between n and n', seeking that value of x which minimizes T. So

$$\frac{\mathrm{d}T}{\mathrm{d}x} = \frac{1}{v}\frac{x}{(x^2+y^2)^{1/2}} - \frac{1}{v'}\frac{(X-x)}{[(X-x)^2+Y^2]^{1/2}} = 0$$

where

$$\frac{x}{(x^2+y^2)^{1/2}} = \sin i$$

and

$$\frac{(X-x)}{[(X-x)^2+Y^2]^{1/2}} = \sin r'$$

But

$$\frac{1}{v} = \frac{n}{c}$$

and

$$\frac{1}{v'} = \frac{n'}{c}$$

Hence

$$n \sin i = n' \sin r'$$

Rays and Wavefronts

Fig. 10.2 showed the ray representation of various wavefronts. In order to reinforce the concept that rays trace the history of wavefronts we consider the examples of a thin lens and a prism.

The Thin Lens

In fig. 10.5 a plane wave in air is incident normally on the plane face of a plano convex glass lens of refractive index n and thickness d at its central

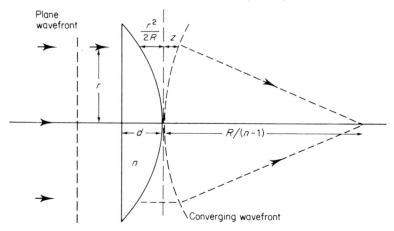

Fig. 10.5. A plane wavefront is normally incident on a plano-convex lens of refractive index n and thickness d at the central axis. The radius of the curved surface $R \gg d$. The wavefront is a surface of constant phase and the optical path length is the same for each section of the wavefront. At a radius r from the central axis the wavefront travels a shorter distance in the denser medium and the lens curves the incident wavefront which converges at a distance $R/(n-1)$ from the lens

axis. Its spherical face has a radius of curvature $R \gg d$. The power of a lens to change the curvature of a wavefront is the inverse of its focal length f. A lens of positive power converges a wavefront, negative power diverges the wavefront.

Simple ray optics gives the power of the plano convex lens as

$$P = \frac{1}{f} = (n-1)\frac{1}{R}$$

but we derive this result from first principles that is, by considering the way in which the lens modifies the wavefront.

At the central axis the wavefront takes a time $t = nd/c$ to traverse the thickness d. At a distance r from the axis the lens is thinner by an amount $r^2/2R$ (using the elementary relation between the sagitta, arc and radius of a circle) so that, in the time $t = nd/c$, points on the wavefront at a distance r from the axis travel a distance

$$(d - r^2/2R)$$

in the lens plus a distance $(r^2/2R + z)$ in air as shown in the figure. Equating the times taken by the two parts of the wave front we have

$$nd/c = (n/c)(d - r^2/2R) + (1/c)(z + r^2/2R)$$

which yields

$$z = (n-1)r^2/2R$$

But this is again the relation between the sagitta z, its arc and a circle of radius $R/(n-1)$ so, in three dimensions, the locus of z is a sphere of radius $R/(n-1)$ and the emerging spherical wavefront converges to a focus at a distance

$$f = R/(n-1)$$

(Problems 10.1, 10.2, 10.3)

The Prism

In fig. 10.6 a section, height y, of a plane wavefront in air is deviated through an angle θ when it is refracted through an isosceles glass prism, base l, vertex angle β and refractive index n. Experiment shows that there is *one*, and only

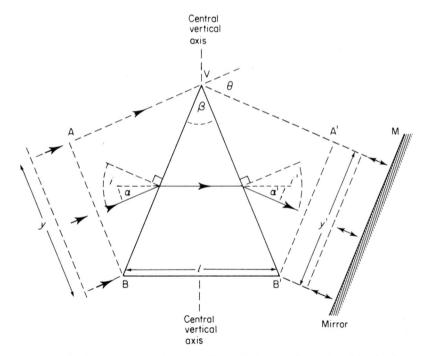

Fig. 10.6. A plane wavefront suffers minimum deviation (θ_{min}) when its passage through a prism is symmetric with respect to the central vertical axis ($\hat{i} = \hat{i}'$). The wavefront obeys the Optical Helmholtz Condition that $ny \tan \alpha$ is a constant where n is the refractive index, y is the width of the wavefront and α is shown. (Here $\hat{\alpha} = \hat{\alpha}'$)

one, value of the incident angle $\hat{i}$ for which the angle of deviation is a minimum $= \theta_{min}$. It is easily shown *using ray optics* that this unique value of $\hat{i}$ requires the passage of the wavefront through the prism to be symmetric about the central vertical axis as shown in the figure so that the incident angle $\hat{i}$ equals the emerging angle $\hat{i}'$. Equating the lengths of the *optical* paths AVA' and BB' ($=nl$) followed by the edges of the wavefront section gives the familiar result

$$\sin \left(\frac{\theta_{min} + \beta}{2} \right) = n \sin \frac{\beta}{2}$$

which is used in the standard experiment to determine n, the refractive index of the prism.

Now there is only *one* value of $\hat{i}$ which produces minimum deviation and this leads us to expect that the passage of the wavefront will be symmetric about the central vertical axis for if a plane mirror (M in the figure) is placed parallel to the emerging wavefront the wavefront is reflected back along its original path, and if $\hat{i} \neq \hat{i}'$ there would be *two* values of incidence, each producing minimum deviation.

However, the real argument for symmetry from a wavefront point of view depends on the optical Helmholtz equation which we shall derive on page 294. This states that the product $ny \tan \alpha$ remains constant as it passes through an optical system irrespective of the local variations of the factors n, y and $\tan \alpha$. Now the wavefront has the same width on entry into and exit from the prism so $y = y'$ and although n changes at the prism faces the initial and final medium for the wavefront is air where $n = 1$.

Hence, from the optical Helmholtz equation $\tan \alpha = \tan \alpha'$ in fig. 10.6. It is evident that as long as its width $y = y'$ the wavefront section will turn through a minimum angle when the physical path length BB' followed by its lower edge is a maximum with respect to AVA', the physical path length of its upper edge.

Ray Optics and Optical Systems

An optical system changes the curvature of a wavefront. It is formed by one or more optical surfaces separating media of different refractive indices. In fig. 10.7 rays from the object point P pass through the optical system to form an image point P'. When the optical surfaces are spherical the line joining P and P', which passes through the centres of curvature of the surfaces, is called the *optical axis*. This axis cuts each optical surface at its *pole*. If the object lies in a plane normal to the optical axis its perfect image lies in a *conjugate* plane normal to the optical axis. Conjugate planes cut the optical axis at conjugate points, e.g. P and P'. In fig. 10.7 the plane at $+\infty$ has a conjugate focal plane cutting the optical axis at the focal point F_0. The plane at $-\infty$ has a conjugate focal plane cutting the optical axis at the focal point F_i.

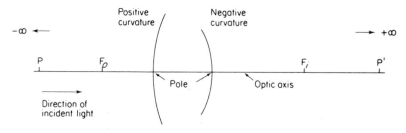

Fig. 10.7. Optical system showing direction of incident light from left to right and optical surfaces of positive and negative curvature. Rays from P pass through P' and this defines P and P' as conjugate points. The conjugate point of F_0 is at $+\infty$, the conjugate point of F_i is at $-\infty$

Paraxial Rays

Perfect geometrical images require perfect plane and spherical optical surfaces and in a real optical system a perfect spherical optical surface is obtained by using only that part of the wavefront close to the optical axis. This means that all angles between the axis and rays are very small. Such rays are called paraxial rays.

Sign Convention

The convention used here involves only the signs of lengths and angles. The direction of incident light is positive and is always taken from left to right. Signs for horizontal and vertical directions are Cartesian. If $AB = l$ then $BA = -l$. The radius of curvature of a surface is measured from its pole to its centre so that, in fig. 10.7, the convex surface presented to the incident light has a positive radius of curvature and the concave surface has a negative radius of curvature.

The Cartesian convention with origin O at the right angles of fig. 10.8 gives the angle between a ray and the optical axis the sign of its tangent.

If the angle between a ray and the axis is α then, for paraxial rays

$$\sin \alpha = \tan \alpha = \alpha$$

and

$$\cos \alpha = 1$$

so that Snell's Law of Refraction

$$n \sin i = n' \sin r'$$

becomes

$$ni = n'r'$$

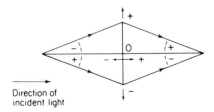

Fig. 10.8. Sign convention for lengths is Cartesian measured from the right angles at O. Angles take the sign of their tangents. O is origin of measurements

Power of a Spherical Surface

In figs. 10.9(a) and (b) a spherical surface separates media of refractive indices n and n'. Any ray through P is refracted to pass through its conjugate point P'. The angles are exaggerated so that the base of h is very close to the pole of the optical surface which is taken as the origin. In fig. 10.9(a) the signs of R, l' and α' are positive with l and α negative. In fig. 10.9(b) R, l, l', α and α' are all positive quantities. In both figures Snell's Law gives

$$ni = n'r'$$

i.e.

$$n(\theta - \alpha) = n'(\theta - \alpha')$$

or

$$n'\alpha' - n\alpha = (n' - n)\theta = \left(\frac{n' - n}{R}\right) h = Ph \tag{10.1}$$

Thus

$$\frac{n'}{l'} - \frac{n}{l} = \frac{n' - n}{R} = P \tag{10.2}$$

where P is the power of the surface. For $n' > n$ the power P is positive and the surface converges the wavefront. For $n' < n$, P is negative and the wavefront diverges. When the radius of curvature R is measured in metres the units of P are *dioptres*.

Magnification by the Spherical Surface

In fig. 10.10 the points QQ' form a conjugate pair, as do PP'. The ray QQ' passes through C the centre of curvature, PQ is the object height y, P' Q' is the image height y' so

$$ni = n'r'$$

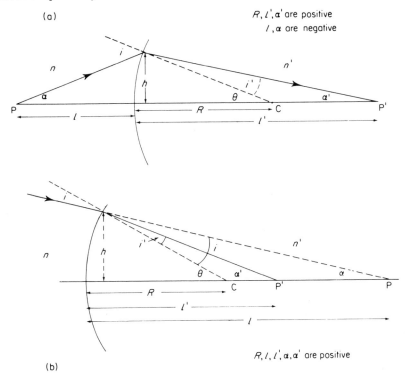

Fig. 10.9. Spherical surface separating media of refractive indices n and n'. Rays from P pass through P'. Snell's Law gives the power of the surface as

$$P = \frac{n'}{l'} - \frac{n}{l} = \frac{n'-n}{R}$$

gives

$$ny/l = n'y'/l'$$

or

$$nyh/l = n'y'h/l'$$

that is

$$ny\alpha = n'y'\alpha'$$

The Transverse Magnification is defined as

$$M_T = y'/y = nl'/n'l.$$

The image y' is inverted so y and y' (and l and l') have opposite signs.

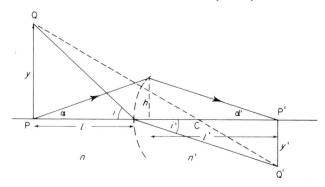

Fig. 10.10. Magnification by a spherical surface. The paraxial form of the Optical Helmholtz Equation is $ny\alpha = n'y'\alpha'$ so Transverse magnification $M_T = y'/y = nl'/ln'$ Angular magnification $M_\alpha = \alpha'/\alpha$. Note that the image is inverted so y and y' (and l and l') have opposite signs

The Angular Magnification is defined as

$$M_\alpha = \alpha'/\alpha$$

Note that

$$M_T = n/n'M_\alpha$$

which, being independent of y, applies to any point on the object so that the object in the plane P is similar to the image in the plane P'.

A series of optical surfaces separating media of refractive indices n, n' n'' yields the expression

$$ny\alpha = n'y'\alpha' = n''y''\alpha''$$

which is the paraxial form of the optical Helmholtz equation quoted on page 286 and derived on page 294.

Power of Two Optically Refracting Surfaces

In fig. 10.11 the refracting surfaces have powers P_1 and P_2 respectively. At the first surface equation (10.1) gives

$$n_1\alpha_1 - n\alpha = P_1h_1$$

and at the second surface

$$n'\alpha' - n_1\alpha_1 = P_2h_2$$

Adding these equations gives

$$n'\alpha' - n\alpha = P_1h_1 + P_2h_2$$

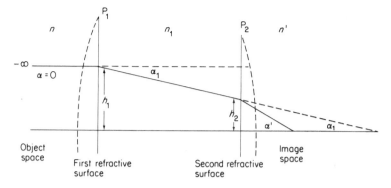

Fig. 10.11. Two optically refracting surfaces of power P_1 and P_2 have a combined power of

$$P = \frac{1}{h_1}(P_1 h_1 + P_2 h_2)$$

If the object is located at $-\infty$ so that $\alpha = 0$ we have

$$n'\alpha' = P_1 h_1 + P_2 h_2$$

or

$$\alpha' = \frac{1}{n'}(P_1 h_1 + P_2 h_2)$$

This gives the same image as a single element of power P if

$$\alpha' = \frac{1}{n'}(P_1 h_1 + P_2 h_2) = \frac{1}{n'}P h_1$$

where

$$P = \frac{1}{h_1}(P_1 h_1 + P_2 h_2) \qquad (10.3)$$

is the total power of the system. *This is our basic equation* and we use it first to find the power of a thin lens in air.

Power of a Thin Lens in Air (fig. 10.12)

Equation (10.2) gives

$$\frac{n'}{l'} - \frac{n}{l} = \frac{n'-n}{R} = P$$

for each surface, so that in fig. 10.12

$$P_1 = (n_1 - 1)/R_1$$

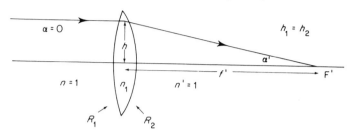

Fig. 10.12. A thin lens of refractive index n_1, and radii of surface curvatures R_1 and R_2 has a power

$$P = (n_1 - 1)\left(\frac{1}{R_1} - \frac{1}{R_2}\right) = \frac{1}{f'}$$

where f' is the focal length. In the figure R_1 is positive and R_2 is negative

and

$$P_2 = (1 - n_1)/R_2$$

From equation (10.3)

$$P = \frac{1}{h_1}(P_1 h_1 + P_2 h_2)$$

with

$$h_1 = h_2$$

we have

$$P = P_1 + P_2$$

so the expression for the thin lens in air with surfaces of power P_1 and P_2 becomes

$$P = \frac{1}{l'} - \frac{1}{l} = (n_1 - 1)\left(\frac{1}{R_1} - \frac{1}{R_2}\right) = \frac{1}{f'}$$

where f' is the focal length.

Applying this result to the plano convex lens of page 284 we have $R_1 = \infty$ and R_2 negative from our sign convention. This gives a positive power which we expect for a converging lens.

Effect of Refractive Index on the Power of a Lens

Suppose, in fig. 10.13, that the object space of the lens remains in air ($n = 1$) but that the image space is a medium of refractive index $n_2' \neq 1$. How does this affect the focal length in the medium n_2'?

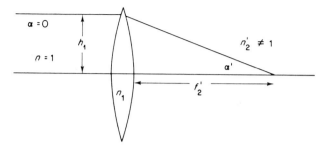

Fig. 10.13. The focal length of a thin lens measured in the medium n'_2 is given by $f'_2 = n'_2 f'$ where f' is the focal length of the lens measured in air

If P is the power of the lens in air we have

$$n'_2 \alpha' - n\alpha = Ph_1$$

and for

$$\alpha = 0$$

we have

$$\alpha' = Ph_1/n'_2 = h_1/n'_2 f'$$

where f' is the focal length in air.
If f'_2 is the focal length in the medium n'_2 then

$$f'_2 \alpha' = h_1$$

so

$$\alpha' = h_1/f'_2 = h_1/n'_2 f'$$

giving

$$f'_2 = n'_2 f'$$

Thus the focal length changes by a factor equal to the refractive index of the medium in which it is measured and the power is affected by the same factor.

If the lens has a medium n_0 in its object space and a medium n_i in its image space then the respective focal lengths f_0 and f_i in these spaces are related by the expression

$$f_i/f_0 = -n_i/n_0$$

where the negative sign shows that f_0 and f_i are measured in opposite directions (f_0 is negative and f_i is positive).

Now, we use this result in deriving the Optical Helmholtz equation.

The Optical Helmholtz Equation

In fig. 10.14 the ray LH_3 from the base of the object, height y, meets the optical axis in the image space at the base L' of the image, height y'. In the similar triangles $F_iL'L_i$ and $F_iH_0H_2$,

$$y'/y = L'L_i/H_0H_2 = F_iL'/F_iH_0 = x'/-f'$$

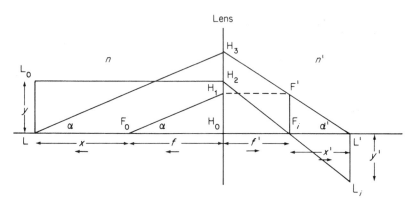

Fig. 10.14. Ray construction through a single lens to prove the Optical Helmholtz Equation

$$ny \tan \alpha = n'y' \tan \alpha'$$

This result is true for systems having any number of optical surfaces. In the figure LH_3 and F_0H_1 are parallel as are H_1F' and H_0F_i

where F_i/H_0 is negative. The negative sign shows that the image is inverted. Now

$$H_0F_0 \tan \alpha = f \tan \alpha = H_0H_1 = F_1F'$$
$$= F_iL' \tan \alpha' = x' \tan \alpha'$$

Therefore

$$x' = f \tan \alpha/\tan \alpha'$$

where f and α are both negative and x' and α' are both positive. This result, together with

$$y'/y = x'/-f'$$

gives

$$y'/y = f \tan \alpha/-f' \tan \alpha'$$

But, from the previous section (page 293)

$$f'/f = -n'/n$$

so

$$y'/y = n \tan \alpha / n' \tan \alpha'$$

Hence

$$ny \tan \alpha = n'y' \tan \alpha'$$

This is the Optical Helmholtz equation which states that the quanity $ny \tan \alpha$ is invariant in passing through the optical system. No restriction is placed on the size of the angles (although we met the paraxial form on page 290). We have derived it here using a single lens, but the result is valid for all optical systems, no matter how complex.

The results of the preceding sections are applied to optical instruments e.g. telescopes and microscopes in the following problems.

(Problems 10.4, 10.5, 10.6, 10.7, 10.8, 10.9, 10.10, 10.11)

Interference and Diffraction

All waves display the phenomena of interference and diffraction which arise from the superposition of more than one wave. At each point of observation within the interference or diffraction pattern the phase difference between any two component waves of the same frequency will depend on the different paths they have followed and the resulting amplitude may be greater or less than that of any single component. Although we speak of separate waves the waves contributing to the interference and diffraction pattern must ultimately derive from the same single source. This avoids random phase effects from separate sources and guarantees coherence. However, even a single source has a finite size and spatial coherence of the light from different parts of the source imposes certain restrictions if interference effects are to be observed. This is discussed in the section on spatial coherence on page 300. The superposition of waves involves the addition of two or more harmonic components with different phases and the basis of our approach is that laid down in the vector addition of fig. 1.11. More formally in the case of diffraction we have shown the equivalence of the Fourier Transform Method on page 267 of Chapter 9.

Interference

Interference effects may be classified in two ways:
(1) Division of wavefront.
(2) Division of amplitude.

(1) *Division of wavefront.* Here the wavefront from a single source passes simultaneously through two or more apertures each of which contributes a wave at the point of superposition. Diffraction also occurs at each aperture.

The difference between interference and diffraction is merely one of scale: in optical diffraction from a narrow slit (or source) the aperture is of the order of the wavelength of the diffracted light. According to Huygens Principle every point on the wavefront in the plane of the slit may be considered as a source of secondary wavelets and the further development of the diffracted wave system may be obtained by superposing these wavelets.

In the interference pattern arising from two or more such narrow slits each slit may be seen as the source of a single wave so the number of superposed components in the final interference pattern equals the number of slits (or sources). This suggests that the complete pattern for more than one slit will display both interference and diffraction effects and we shall see that this is indeed the case.

(2) *Division of amplitude.* Here a beam of light or ray is reflected and transmitted at a boundary between media of different refractive indices. The incident, reflected and transmitted components form separate waves and follow different optical paths. They interfere when they are recombined.

Division of Wavefront

Interference between waves from two slits or sources. In fig. 10.15 let S_1 and S_2 be two equal sources separated by a distance f, each generating a wave of angular frequency ω and amplitude a. At a point P sufficiently distant from S_1 and S_2 only plane wavefronts arrive with displacements

$$y_1 = a \sin (\omega t - kx_1) \quad \text{from } S_1$$

and

$$y_2 = a \sin (\omega t - kx_2) \quad \text{from } S_2$$

so that the phase difference between the two signals at P is given by

$$\delta = k(x_2 - x_1) = \frac{2\pi}{\lambda}(x_2 - x_1)$$

This phase difference δ, which arises from the path difference $x_2 - x_1$, depends only on x_1, x_2 and the wavelength λ and not on any variation in the source behaviour. This requires that there shall be no sudden changes of phase in the signal generated at either source—such sources are called *coherent*.

The superposition of displacements at P gives a resultant

$$R = y_1 + y_2 = a[\sin (\omega t - kx_1) + \sin (\omega t - kx_2)]$$

and an intensity magnitude

$$I = R^2 = 2a^2(1 + \cos \delta)$$

$$= 4a^2 \cos^2 \frac{\delta}{2}$$

(the time dependent term is not included).

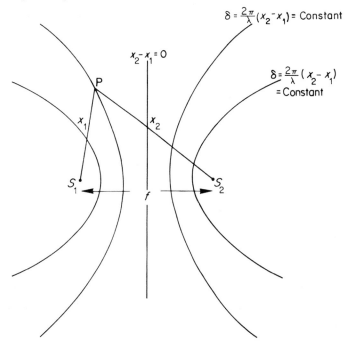

Fig. 10.15. Interference at P between waves from equal sources S_1 and S_2, separation f, depends only on the path difference $x_2 - x_1$. Loci of points with constant phase difference $\delta = (2\pi/\lambda)(x_2 - x_1)$ are the family of hyperbolas with S_1 and S_2 as foci

When

$$\cos\frac{\delta}{2} = 1$$

the intensity is a maximum,

$$I = 4a^2$$

and the component displacements reinforce each other to give *constructive interference*. This occurs when

$$\frac{\delta}{2} = \frac{\pi}{\lambda}(x_2 - x_1) = n\pi$$

that is, when the path difference

$$x_2 - x_1 = n\lambda$$

When

$$\cos\frac{\delta}{2} = 0$$

the intensity is zero and the components cancel to give *destructive interference*. This requires that

$$\frac{\delta}{2} = (2n+1)\frac{\pi}{2} = \frac{\pi}{\lambda}(x_2 - x_1)$$

or, the path difference

$$x_2 - x_1 = (n + \tfrac{1}{2})\lambda$$

The loci or sets of points for which $x_2 - x_1$ (or δ) is constant are shown in fig. 10.1 to form hyperbolas about the foci S_1 and S_2 (in three dimensions the loci would be the hyperbolic surfaces of revolution).

Interference from Two Equal Sources of Separation f

(a) *Separation $f \gg \lambda$. Young's Slit Experiment*

One of the best known methods for producing optical interference effects is the Young's slit experiment. Here the two coherent sources, fig. 10.16, are two identical slits S_1 and S_2 illuminated by a monochromatic wave system from a single source equidistant from S_1 and S_2. The observation point P lies on a screen which is set at a distance l from the plane of the slits.

The intensity at P is given by

$$I = R^2 = 4a^2 \cos^2 \frac{\delta}{2}$$

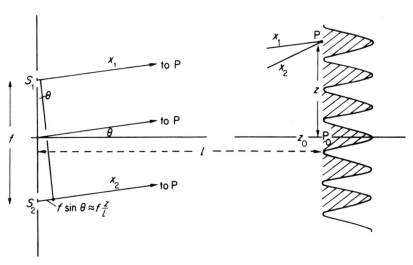

Fig. 10.16. Waves from equal sources S_1 and S_2 interfere at P with phase difference $\delta = (2\pi/\lambda)(x_2 - x_1) = (2\pi/\lambda) f \sin \theta \approx (2\pi/\lambda) f(z/l)$. The distance $l \gg z$ and f so S_1 P and S_2 P are effectively parallel. Interference fringes of intensity $I = I_0 \cos^2 \delta/2$ are formed in the plane PP_0

and the distances $PP_0 = z$ and slit separation f are both very much less than l (experimentally $\approx 10^{-3} l$). This is indicated by the break in the lines x_1 and x_2 in fig. 10.16 where S_1P and S_2P may be considered as sufficiently parallel for the path difference to be written as

$$x_2 - x_1 = f \sin \theta = f \frac{z}{l}$$

to a very close approximation.

Thus

$$\delta = \frac{2\pi}{\lambda}(x_2 - x_1) = \frac{2\pi}{\lambda} f \sin \theta = \frac{2\pi}{\lambda} f \frac{z}{l}$$

If

$$I = 4a^2 \cos^2 \frac{\delta}{2}$$

then

$$I = I_0 = 4a^2 \quad \text{when } \cos \frac{\delta}{2} = 1$$

that is, when the path difference

$$f \frac{z}{l} = 0, \qquad \pm\lambda, \qquad \pm 2\lambda, \qquad \ldots \pm n\lambda$$

and

$$I = 0 \quad \text{when } \cos \frac{\delta}{2} = 0$$

that is, when

$$f \frac{z}{l} = \pm \frac{\lambda}{2}, \qquad \pm \frac{3\lambda}{2}, \qquad \pm (n + \tfrac{1}{2})\lambda$$

Taking the point P_0 as $z = 0$, the variation of intensity with z on the screen $P_0 P$ will be that of fig. 10.16, a series of alternating straight bright and dark fringes parallel to the slit directions, the bright fringes having $I = 4a^2$ whenever $z = n\lambda l/f$ and the dark fringes $I = 0$, occurring when $z = (n + \tfrac{1}{2})\lambda l/f$, n being called the *order of interference* of the fringes. The zero order $n = 0$ at the point P_0 is the central bright fringe. The distance on the screen between two bright fringes of orders n and $n + 1$ is given by

$$z_{n+1} - z_n = [(n + 1) - n]\frac{\lambda l}{f} = \frac{\lambda l}{f}$$

which is also the physical separation between two consecutive dark fringes. The spacing between the fringes is therefore constant and independent of n, and a measurement of the spacing, l and f determines λ.

The intensity distribution curve (fig. 10.17) shows that when the two wave trains arrive at P exactly out of phase they interfere destructively and the resulting intensity or energy flux is zero. Energy conservation requires that the energy must be redistributed, and that lost at zero intensity is found in the intensity peaks. The average value of $\cos^2 \delta/2$ is $\frac{1}{2}$, and the dotted line at $I = 2a^2$ is the average intensity value over the interference system which is equal to the sum of the separate intensities from each slit.

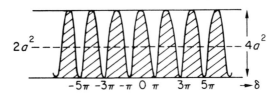

Fig. 10.17. Intensity of interference fringes is proportional to $\cos^2 \delta/2$, where δ is the phase difference between the interfering waves. The energy which is lost in destructive interference (minima) is redistributed into regions of constructive interference (maxima)

There are two important points to remember about the intensity interference fringes when discussing diffraction phenomena; these are

    (a) the intensity varies with $\cos^2 \delta/2$,

and (b) the maxima occur for path differences of zero or integral numbers of the wavelength, whilst the minima represent path differences of odd numbers of the half-wavelength.

Spatial coherence. In the preceding section nothing has been said about the size of the source producing the plane wave which falls on S_1 and S_2. If this source is an ideal *point* source A equidistant from S_1 and S_2, fig. 10.18, then a single set of $\cos^2$ fringes is produced. But every source has a finite size, given by AB in fig. 10.18, and each point on the line source AB produces its own set of interference fringes in the plane PP_0; the eye observing the sum of their intensities.

If the solid curve $A'C$ is the intensity distribution for the point A of the source and the broken curves up to B' represent the corresponding fringes for points along AB the resulting intensity curve is DE. Unless $A'B'$ extends to C the variations of DE will be seen as faint interference bands. These intensity variations were quantified by Michelson, who defined the

$$\text{Visibility} = \frac{I_{\max} - I_{\min}}{I_{\max} + I_{\min}}$$

The $\cos^2$ fringes from a point source obviously have a visibility of unity because the minimum intensity $I_{\min} = 0$.

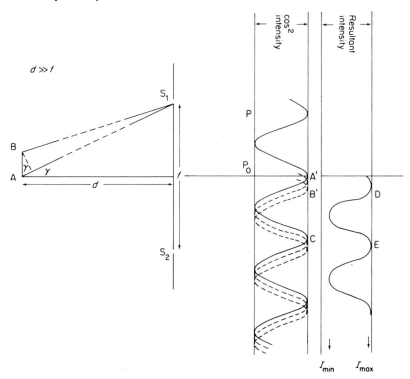

Fig. 10.18. The point source A produces the $\cos^2$ interference fringes represented by the solid curve A′C. Other points on the line source AB produce $\cos^2$ fringes (the displaced broken curves B′) and the observed total intensity is the curve DE. When the points on AB extend A′B′ to C the fringes disappear and the field is uniformly illuminated

When A′B′ of fig. 10.18 = A′C, the point source fringe separation (or a multiple of it) the field is uniformly illuminated, fringe visibility = 0 and the fringes disappear.

This occurs when the path difference

$$AS_2 - BS_1 \approx AB \sin \gamma = \lambda \quad \text{where } AS_2 = AS_1.$$

Thus the requirement for fringes of good visibility imposes a limit on the finite size of the source. Light from points on the source must be *spatially coherent* in the sense that AB sin $\gamma \ll \lambda$ in fig. 10.18.

But for $f \ll d$,

$$\sin \gamma \approx f/2d$$

so the coherence condition becomes

$$\sin \gamma = f/2d \ll \lambda/AB$$

or

$$AB/d \ll \lambda/f$$

where AB/d measures the angle subtended by the source at the plane S_1S_2. *Spatial coherence* therefore requires that the angle subtended by the source

$$\ll \lambda/f$$

where f is the linear size of the diffracting system. (Note also that λ/f measures $\theta(\sim z/l)$ the angular separation of the fringes in fig. 10.16.)

As an example of spatial coherence we may consider the production of Young's interference fringes using the sun as a source.

The sun subtends an angle of 0.018 radians at the earth and if we accept the approximation

$$\frac{AB}{d} \ll \frac{\lambda}{f} \approx \frac{\lambda}{4f}$$

with $\lambda = 0.5$ microns,

we have

$$f \sim \frac{0 \cdot 5}{4(0 \cdot 018)} \sim 14 \text{ microns.}$$

This small value of slit separation is required to meet the spatial coherence condition.

(b) *Separation* $f \ll \lambda$ $(kf \ll 1$ *where* $k = 2\pi/\lambda)$

If there is a zero phase difference between the signals leaving the sources S_1 and S_2 of fig. 10.16 then the intensity at some distant point P may be written

$$I = 4a^2 \cos^2 \frac{\delta}{2} = 4I_s \cos^2 \frac{kf \sin \theta}{2} \approx 4I_s$$

where the path difference $S_2P - S_1P = f \sin \theta$ and $I_s = a^2$ is the intensity from each source.

We note that, since $f \ll \lambda$ $(kf \ll 1)$, the intensity has a very small θ dependence and the two sources may be effectively replaced by a single source of amplitude 2a.

Dipole Radiation $(f \ll \lambda)$

Suppose however that the signals leaving the sources S_1 and S_2 are anti phase so that their total phase difference at some distant point P is

$$\delta = (\delta_0 + kf \sin \theta)$$

where $\delta_0 = \pi$ is the phase difference introduced at source.

The intensity at P is given by

$$I = 4I_s \cos^2 \frac{\delta}{2} = 4I_s \cos^2 \left(\frac{\pi}{2} + \frac{kf \sin \theta}{2} \right)$$

$$= 4I_s \sin^2 \left(\frac{kf \sin \theta}{2} \right)$$

$$\approx I_s (kf \sin \theta)^2$$

because

$$kf \ll 1$$

Two anti phase sources of this kind constitute a *dipole* whose radiation intensity $I \ll I_s$ the radiation from a single source, when $kf \ll 1$. The efficiency of radiation is seen to depend on the product kf and, for a fixed separation f the dipole is a less efficient radiator at low frequencies (small k) than at higher frequencies. Fig. 10.19 shows the radiation intensity I plotted against the

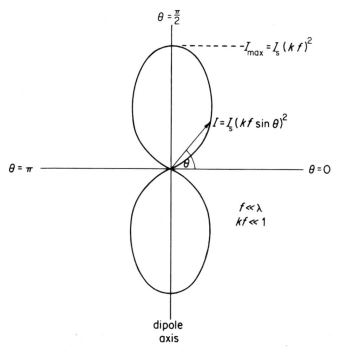

Fig. 10.19. Intensity I versus direction θ for interference pattern between waves from two equal sources, π radians out of phase (dipole) with separation $f \ll \lambda$. The dipole axis is along the direction $\theta = \pm \pi/2$

polar angle θ and we see that for the dipole axis along the direction $\theta = \pi/2$, completely destructive interference occurs only on the perpendicular axis $\theta = 0$ and $\theta = \pi$. There is no direction (value of θ) giving completely constructive interference. The highest value of the radiated intensity occurs along the axis $\theta = \pi/2$ and $\theta = 3\pi/2$ but even this is only

$$I = (kf)^2 I_s$$

where

$$kf \ll 1$$

The directional properties of a radiating dipole are incorporated in the design of transmitting aerials. In acoustics a loudspeaker may be considered as a multi-dipole source, the face of the loudspeaker generating compression waves whilst its rear propagates rarefactions. Acoustic reflections from surrounding walls give rise to undesirable interference effects which are avoided by enclosing the speaker in a cabinet. Bass reflex or phase inverter cabinets incorporate a vent on the same side as the speaker face at an acoustic distance of half a wavelength from the rear of the speaker. The vent thus acts as a second source *in phase* with the speaker face and radiation is improved.

(Problems 10.12, 10.13, 10.14, 10.15, 10.16)

Interference from Linear Array of N Equal Sources

Fig. 10.20 shows a linear array of N equal sources with constant separation f generating signals which are all in phase ($\delta_0 = 0$). At a distant point P in a direction θ from the sources the phase difference between the signals from two successive sources is given by

$$\delta = \frac{2\pi}{\lambda} f \sin \theta$$

and the resultant at P is found by superposing the equal contributions from each source with the constant phase difference δ between successive contributions.

But we found from fig. 1.11 that the resultant of such a superposition was given by

$$R = a \frac{\sin (N\delta/2)}{\sin (\delta/2)}$$

where a is the signal amplitude at each source, so the intensity may be written

$$I = R^2 = a^2 \frac{\sin^2 (N\delta/2)}{\sin^2 (\delta/2)} = I_s \frac{\sin^2 (N\pi f \sin \theta/\lambda)}{\sin^2 (\pi f \sin \theta/\lambda)}$$

$$= I_s \frac{\sin^2 N\beta}{\sin^2 \beta}$$

where I_s is the intensity from each source and $\beta = \pi f \sin \theta/\lambda$.

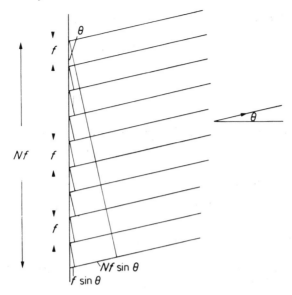

Fig. 10.20. Linear array of N equal sources separation f radiating in a direction θ to a distant point P. The resulting amplitude at P (see fig. 1.11) is given by

$$R = a[\sin N(\delta/2)/\sin(\delta/2)]$$

where a is the amplitude from each source and

$$\delta = (2\pi/\lambda)f \sin\theta$$

is the common phase difference between successive sources

If we take the case of $N = 2$ then

$$I = I_s \frac{\sin^2 2\beta}{\sin^2 \beta} = 4I_s \cos^2 \beta = 4I_s \cos^2 \frac{\delta}{2}$$

which gives us the Young's Slit Interference pattern.

We can follow the intensity pattern for N sources by considering the behaviour of the term $\sin^2 N\beta / \sin^2 \beta$.

We see that when

$$\beta = \frac{\pi}{\lambda} \sin\theta = 0 \pm \pi \pm 2\pi, \text{ etc.}$$

i.e. when

$$f \sin\theta = 0, \pm\lambda, \pm2\lambda \ldots \pm n\lambda$$

constructive interference of the order n takes place, and

$$\frac{\sin^2 N\beta}{\sin^2 \beta} \to \frac{N^2 \beta^2}{\beta^2} \to N^2$$

giving

$$I = N^2 I_s$$

that is, a very strong intensity at the *Principal Maximum* condition of

$$f \sin \theta = n\lambda$$

We can display the behaviour of the $\sin^2 N\beta / \sin^2 \beta$ term as follows

Numerator $\sin^2 N\beta$ is zero for $N\beta \to 0\pi \ldots N\pi \ldots 2N\pi$

$$\downarrow \qquad \downarrow \qquad \downarrow$$

Denominator $\sin^2 \beta$ is zero for $\beta \to 0 \ldots \pi \ldots 2\pi$

The coincidence of zeros for both numerator and denominator determine the Principal Maxima with the factor N^2 in the intensity, i.e. whenever $f \sin \theta = n\lambda$.

Between these principal maxima are $N - 1$ points of zero intensity which occur whenever the numerator $\sin^2 N\beta = 0$ but where $\sin^2 \beta$ remains finite. These occur when

$$f \sin \theta = \frac{\lambda}{N}, \qquad \frac{2\lambda}{N} \ldots (n-1)\frac{\lambda}{N}$$

The $N - 2$ subsidiary maxima which occur between the principal maxima have much lower intensities because none of them contains the factor N^2. Fig. 10.21 shows the intensity curves for $N = 2, 4, 8$ and $N \to \infty$.

Two scales are given on the horizontal axis. One shows how the maxima occur at the order of interference $n = f \sin \theta / \lambda$. The other, using units of $\sin \theta$ as the ordinate displays two features. It shows that the separation between the principal maxima in units of $\sin \theta$ is λ / f and that the width of half the base of the principal maxima in these units is λ / Nf (the same value as the width of the base of subsidiary maxima). As N increases not only does the principal intensity increase as N^2 but the width of the principal maximum becomes very small.

As N becomes very large, the interference pattern becomes highly directional, very sharply defined peaks of high intensity occurring whenever $\sin \theta$ changes by λ / f.

The directional properties of such a linear array are widely used in both transmitting and receiving aerials and the polar plot for $N = 4$ (fig. 10.22) displays these features. For N large, such an array, used as a receiver, forms the basis of a radio telescope where the receivers (sources) are set at a constant (but adjustable) separation f and tuned to receive a fixed wavelength. Each receiver takes the form of a parabolic reflector, the axes of which are kept

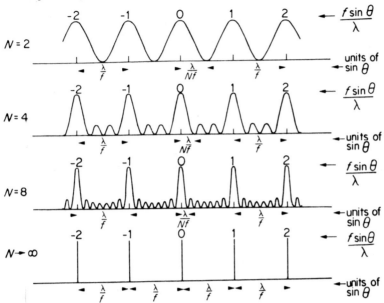

Fig. 10.21. Intensity of interference patterns from linear arrays of N equal sources of separation f. The horizontal axis in units of $f\sin\theta/\lambda$ gives the spectral order n of interference. The axis in units of $\sin\theta$ shows that the separation between principal maxima is given by $\sin\theta = \lambda/f$ and the half-width of the principal maximum is given by $\sin\theta = \lambda/Nf$

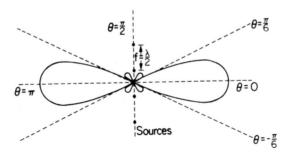

Fig. 10.22. Polar plot of the intensity of the interference pattern from a linear array of four sources with common separation $f = \lambda/2$. Note that the half-width of the principal maximum is $\theta = \pi/6$ satisfying the relation $\sin\theta = \lambda/Nf$ and that the separation between principal maxima satisfies the relation that the change in $\sin\theta = \lambda/f$

parallel as the reflectors are oriented in different directions. The angular separation between the directions of incidence for which the received signal is a maximum is given by sin $\theta = \lambda/f$.

(Problems 10.17, 10.18)

Division of Amplitude

We now consider interference effects produced by division of amplitude. In fig. 10.23 a ray of monochromatic light of wavelength λ in air is incident at an angle i on a plane parallel slab of material thickness t and refractive index

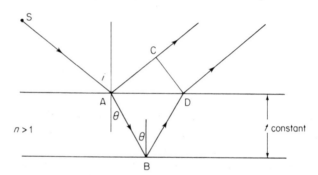

Fig. 10.23. Fringes of constant inclination. Interference fringes formed at infinity by division of amplitude when the material thickness t is constant. The mth order bright fringe is a circle centred at S and occurs for the constant θ value in $2nt \cos \theta = (m + \frac{1}{2})\lambda$

$n > 1$. It suffers partial reflexion and transmission at the upper surface, some of the transmitted light is reflected at the lower surface and emerges parallel to the first reflexion with a phase difference determined by the extra optical path it has travelled in the material. These parallel beams meet and interfere at infinity but they may be brought to focus by a lens. Their optical path difference is seen to be

$$n(AB + BD) - AC = 2nAB - AC$$

$$= 2nt/\cos \theta - 2t \tan \theta \sin i$$

$$= \frac{2nt}{\cos \theta}(1 - \sin^2 \theta) = 2nt \cos \theta$$

(because $\sin i = n \sin \theta$).

This path difference introduces a phase difference

$$\Delta\phi = \frac{2\pi}{\lambda} 2nt \cos\theta$$

but an additional phase change of π radians occurs at the upper surface. Thus if $2nt \cos\theta = m\lambda$ (m an integer) the two beams are antiphase and cancel to give zero intensity, a minimum of interference. If $2nt \cos\theta = (m + \frac{1}{2})\lambda$ the amplitudes will reinforce to give an interference maximum.

Since t is constant the locus of each interference fringe is determined by a constant value of θ which depends on a constant angle i. This gives a circular fringe centred on S. An extended source produces a range of constant θ values at one viewing position so the complete pattern is obviously a set of concentric circular fringes centred on S and formed at infinity. They are fringes of *equal inclination* and are called Haidinger fringes. They are observed to high orders of interference, that is values of m, so that t may be relatively large.

When the thickness t is not constant and the faces of the slab form a wedge, fig. 10.24(a) and (b) the interfering rays are not parallel but meet at points (real or virtual) near the wedge. The resulting interference fringes are localized near the wedge and are almost parallel to the thin end of the wedge. When observations are made at or near the normal to the wedge $\cos\theta \sim 1$ and changes slowly in this region so that $2nt \cos\theta \approx 2nt$. The condition for bright fringes then becomes

$$2nt = (m + \frac{1}{2})\lambda$$

and each fringe locates a particular value of the thickness t of the wedge and this defines the pattern as *fringes of equal thickness*. As the value of m increases to $m + 1$ the thickness of the wedge increases by $\lambda/2n$ so the fringes allow measurements to be made to within a fraction of a wavelength and are of great practical importance.

The spectral colours of a thin film of oil floating on water are fringes of constant thickness. Each frequency component of white light produces an interference fringe at that film thickness appropriate to its own particular wavelength.

In the laboratory the most familiar fringes of constant thickness are Newton's Rings.

Newton's Rings

Here the wedge of varying thickness is the air gap between two spherical surfaces of different curvature. A constant value of t yields a circular fringe and the pattern is one of concentric fringes alternately dark and bright. The simplest example, fig. 10.25, is a plano convex lens resting on a plane reflecting surface where the system is illuminated from above using a partially reflecting glass plate tilted at 45°. Each downward ray is partially reflected at each surface of the lens and at the plane surface. Interference takes place between

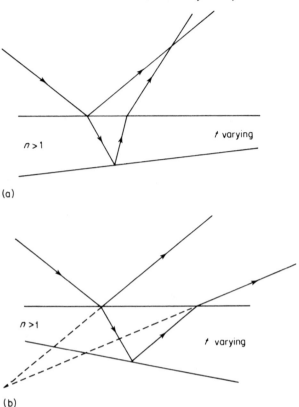

t varying

$n > 1$

(a)

$n > 1$

t varying

(b)

Fig. 10.24. Fringes of constant thickness. When the thickness t of the material is not constant the fringes are localized where the interfering beams meet (a) in a real position and (b) in a virtual position. These fringes are almost parallel to the line where $t = 0$ and each fringe defines a locus of constant t

the light beams reflected at each surface of the air gap. At the lower (air to glass) surface of the gap there is a π radians phase change upon reflexion and the centre of the interference fringe pattern, at the point of contact, is dark. Moving out from the centre, successive rings are light and dark as the air gap thickness increases in units of $\lambda/2$. If R is the radius of curvature of the spherical face of the lens, the thickness t of the air gap at a radius r from the centre is given approximately by $t \approx r^2/2R$. In the mth order of interference a bright ring requires

$$2t = (m + \tfrac{1}{2})\lambda = r^2/R$$

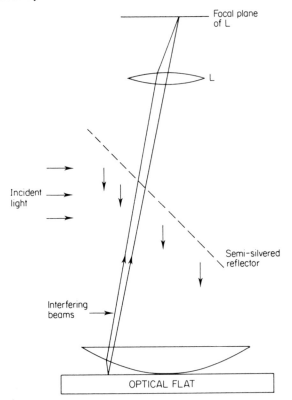

Fig. 10.25. Newton's rings of interference formed by an air film of varying thickness between the lens and the optical flat. The fringes are circular, each fringe defining a constant value of the air film thickness

and because $t \propto r^2$ the fringes become more crowded with increasing r. Rings may be observed with very simple equipment and good quality apparatus can produce fringes for $m > 100$.

(Problem 10.19)

Michelson's Spectral Interferometer

This instrument can produce both types of interference fringes, that is, *circular fringes of equal inclination at infinity* and *localized fringes of equal thickness*. At the end of the nineteenth century it was one of the most important instruments for measuring the structure of spectral lines.

As shown in fig. 10.26 it consists of two identical plane parallel glass plates G_1 and G_2 and two highly reflecting plane mirrors M_1 and M_2. G_1 has a

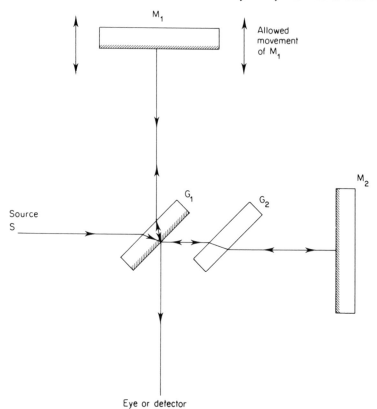

Fig. 10.26. Michelson's Spectral Interferometer. The beam from source S splits at the back face of G_1, and the two parts are reflected at mirrors M_1 and M_2 to recombine and interfere at the eye or detector. G_2 is not necessary with monochromatic light but is required to produce fringes when S is a white light source

partially silvered back face, G_2 does not. In the figure G_1 and G_2 are parallel and M_1 and M_2 are perpendicular. Slow, accurately monitored motion of M_1 is allowed in the direction of the arrows but the mounting of M_2 is fixed although the angle of the mirror plane may be tilted so that M_1 and M_2 are no longer perpendicular.

The incident beam from an extended source divides at the back face of G_1. A part of it is reflected back through G_1 to M_1 where it is returned through G_1 into the eye or detector. The remainder of the incident beam reaches M_2 via G_2 and returns through G_2 to be reflected at the back face of G_1 into the eye or detector where it interferes with the beam from the M_1 arm of the interferometer. The presence of G_2 assures that each of the two interfering beams has the same optical path in glass. This condition is not essential for fringes with monochromatic light but it is required with a white light source where dispersion in glass becomes important.

An observer at the detector looking into G_1 will see M_1, a reflected image of M_2 (M_2' say) and the images S_1 and S_2' of the source provided by M_1 and M_2. This may be represented by the linear configuration of fig. 10.27 which shows how interference takes place and what type of fringes are produced.

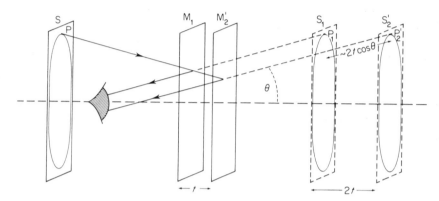

Fig. 10.27. Linear configuration to show fringe formation by a Michelson interferometer. A ray from point P on the extended source S reflects at M_1, and appears to come from P_1 in the reflected plane S_1. The ray is reflected from M_2 (shown here as M_2') and appears to come from P_2' in the reflected plane S_2'. The path difference at the detector between the interfering beams is effectively $2t \cos \theta$ where t is the difference between the path lengths from the source S to the separate mirrors M_1 and M_2

When the optical paths in the interferometer arms are equal and M_1 and M_2 are perpendicular the planes of M_1 and the image M_2' are coincident. However a small optical path difference t between the arms becomes a difference of $2t$ between the mirrored images of the source as shown in fig. 10.27. The divided ray from a single point P on the extended source is reflected at M_1 and M_2 (shown as M_2') but these reflexions appear to come from P_1 and P_2' in the image planes of the mirrors. The path difference between the rays from P_1 and P_2' is evidently $2t \cos \theta$. When $2t \cos \theta = m\lambda$ a maximum of interference occurs and for constant θ the interference fringe is a circle. The extended source produces a range of constant θ values and a pattern of concentric circular fringes of constant inclination.

If the path difference t is very small and the plane of M_2 is now tilted, a wedge is formed and straight localized fringes may be observed at the narrowest part of the wedge. As the wedge thickens the fringes begin to curve because the path difference becomes more strongly dependent upon the angle of observation. These curved fringes are always convex towards the thin end of the wedge.

The Structure of Spectral Lines

The discussion on spatial coherence, page 300, showed that two close identical sources emitting the same wavelength λ produced interference fringe systems slightly displaced from each other, fig. 10.18.

The same effect is produced by a *single* source, such as sodium, emitting two wavelengths, λ and $\lambda - \Delta\lambda$ so that the maxima and minima of the $\cos^2$ fringes for λ are slightly displaced from those for $\lambda - \Delta\lambda$. This displacement increases with the order of interference m until a value of m is reached when the maximum for λ coincides with a minimum for $\lambda - \Delta\lambda$ and the fringes disappear as their visibility is reduced to zero.

In 1862, Fizeau, using a sodium source to produce Newton's Rings, found that the fringes disappeared at the order $m = 490$ but returned to maximum visibility at $m = 980$. He correctly identified the presence of two components in the spectral line.

The visibility

$$(I_{max} - I_{min})/(I_{max} + I_{min})$$

equals zero when

$$m\lambda = (m + \tfrac{1}{2})(\lambda - \Delta\lambda)$$

and for $\lambda = 0\cdot5893$ microns and $m = 490$ we have $\Delta\lambda = 0\cdot0006$ microns, (6 Å), which are the accepted values for the D lines of the sodium doublet.

Using his spectral interferometer, Michelson extended this work between the years 1890 and 1900, plotting the visibility of various fringe systems and building a mechanical harmonic analyser into which he fed different component frequencies in an attempt to reproduce his visibility curves. The sodium doublet with angular frequency components ω and $\omega + \Delta\omega$ produced a visibility curve similar to that of figs. 1.7 and 3.4 and was easy to interpret. More complicated visibility patterns were not easy to reproduce and the modern method of Fourier Transform spectroscopy reverses the procedure by extracting the frequency components from the observed pattern.

Michelson did however confirm that the cadmium red line, $\lambda = 0\cdot6438$ microns was highly monochromatic. The visibility had still to reach a minimum when the path difference in his interferometer arms was $0\cdot2$ metres.

Fabry–Perot Interferometer

The interference fringes produced by division of amplitude which we have discussed so far have been observed as reflected light and have been produced by only two interfering beams. We now consider fringes which are observed in transmission and which require multiple reflections. They are fringes of constant inclination formed in a pattern of concentric circles by the Fabry–

Perot interferometer. The fringes are particularly narrow and sharply defined so that a beam consisting of two wavelengths λ and $\lambda - \Delta\lambda$ forms two patterns of rings which are easily separated for small $\Delta\lambda$. This instrument therefore has an extremely high resolving power. The main component of the interferometer is an etalon fig. 10.28 which consists of two plane parallel glass plates with identical highly reflecting inner surfaces S_1 and S_2 which are separated by a distance d.

Suppose a monochromatic beam of unit amplitude, angular frequency ω and wavelength (in air) of λ strikes the surface S_1 as shown. A fraction t of

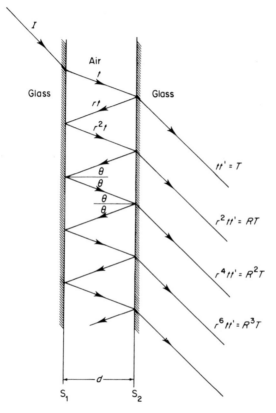

Fig. 10.28. S_1 and S_2 are the highly reflecting inner surfaces of a Fabry–Perot etalon with a constant air gap thickness d. Multiple reflections produce parallel interfering beams with amplitudes T, RT, R^2T, etc. each beam having a phase difference

$$\delta = 4\pi d \cos \theta/\lambda$$

with respect to its neighbour

this beam is transmitted in passing from glass to air. At S_2 a further fraction t' is transmitted in passing from air to glass to give an emerging beam of amplitude $tt' = T$. The reflection coefficient at the air–S_1 and air–S_2 surfaces is r so each subsequent emerging beam is parallel but has an amplitude factor $r^2 = R$ with respect to its predecessor. Other reflection and transmission losses are common to all beams and do not affect the analysis. Each emerging beam has a phase lag $\delta = 4\pi d \cos \theta / \lambda$ with respect to its predecessor and these parallel beams interfere when they are brought to focus via a lens.

The vector sum of the transmitted interfering amplitudes together with their appropriate phases may be written

$$A = T\, e^{i\omega t} + TR\, e^{i(\omega t - \delta)} + TR^2\, e^{i(\omega t - 2\delta)} \dots$$

$$= T\, e^{i\omega t}[1 + R\, e^{-i\delta} + R^2\, e^{-i2\delta} \dots$$

which is an infinite geometric progression with the sum

$$A = T\, e^{i\omega t}/(1 - R\, e^{-i\delta})$$

This has a complex conjugate

$$A^* = T\, e^{-i\omega t}/(1 - R\, e^{i\delta})$$

If the incident unit intensity is I_0 the fraction of this intensity in the transmitted beam may be written

$$\frac{I_t}{I_0} = \frac{AA^*}{I_0} = \frac{T^2}{(1 - R\, e^{-i\delta})(1 - R\, e^{i\delta})} = \frac{T^2}{(1 + R^2 - 2R \cos \delta)}$$

or, with

$$\cos \delta = 1 - 2 \sin^2 \delta/2$$

as

$$\frac{I_t}{I_0} = \frac{T^2}{(1 - R)^2 + 4R \sin^2 \delta/2} = \frac{T^2}{(1 - R)^2} \frac{1}{1 + [4R \sin^2 \delta/2/(1 - R)^2]}$$

But the factor $T^2/(1 - R)^2$ is a constant, written C so

$$\frac{I_t}{I_0} = C \cdot \frac{1}{1 + [4R \sin^2 \delta/2/(1 - R)^2]}$$

Writing $CI_0 = I_{max}$, the graph of I_t versus δ in fig. 10.29 shows that as the reflection coefficient of the inner surfaces is increased, the interference fringes become narrow and more sharply defined. Values of $R > 0.9$ may be reached using the special techniques of multilayer dielectric coating. In one of these techniques a glass plate is coated with alternate layers of high and low refractive index materials so that each boundary presents a large change of refractive index and hence a large reflection. If the *optical* thickness of each layer is $\lambda/4$ the emerging beams are all in phase and the reflected intensity is high.

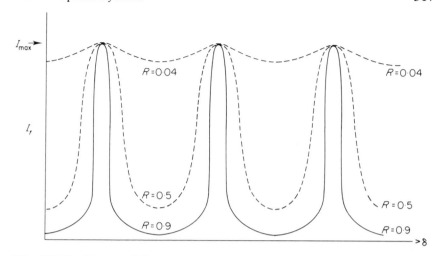

Fig. 10.29. Observed intensity of fringes produced by a Fabry–Perot interferometer. Transmitted intensity I_t versus δ. $R = r^2$ where r is the reflexion coefficient of the inner surfaces of the etalon. As R increases the fringes become narrower and more sharply defined

Resolving Power of the Fabry–Perot Interferometer

Fig. 10.29 shows that a value of $R = 0 \cdot 9$ produces such narrow and sharply defined fringes that if the incident beam has two components λ and $\lambda - \Delta\lambda$ the two sets of fringes should be easily separated. The criterion for separation depends on the shape of the fringes: the diffraction grating of page 328 uses the Rayleigh criterion, but the fringes here are so sharp that they are resolved at a much smaller separation than that required by Rayleigh.

Here the fringes of the two wavelengths may be resolved when they cross at half their maximum intensities, that is, at $I_t = I_{max}/2$ in fig. 10.30.

Using the expression

$$I_t = I_{max} \cdot \cfrac{1}{1 + \cfrac{4R \sin^2 \delta/2}{(1-R)^2}}$$

we see that $I_t = I_{max}$ when $\delta = 0$ and $I_t = I_{max}/2$ when the factor

$$4R \sin^2 \delta/2/(1-R)^2 = 1$$

The fringes are so narrow that they are visible only for very small values of δ so we may replace $\sin \delta/2$ by $\delta/2$ in the expression

$$4R \sin^2 \delta/2/(1-R)^2 = 1$$

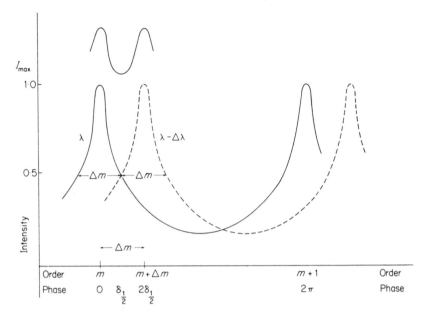

Fig. 10.30. Fabry–Perot interference fringes for two wavelengths λ and $\lambda - \Delta\lambda$ are resolved at order m when they cross at half their maximum intensity. Moving from order m to $m+1$ changes the phase δ by 2π radians and the full 'half-value' width of each maximum is given by $\Delta m = 2\delta_{1/2}$ which is also the separation between the maxima of λ and $\lambda - \Delta\lambda$ when the fringes are just resolved

to give the value

$$\delta_{1/2} = \frac{(1-R)}{R^{1/2}}$$

as that phase departure from the maximum, $\delta = 0$, which produces the intensity $I_t = I_{max}/2$ for wavelength λ. Our criterion for resolution means, therefore, that the maximum intensity for $\lambda - \Delta\lambda$ is removed an extra amount $\delta_{1/2}$ along the phase axis of fig. 10.30. This axis also shows the order of interference m at which the wavelengths are resolved, together with the order $m+1$ which represents a phase shift of $\delta = 2\pi$ along the phase axis.

In the mth order of interference we have

$$2d \cos\theta = m\lambda$$

and for fringes of equal inclination (θ constant), logarithmic differentiation gives

$$\lambda/\Delta\lambda = -m/\Delta m$$

Now $\Delta m = 1$ represents a phase change of $\delta = 2\pi$ and the phase difference of $2.\delta_{1/2}$ which just resolves the two wavelengths corresponds to a change of order

$$\Delta m = 2.\delta_{1/2}/2\pi$$

Thus the resolving power, defined as

$$\frac{\lambda}{\Delta\lambda} = \left|\frac{m}{\Delta m}\right| = \frac{m\pi}{\delta_{1/2}} = \frac{m\pi R^{1/2}}{(1-R)}$$

The equivalent expression for the resolving power in the mth order for a diffraction grating of N lines (interfering beams) is shown on page 328 to be

$$\frac{\lambda}{\Delta\lambda} = mN$$

so we may express

$$N' = \pi R^{1/2}/(1-R)$$

as the effective number of interfering beams in the Fabry–Perot interferometer.

This quantity N' is called the *finesse* of the etalon and is a measure of its quality. We see that

$$N' = \frac{2\pi}{2\delta_{1/2}} = \frac{1}{\Delta m} = \frac{\text{separation between orders } m \text{ and } m + 1}{\text{'half value' width of } m\text{th order}}$$

Thus, using one wavelength only, the ratio of the separation between successive fringes to the narrowness of each fringe measures the quality of the etalon. A typical value of $N' \sim 30$.

Free Spectral Range

There is a limit to the wavelength difference $\Delta\lambda$ which can be resolved with the Fabry–Perot interferometer. This limit is reached when $\Delta\lambda$ is such that the circular fringe for λ in the mth order coincides with that for $\lambda - \Delta\lambda$ in the $m + 1$th order. The pattern then loses its unique definition and this value of $\Delta\lambda$ is called the *free spectral range*.

From the preceding section we have the expression

$$\frac{\lambda}{\Delta\lambda} = -\frac{m}{\Delta m}$$

and in the limit when $\Delta\lambda$ represents the free spectral range then

$$\Delta m = 1$$

and

$$\Delta\lambda = -\lambda/m$$

But $m\lambda = 2d$ when $\theta \approx 0$ so the free spectral range

$$\Delta\lambda = -\lambda^2/2d$$

Typically $d \sim 10^{-2}$m and for λ (cadmium red) $= 0.6438$ microns we have, from $2d = m\lambda$, a value of

$$m \approx 3 \times 10^4$$

Now the resolving power

$$\frac{\lambda}{\Delta\lambda} = mN'$$

so, for

$$N' \approx 30$$

the resolving power can be as high as 1 part in 10^6.

Diffraction

Diffraction is classified as Fraunhofer or Fresnel. In Fraunhofer diffraction the pattern is formed at such a distance from the diffracting system that the waves generating the pattern may be considered as plane. A Fresnel diffraction pattern is formed so close to the diffracting system that the waves generating the pattern still retain their curved characteristics.

Fraunhofer Diffraction

The single narrow slit. Earlier in this chapter it was stated that the difference between interference and diffraction is merely one of scale and not of physical behaviour.

Suppose we contract the scale of the N equal sources separation f of fig. 10.20 until the separation between the first and the last source, originally Nf, becomes equal to a distance d where d is now assumed to be the width of a narrow slit on which falls a monochromatic wavefront of wavelength λ where $d \sim \lambda$. Each of the large number N equal sources may now be considered as the origin of secondary wavelets generated across the plane of the slit on the basis of Huygens' Principle to form a system of waves diffracted in all directions.

When these diffracted waves are focused on a screen as shown in fig. 10.31 the intensity distribution of the diffracted waves may be found in terms of the aperture of the slit, the wavelength λ and the angle of diffraction θ. In fig. 10.31 a plane light wave falls normally on the slit aperture of width d and the waves diffracted at an angle θ are brought to focus at a point P on the screen P P_0. The point P is sufficiently distant from the slit for all wavefronts reaching it to be plane and we limit our discussion to *Fraunhofer Diffraction*.

Finding the amplitude of the light at P is the simple problem of superposing all the small contributions from the N equal sources in the plane of the slit,

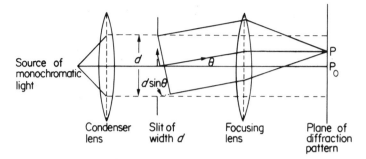

Fig. 10.31. A monochromatic wave normally incident on a narrow slit of width d is diffracted through an angle θ and the light in this direction is focused at a point P. The amplitude at P is the superposition of all the secondary waves in the plane of the slit with their appropriate phases. The extreme phase difference from contributing waves at opposite edges of the slit is $\phi = 2\pi d \sin \theta / \lambda = 2\alpha$

taking into account the phase differences which arise from the variation in path length from P to these different sources. We have already solved this problem several times. In Chapter 9 we took it as an example of the Fourier Transform method but here we reapply the result already used in this chapter on page 304, namely that the intensity at P is given by

$$I = I_s \frac{\sin^2 N\beta}{\sin^2 \beta} \quad \text{where } N\beta = \frac{\pi}{\lambda} Nf \sin \theta$$

is half the phase difference between the contributions from the first and last sources. But now $Nf = d$ the slit width, and if we replace β by α where $\alpha = (\pi/\lambda)d \sin \theta$ is now half the phase difference between the contributions from the opposite edges of the slit, the intensity of the diffracted light at P is given by

$$I = I_s \frac{\sin^2 (\pi/\lambda)d \sin \theta}{\sin^2 (\pi/\lambda N)d \sin \theta} = I_s \frac{\sin^2 \alpha}{\sin^2 (\alpha/N)}$$

For N large

$$\sin^2 \frac{\alpha}{N} \to \left(\frac{\alpha}{N}\right)^2$$

and we have

$$I = N^2 I_s \frac{\sin^2 \alpha}{\alpha^2} = I_0 \frac{\sin^2 \alpha}{\alpha^2}$$

(recall that in the Fourier Transform derivation on page 269,

$$I_0 = \frac{d^2 h^2}{4\pi^2}$$

where h was the amplitude from each source).

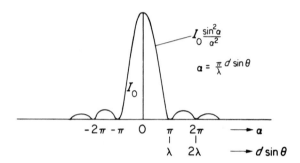

Fig. 10.32. Diffraction pattern from a single narrow slit of width d has an intensity $I = I_0 \sin^2 \alpha/\alpha^2$
where $\alpha = \pi\, d \sin \theta/\lambda$

Plotting $I = I_0(\sin^2 \alpha/\alpha^2)$ with $\alpha = (\pi/\lambda)d \sin \theta$ in fig. 10.32 we see that its pattern is symmetrical about the value

$$\alpha = \theta = 0$$

where $I = I_0$ because $\sin \alpha/\alpha \to 1$ as $\alpha \to 0$. The intensity $I = 0$ whenever $\sin \alpha = 0$ that is, whenever α is a multiple of π or

$$\alpha = \frac{\pi}{\lambda}d \sin \theta = \pm\pi \pm 2\pi \pm 3\pi, \text{ etc}$$

giving

$$d \sin \theta = \pm\lambda \pm 2\lambda \pm 3\lambda, \text{ etc}$$

This condition for *diffraction minima* is the same as that for *interference maxima* between two slits of separation d, and this is important when we consider the problem of light transmission through more than one slit.

The intensity distribution maxima occur whenever the factor $\sin^2 \alpha/\alpha^2$ has a maximum, that is, when

$$\frac{d}{d\alpha}\left(\frac{\sin \alpha}{\alpha}\right)^2 = \frac{d}{d\alpha}\left(\frac{\sin \alpha}{\alpha}\right) = 0$$

or

$$\frac{\cos \alpha}{\alpha} - \frac{\sin \alpha}{\alpha^2} = 0$$

This occurs whenever $\alpha = \tan \alpha$, and fig. 10.33 shows that the roots of this equation are closely approximated by $\alpha = \pm 3\pi/2$, $\pm 5\pi/2$, etc. (see problem at end of chapter on exact values).

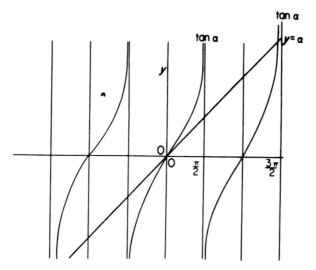

Fig. 10.33. Position of principal and subsidiary maxima of single slit diffraction pattern is given by the intersections of $y = \alpha$ and $y = \tan \alpha$

The table below shows the relative intensities of the subsidiary maxima with respect to the principal maximum I_0.

α	$\dfrac{\sin^2 \alpha}{\alpha^2}$	$\dfrac{I_0 \sin^2 \alpha}{\alpha^2}$
0	1	I_0
$\dfrac{3\pi}{2}$	$\dfrac{4}{9\pi^2}$	$\dfrac{I_0}{22 \cdot 2}$
$\dfrac{5\pi}{2}$	$\dfrac{4}{25\pi^2}$	$\dfrac{I_0}{61 \cdot 7}$
$\dfrac{7\pi}{2}$	$\dfrac{4}{49\pi^2}$	$\dfrac{I_0}{121}$

The rapid decrease in intensity as we move from the centre of the pattern explains why only the first two or three subsidiary maxima are normally visible.

Scale of the Intensity Distribution

The width of the principal maximum is governed by the condition $d \sin \theta = \pm\lambda$. A constant wavelength λ means that a decrease in the slit width d will increase the value of $\sin \theta$ and will widen the principal maximum and the separation between subsidiary maxima. The narrower the slit the wider the diffraction pattern, that is, in terms of a Fourier transform the narrower the pulse in x-space the greater the region in k- or wave number space required to represent it.

(Problems 10.20, 10.21)

Intensity Distribution for Interference with Diffraction from *N* Identical Slits

The extension of the analysis from the example of one slit to that of N equal slits of width d and common spacing f, fig. 10.34, is very simple.

To obtain the expression for the intensity at a point P of diffracted light from a single slit we considered the contributions from the multiple equal sources across the plane of the slit.

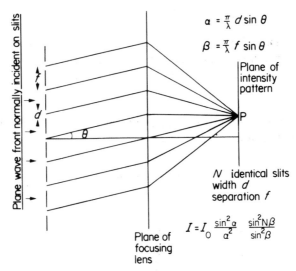

Fig. 10.34. Intensity distribution for diffraction by N equal slits is

$$I = I_0 \frac{\sin^2 \alpha}{\alpha^2} \frac{\sin^2 N\beta}{\sin^2 \beta}$$

the product of the diffraction intensity for one slit, $I_0 \sin^2 \alpha / \alpha^2$ and the interference intensity between N sources $\sin^2 N\beta / \sin^2 \beta$, where $\alpha = (\pi/\lambda)d \sin \theta$ and $\beta = (\pi/\lambda)f \sin \theta$

We obtained the result

$$I = I_0 \frac{\sin^2 \alpha}{\alpha^2}$$

by contracting the original linear array of N sources of spacing f on page 304. If we expand the system again to recover the linear array, where each source is *now* a slit giving us the diffraction contribution

$$I_s = I_0 \frac{\sin^2 \alpha}{\alpha^2}$$

we need only insert this value at I_s in the original expression for the interference intensity,

$$I = I_s \frac{\sin^2 N\beta}{\sin^2 \beta}$$

on page 304 where

$$\beta = \frac{\pi}{\lambda} f \sin \theta$$

to obtain, for the intensity at P in fig. 10.34, the value

$$I = I_0 \frac{\sin^2 \alpha}{\alpha^2} \frac{\sin^2 N\beta}{\sin^2 \beta}$$

where

$$\alpha = \frac{\pi}{\lambda} d \sin \theta$$

Note that this expression combines the diffraction term $\sin^2 \alpha / \alpha^2$ for each slit (source) and the interference term $\sin^2 N\beta / \sin^2 \beta$ from N sources (which confirms what we expected from the opening paragraphs on interference). The diffraction pattern for any number of slits will always have an envelope

$$\frac{\sin^2 \alpha}{\alpha^2} \quad \text{(single slit diffraction)}$$

modifying the intensity of the multiple slit (source) interference pattern

$$\frac{\sin^2 N\beta}{\sin^2 \beta}$$

Fraunhofer Diffraction for Two Equal Slits ($N = 2$)

When $N = 2$ the factor

$$\frac{\sin^2 N\beta}{\sin^2 \beta} = 4 \cos^2 \beta$$

so that the intensity

$$I = 4I_0 \frac{\sin^2 \alpha}{\alpha^2} \cos^2 \beta$$

the factor 4 arising from N^2 whilst the $\cos^2 \beta$ term is familiar from the double source interference discussion. The intensity distribution for $N = 2$, $f = 2d$, is shown in fig. 10.35. The intensity is zero at the diffraction minima when $d \sin \theta = n\lambda$. It is also zero at the interference minima when $f \sin \theta = (n + \frac{1}{2})\lambda$.

At some value of θ an **interference maximum** occurs for $f \sin \theta = n\lambda$ at the same position as a **diffraction minimum** occurs for $d \sin \theta = m\lambda$.

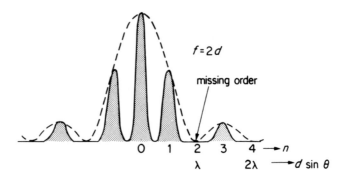

Fig. 10.35. Diffraction pattern for two equal slits, show-ing interference fringes modified by the envelope of a single slit diffraction pattern. Whenever diffraction minima coincide with interference maxima a fringe is suppressed to give a 'missing order' of interference

In this case the diffraction minimum suppresses the interference maximum and the order n of interference is called a *missing order*.

The value of n depends upon the ratio of the slit spacing to the slit width for

$$\frac{n\lambda}{m\lambda} = \frac{f \sin \theta}{d \sin \theta}$$

i.e.

$$\frac{n}{m} = \frac{f}{d} = \frac{\beta}{\alpha}$$

Thus, if

$$\frac{f}{d} = 2$$

the missing orders will be $n = 2, 4, 6, 8$ etc. for $m = 1, 2, 3, 4$, etc.

The ratio

$$\frac{f}{d} = \frac{\beta}{\alpha}$$

governs the scale of the diffraction pattern since this determines the number of interference fringes between diffraction minima and the scale of the diffraction envelope is governed by α.

(Problem 10.22)

Transmission Diffraction Grating (*N* Large)

A large number N of equivalent slits forms a transmission diffraction grating where the common separation f between successive slits is called the *grating space*.

Again, in the expression for the intensity

$$I = I_0 \frac{\sin^2 \alpha}{\alpha^2} \frac{\sin^2 N\beta}{\sin^2 \beta}$$

the pattern lies under the single slit diffraction term (fig. 10.36).

$$\frac{\sin^2 \alpha}{\alpha^2}$$

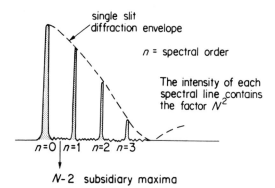

single slit diffraction envelope

n = spectral order

The intensity of each spectral line contains the factor N^2

$n=0$ | $n=1$ $n=2$ $n=3$

$N-2$ subsidiary maxima

Fig. 10.36. Spectral line of a given wavelength produced by a diffraction grating loses intensity with increasing order n as it is modified by the single slit diffraction envelope. At the principal maxima each spectral line has an intensity factor N^2 where N is the number of lines in the grating

The principal interference maxima occur at

$$f \sin \theta = n\lambda$$

having the factor N^2 in their intensity and these are observed as *spectral lines* of order n. We see, however, that the intensities of the spectral lines of a given wavelength decrease with increasing spectral order because of the modifying $\sin^2 \alpha / \alpha^2$ envelope.

Resolving Power of Diffraction Grating

The importance of the diffraction grating as an optical instrument lies in its ability to resolve the spectral lines of two wavelengths which are too close to be separated by the naked eye. If these two wavelengths are λ and $\lambda + d\lambda$ where $d\lambda/\lambda$ is very small the *Resolving Power* for any optical instrument is given by the ratio $\lambda/d\lambda$.

Two such lines are just resolved, according to Rayleigh's Criterion, when the maximum of one falls upon the first minimum of the other. If the lines are closer than this their separate intensities cannot be distinguished.

If we recall that the spectral lines are the principal maxima of the interference pattern from many slits we may display Rayleigh's Criterion in fig. 10.37 where the nth order spectral lines of the two wavelengths are plotted on an axis measured in units of $\sin \theta$. We have already seen in fig. 10.21 that the half width of the spectral lines (principal maxima) measured in such units is given by λ/Nf where N is now the number of grating lines (slits) and f is the grating space. In fig. 10.37 the nth order of wavelength λ occurs when

$$f \sin \theta = n\lambda$$

whilst the nth order for $\lambda + d\lambda$ satisfies the condition

$$f[\sin \theta + \Delta(\sin \theta)] = n(\lambda + d\lambda)$$

so that

$$f\Delta(\sin \theta) = n \ d\lambda$$

Rayleigh's Criterion requires that the fractional change

$$\Delta(\sin \theta) = \frac{\lambda}{Nf}$$

so that

$$f\Delta(\sin \theta) = n \ d\lambda = \frac{\lambda}{N}$$

Hence the Resolving Power of the diffraction grating in the nth order is given by

$$\frac{\lambda}{d\lambda} = Nn$$

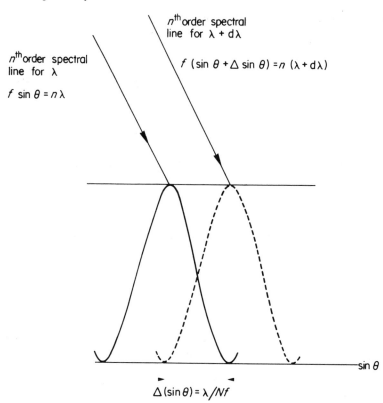

n^{th} order spectral
line for $\lambda + d\lambda$

n^{th} order spectral
line for λ

$f \sin \theta = n \lambda$

$f (\sin \theta + \Delta \sin \theta) = n (\lambda + d\lambda)$

$\overline{} \sin \theta$

$\Delta (\sin \theta) = \lambda / Nf$

Fig. 10.37. Rayleigh's criterion states that the two wavelengths λ and $\lambda + d\lambda$ are just resolved in the nth spectral order when the maximum of one line falls upon the first minimum of the other as shown. This separation, in units of $\sin \theta$, is given by λ / Nf where N is the number of diffraction lines in the grating and f is the grating space. This leads to the result that the resolving power of the grating $\lambda / d\lambda = nN$

Note that the Resolving Power increases with the number of grating lines N and the spectral order n. A limitation is placed on the useful range of n by the decrease of intensity with increasing n due to the modifying diffraction envelope

$$\frac{\sin^2 \alpha}{\alpha^2} \quad \text{(fig. 10.36)}$$

Resolving Power in Terms of the Bandwidth Theorem

A spectral line in the nth order is formed when $f \sin \theta = n \lambda$ where $f \sin \theta$ is the path difference between light coming from two successive slits in the grating. The extreme path difference between light coming from opposite ends of the grating of N lines is therefore given by

$$Nf \sin \theta = Nn\lambda$$

and the time difference between signals travelling these extreme paths is

$$\Delta t = \frac{Nn\lambda}{c}$$

where c is the velocity of light.

The light frequency $\nu = c/\lambda$ has a resolvable differential change

$$|\Delta \nu| = c\frac{|\Delta \lambda|}{\lambda^2} = \frac{c}{Nn\lambda}$$

because $\Delta\lambda/\lambda = 1/Nn$ (from the inverse of the Resolving Power).

Hence

$$\Delta \nu = \frac{c}{Nn\lambda} = \frac{1}{\Delta t}$$

or $\Delta\nu\,\Delta t = 1$ (the Bandwidth Theorem).

Thus the frequency difference which can be resolved is the inverse of the time difference between signals following the extreme paths

$$(\Delta\nu\,\Delta t = 1 \quad \text{is equivalent of course to } \Delta\omega\,\Delta t = 2\pi)$$

If we now write the extreme path difference as

$$Nn\lambda = \Delta x$$

we have, from the inverse of the Resolving Power, that

$$\frac{\Delta\lambda}{\lambda} = \frac{1}{Nn}$$

so

$$\frac{|\Delta\lambda|}{\lambda^2} = \Delta\!\left(\frac{1}{\lambda}\right) = \frac{\Delta k}{2\pi} = \frac{1}{Nn\lambda} = \frac{1}{\Delta x}$$

where the wave number $k = 2\pi/\lambda$.

Hence we also have

$$\Delta x\,\Delta k = 2\pi$$

where Δk is a measure of the resolvable wavelength difference expressed in terms of the difference Δx between the extreme paths.

On pages 65 and 67 we discussed the quality factor Q of an oscillatory system. Note that the resolving power may be considered as the Q of an instrument for

$$\frac{\lambda}{\Delta\lambda} = \left|\frac{\nu}{\Delta\nu}\right| = \frac{\omega}{\Delta\omega} = Q$$

(Problems 10.23, 10.24, 10.25, 10.26)

Fraunhofer Diffraction from a Rectangular Aperture

The value of the Fourier transform method of Chapter 9 becomes apparent when we consider plane wave diffraction from an aperture which is finite in two dimensions.

Although Chapter 9 carried through the transform analysis for the case of only one variable it is equally applicable to functions of more than one variable. In two dimensions, the function $f(x)$ becomes the function $f(x, y)$, giving a transform $F(k_x, k_y)$ where the subscripts give the directions with which the wave numbers are associated.

In fig. 10.38 a plane wave front is diffracted as it passes through the rectangular aperture of dimensions d in the x-direction and b in the y-direction. The vector $\mathbf{k}$, which is normal to the diffracted wave front, has direction cosines l and m with respect to the x- and y-axes respectively. This wavefront is brought to a focus at point P, and the amplitude at P is the superposition of the contributions from all points (x, y) in the aperture with their appropriate phases.

A typical point (x, y) in the aperture may be denoted by the vector $\mathbf{r}$; the difference in phase between the contribution from this point and the central

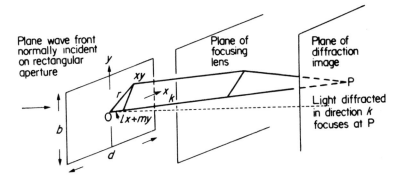

Fig. 10.38. Plane waves of monochromatic light incident normally on a rectangular aperture are diffracted in a direction $\mathbf{k}$. All light in this direction is brought to focus at P in the image plane. The amplitude at P is the superposition of contributions from all the typical points, x, y in the aperture plane with their appropriate phase relationships

point O of the aperture is, of course, $(2\pi/\lambda)$ (path difference). But the path difference is merely the projection of the vector $\mathbf{r}$ upon the vector $\mathbf{k}$, and the phase difference is $\mathbf{k} \cdot \mathbf{r} = (2\pi/\lambda)(lx + my)$, where $lx + my$ is the projection of $\mathbf{r}$ on $\mathbf{k}$.

If we write

$$\frac{2\pi l}{\lambda} = k_x \quad \text{and} \quad \frac{2\pi m}{\lambda} = k_y$$

we have the Fourier transform in two dimensions

$$F(k_x, k_y) = \frac{1}{(2\pi)^2} \int_{-\infty}^{\infty} \int_{-\infty}^{\infty} f(x, y) \, e^{-i(k_x x + k_y y)} \, dx \, dy$$

where $f(x, y)$ is the amplitude of the small contributions from the points in the aperture.

Taking $f(x, y)$ equal to a constant a, we have $F(k_x, k_y)$ the amplitude in k-space at P

$$= \frac{a}{(2\pi)^2} \int_{-d/2}^{+d/2} \int_{-b/2}^{+b/2} e^{-ik_x x} \, e^{-ik_y y} \, dx \, dy$$

$$= \frac{a}{4\pi^2} bd \frac{\sin \alpha}{\alpha} \frac{\sin \beta}{\beta}$$

where

$$\alpha = \frac{\pi l d}{\lambda} = \frac{k_x d}{2}$$

and

$$\beta = \frac{\pi m b}{\lambda} = \frac{k_y b}{2}$$

Physically the integration with respect to y evaluates the contribution of a strip of the aperture along the y direction, and integrating with respect to x then adds the contributions of all these strips with their appropriate phase relationships.

The intensity distribution of the rectangular aperture is given by

$$I = I_0 \frac{\sin^2 \alpha}{\alpha^2} \frac{\sin^2 \beta}{\beta^2}$$

and relative intensities of the subsidiary maxima depend upon the product of the two diffraction terms $\sin^2 \alpha / \alpha^2$ and $\sin^2 \beta / \beta^2$.

These relative values will therefore be numerically equal to the product of any two terms of the series

$$\frac{4}{9\pi^2}, \quad \frac{4}{25\pi^2}, \quad \frac{4}{49\pi^2}, \quad \text{etc.}$$

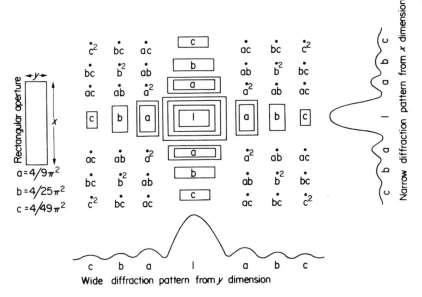

Fig. 10.39. The distribution of intensity in the diffraction pattern from a rectangular aperture is seen as the product of two single-slit diffraction patterns, a wide diffraction pattern from the narrow dimension of the slit and a narrow diffraction pattern from the wide dimension of the slit. This 'rotates' the diffraction pattern through 90° with respect to the aperture

The diffraction pattern from such an aperture together with a plan showing the relative intensities is given in fig. 10.39.

Fraunhofer Diffraction from a Circular Aperture

Diffraction through a circular aperture presents another two-dimensional problem to which the Fourier transform technique may be applied.

As in the case of the rectangular aperture, the diffracted plane wave propagates in a direction k with direction cosines l and m with respect to the x- and y-axes (fig. 10.40a). The circular aperture has a radius d and any point in it is specified by polar coordinates (r, θ) where $x = r \cos \theta$ and $y = r \sin \theta$. This plane wave front in direction k is focused at a point P in the plane of the diffraction pattern and the amplitude at P is the superposition of the contributions from all points (r, θ) in the aperture with their appropriate phase relationships. The phase difference between the contribution from a point defined (x, y) and that from the central point of the aperture is

$$\frac{2\pi}{\lambda}(\text{path difference}) = \frac{2\pi}{\lambda}(lx + my) = k_x x + k_y y$$

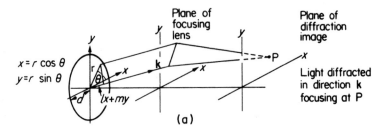

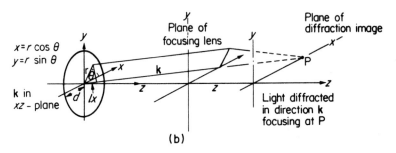

Fig. 10.40. (a) A plane monochromatic wave diffracted in a direction **k** from a circular aperture is focused at a point P in the image plane. Contributions from all points x, y in the aperture superpose at P with appropriate phase relationships. (b) The direction **k** of (a) is chosen to lie wholly in the xz-plane to simplify the analysis. No generality is lost because of circular symmetry. The variation of the amplitude of diffracted light along any one radius determines the complete pattern

as with the rectangular aperture, so that the Fourier transform becomes

$$F(k_x k_y) = \frac{1}{(2\pi)^2} \int_{-\infty}^{\infty} \int_{-\infty}^{\infty} f(x, y) \, e^{-i(k_x x + k_y y)} \, \mathrm{d}x \, \mathrm{d}y$$

If we use polar coordinates, $f(x, y)$ becomes $f(r, \theta)$ and $\mathrm{d}x \, \mathrm{d}y$ becomes $r \, \mathrm{d}r \, \mathrm{d}\theta$, where the limits of θ are from 0 to 2π. Moreover, because of the circular symmetry we may simplify the problem. The amplitude or intensity distribution along any radius of the diffraction pattern is sufficient to define the whole of the pattern, and we may choose this single radial direction conveniently by restricting **k** to lie wholly in the xz plane (fig. 10.40b) so that $m = k_y = 0$ and the phase difference is simply

$$\frac{2\pi}{\lambda} l x = k_x x = k_x r \cos \theta$$

Assuming that $f(r, \theta)$ is a constant amplitude a at all points in the circular aperture, the transform becomes

$$F(k_x) = \frac{a}{2\pi} \int_0^{2\pi} d\theta \int_0^d e^{-ik_x r \cos\theta} r \, dr$$

This can be integrated by parts with respect to r and then term by term in a power series for $\cos\theta$, but the result is well known and conveniently expressed in terms of a *Bessel function* as

$$F(k_x) = \frac{ad}{k_x} J_1(k_x d)$$

where $J_1(k_x d)$ is called a Bessel function of the first order.

Bessel functions are series expansions which are analogous to sine and cosine functions. Where sines and cosines are those functions which satisfy rectangular boundary conditions defined in Cartesian coordinates, Bessel functions satisfy circular or cylindrical boundary conditions requiring polar coordinates.

Standing waves on a circular membrane, e.g. a drum, would require definition in terms of Bessel functions.

The Bessel function of order n is written

$$J_n(x) = \frac{x^n}{2^n n!}\left(1 - \frac{x^2}{2 \cdot 2n + 2} + \frac{x^4}{2 \cdot 4 \cdot 2n + 2 \cdot 2n + 4} \cdots \right)$$

so that

$$J_1(x) = \frac{x}{2} - \frac{x^3}{2^2 4} + \frac{x^5}{2^2 4^2 6} - \frac{x^7}{2^2 4^2 6^2 8}$$

The expression $a^2 d^4 [J_1(k_x d)/k_x d]^2$, which measures the intensity along any radius of the diffraction pattern due to a circular aperture is normalized and plotted in fig. 10.41.

$J_1(k_x d)$ has an infinite number of zeros, and the diffraction pattern is formed by an infinite number of light and dark concentric rings. The first dark band will occur at the first zero of $J_1(k_x d)$ which is given by $k_x d = 1 \cdot 219\pi$.

However,

$$k_x d = \frac{2\pi}{\lambda} l d = \frac{2\pi}{\lambda} d \sin\theta'$$

where θ' is the angle between the vector **k** and the z-axis and defines the angle of diffraction. The first minimum therefore occurs at $d \sin\theta' = 0 \cdot 61\lambda$ and the next minimum at $d \sin\theta' = 1 \cdot 16\lambda$.

If the aperture were square with a side length $2d$ (the diameter of the circle) the first dark fringe would be at $d \sin\theta' = 0 \cdot 5\lambda$ and the second at $d \sin\theta' = \lambda$.

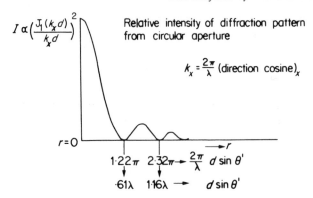

Fig. 10.41. Intensity of the diffraction pattern from a circular aperture of radius d versus r, the radius of the pattern. The intensity is proportional to $[J_1(k_xd)/k_xd]^2$, where J_1 is Bessel's function of order 1. The pattern consists of a central circular principal maximum surrounded by a series of concentric rings of minima and subsidiary maxima of rapidly diminishing intensity

As the radius of the circular aperture is reduced the value of θ' for the first minimum is increased and the whole pattern expands. This reminds us that a reduction of the pulse in x-space requires an increase in wave number or k-space to represent it.

Fresnel Diffraction

The Straight Edge and Slit

Our discussion of Fraunhofer diffraction considered a plane wave normally incident upon a slit in a plane screen so that waves at each point in the plane of the slit were in phase. Each point in the plane became the source of a new wavefront and the superposition of these wavefronts generated a diffraction pattern. At a sufficient distance from the slit the superposed wavefronts were plane and this defined the condition for Fraunhofer diffraction. Its pattern followed from summing the contributions from these waves together with their relative phases and on page 21 we saw that these formed an arc of constant length. When the contributions were all in phase the arc was a straight line but as the relative phases increased the arc curved to form *closed* circles of decreasing radii. The length of the chord joining the ends of the arc measured the resulting amplitude of the superposition and the square of that length measured the light intensity within the pattern.

Nearer the slit where the superposed wavefronts are not yet plane but retain their curved character the diffraction pattern is that of Fresnel. There

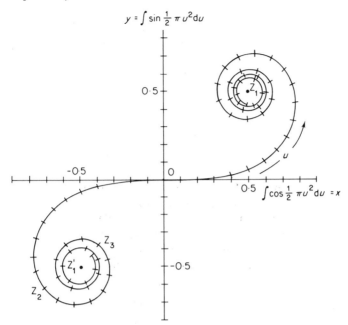

$$y = \int \sin \tfrac{1}{2}\pi u^2 du$$

$$\int \cos \tfrac{1}{2}\pi u^2 du = x$$

Fig. 10.42. Cornu spiral associated with Fresnel diffraction. The spiral in the first quadrant represents the contribution from the upper half of an infinite plane wavefront above an infinite straight edge. The third quadrant spiral results from the downward withdrawal of the straight edge. The width of the wavefront contributing to the diffraction pattern is correlated with the length u along the spiral. The upper half of the wavefront above the straight edge contributes an intensity. $(OZ_1)^2$ that is, the square of the length of the chord from the origin to the spiral eye. This intensity is $0 \cdot 25$ of the intensity $(Z_1 Z_1')^2$ due to the whole wavefront

is no sharp division between Fresnel and Fraunhofer diffraction, the pattern changes continuously from Fresnel to Fraunhofer as the distance from the slit increases.

The Fresnel pattern is determined by a procedure exactly similar to that in Fraunhofer diffraction, an arc of constant length is obtained but now it convolutes around the arms of a pair of joined spirals, fig. 10.42, and not around closed circles.

An understanding of Fresnel diffraction is most easily gained by first considering, not the slit, but a straight edge formed by covering the lower half of the incident plane wavefront with an infinite plane screen. The undisturbed upper half of the wavefront will contribute one half of the total spiral pattern, that part in the first quadrant.

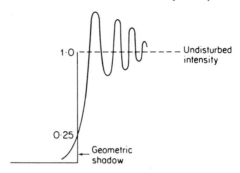

Fig. 10.43. Fresnel diffraction pattern from a straight edge. Light is found within the geometric shadow and fringes of varying intensity form the observed pattern. The intensity at the geometric shadow is 0·25 of that due to the undisturbed wavefront

The Fresnel diffraction pattern from a straight edge, fig. 10.43, has several significant features. In the first place light is found beyond the geometric shadow; this confirms its wave nature and requires a Huygens wavelet to contribute to points not directly ahead of it (see the discussion on page 278). Also, near the edge there are fringes of intensity greater and less than that of the normal undisturbed intensity (taken here as unity). On this scale the intensity at the geometric shadow is exactly 0·25.

To explain the origin of this pattern we consider the point O at the straight edge of fig. 10.44 and the point P directly ahead of O. The line OP defines the geometric shadow. Below O the screen cuts off the wavefront. The phase difference between the contributions to the disturbance at P from O and from a point H, height h above O is given by

$$\Delta(h) = \frac{2\pi}{\lambda}(\mathrm{HP} - \mathrm{OP}) \approx \frac{2\pi}{\lambda}\frac{1}{2}\frac{h^2}{l}$$

where OP = l and higher powers of h^2/l^2 are neglected.

We now divide the wavefront above O into strips which are parallel to the infinite straight edge and we call these strips 'half period zones'. This name derives from the fact that the width of each strip is chosen so that the contributions to the disturbance at P from the lower and upper edges of a given strip differ in phase by π radians.

Since the phase $\Delta(h) \propto h^2$ we shall not expect these strips or half period zones to be of equal width and fig. 10.45 shows how the width of each strip decreases as h increases. The total contribution from a strip will depend upon its area, that is, upon its width. The amplitude and phase of the contribution

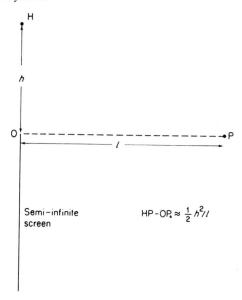

Fig. 10.44. Line OP normal to the straight edge defines the geometric shadow. The wavefront at height h above O makes a contribution to the disturbance at P which has a phase lag of $\pi h^2/\lambda l$ with respect to that from O. The total disturbance at P is the vector sum (amplitude and phase) of all contributions from the wavefront section above O

at P from a narrow strip of width dh at a height h above O may be written as $(dh)\, e^{i\Delta}$ where $\Delta = \pi h^2/\lambda l$.

This contribution may be resolved into two perpendicular components

$$dx = dh \cos \Delta$$

and

$$dy = dh \sin \Delta$$

If we now plot the vector sum of these contributions the total disturbance at P from that section of the wavefront measured from O to a height h will have the component values $x = \int dx$ and $y = \int dy$. These integrals are usually expressed in terms of the dimensionless variable $u = h(2/\lambda l)^{1/2}$ the physical significance of which we shall see shortly.

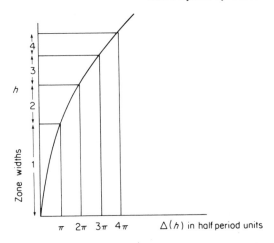

Fig. 10.45. Variation of the width of each half period zone with height h above the straight edge

We then have $\Delta = \pi u^2/2$ and $dh = (\lambda l/2)^{1/2} du$ and the integrals become

$$x = \int dx = \int_0^u \cos (\pi u^2/2)\, du$$

and

$$y = \int dy = \int_0^u \sin (\pi u^2/2)\, du$$

These integrals are called Fresnels Integrals and the locus of the coordinates x and y with variation of u (that is of h) is the spiral in the first quadrant of fig. 10.42. The complete figure is known as Cornu's spiral.

As h, the width of the contributing wavefront above the straight edge, increases, we measure the increasing length u from 0 along the curve of the spiral in the first quadrant until, as h and $u \to \infty$ we reach Z_1 the centre of the spiral eye with coordinate $x = \frac{1}{2}$, $y = \frac{1}{2}$.

The tangent to the spiral curve is

$$\frac{dy}{dx} = \tan \frac{\pi u^2}{2}$$

and this is zero when the phase

$$\Delta(h) = \pi h^2/\lambda l = \pi u^2/2 = m\pi$$

where m is an integer so that $u = \sqrt{(2m)}$ relates u, the distance measured along the spiral to m the number of half period zones contributing to the disturbance at P. The total intensity at P due to all the half period zones above the straight edge is given by the square of the length of the 'chord' OZ_1. This is the intensity at the geometric shadow.

Suppose now that we keep P fixed as we slowly withdraw the screen vertically downwards from O. This begins to reveal contributions to P from the lower part of the wavefront that is, the part which contributes to the Cornu spiral in the third quadrant. The length u now includes not only the whole of the upper spiral arm but an increasing part of the lower spiral until, when u has extended to Z_2 the 'chord' Z_1Z_2 has its maximum value and this corresponds to the fringe of maximum intensity nearest the straight edge. Further withdrawal of the screen lengthens u to the position Z_3 which corresponds to the first minimum of the fringe pattern and the convolutions of an increasing length u around the spiral eye will produce further intensity oscillations of decreasing magnitude until, with the final removal of the screen, u is now the total length of the spiral and the square of the 'chord' length Z_1Z_1' gives the undisturbed intensity of unit value. But $Z_1Z_1' = 2Z_1O$ so that the undisturbed intensity $(Z_1Z_1')^2$ is a factor of four greater than $(Z_1O)^2$ the intensity at the geometric shadow.

The Fresnel diffraction pattern from a slit may now be seen as that due to a fixed height h of the wavefront equal to that of the slit width. This defines a fixed length u of the spiral between the end points of which the 'chord' is drawn and its length measured and squared to give the intensity. At a given distance from the slit the intensity at a point P in the diffraction pattern will correlate with the precise location of the fixed length u along the spiral. At the centre of the pattern P is symmetric with respect to the upper and lower edges of the slit and the fixed length u is centred about O fig. 10.46. As P moves across the pattern towards the geometric shadow the length u moves around the convolutions of the spiral. In the geometric shadow this length is located entirely within the first or third quadrant of the spiral and the magnitude of the 'chord' between its ends is less than OZ_1. When the slit is wide enough to produce the central minimum of the diffraction pattern in fig. 10.47 the length u is centred at O with its ends at Z_3 and Z_4 in fig. 10.46.

Circular aperture (Fresnel diffraction)

In this case the half period zones become annuli of decreasing width. If r_n is the mean radius of the half period zone whose phase lag is $n\pi$ with respect to the contribution from the central ring the path difference in fig. 10.48 is given by

$$NP - OP = \Delta = n\lambda/2 = \tfrac{1}{2}r_n^2/l$$

Unlike the rectangular example of the straight edge where the area of the half period zone was proportional to its width dh each zone here has the same area equal to $\pi\lambda l$.

Each zone thus contributes equally to the disturbance at P except for a factor arising from the rigorous Kirchhoff theory which, until now, we have

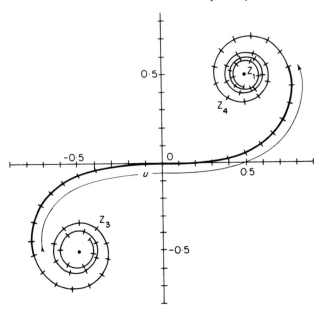

Fig. 10.46. The slit width h defines a fixed length u of the spiral. The intensity at a point P in the diffraction pattern is correlated with the precise location of u on the spiral. When P is at the centre of the pattern u is centred on O and moves along the spiral as P moves towards the geometric shadow. Within the geometric shadow the chord joining the ends of u is less than OZ_1

been able to ignore. This is the so called obliquity factor $\cos \chi$ where χ is shown in the figure. This factor is negligible for small values of n but its effect is to reduce a zone contribution as n increases. A large circular aperture with many zones produces, in the limit, an undisturbed normal intensity on the axis and from fig. 10.49 where we show the magnitude and phase from successive half zones we see that the sum of these vectors which represents the amplitude produced by an undisturbed wave is only half of that from the innermost zone.

It is evident that if alternate zones transmit no light then the contributions from the remaining zones would all be in phase and combine to produce a high intensity at P similar to the focusing effect of a lens. Such circular 'zone plates' are made by blacking out the appropriate areas of a glass slide, fig. 10.50. A further refinement increases the intensity still more. If the alternate zone areas are not blacked out but become areas where the *optical* thickness of the glass is reduced, via etching, by $\lambda/2$ the light transmitted through these zones is advanced in phase by π radians so that the contributions from all the zones are now in phase.

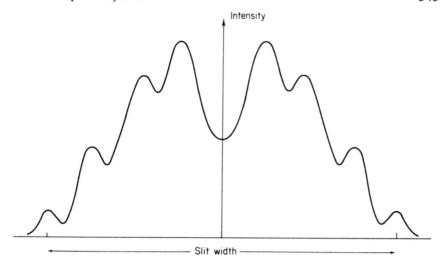

Intensity

Slit width

Fig. 10.47. Fresnel diffraction pattern from a slit which is wide enough for the spiral length u to be centred at O and to end on points Z_3 and Z_4 of fig. 10.46. This produces the intensity minimum at the centre of the pattern

Holography

Why is it that when we observe an object we see it in three dimensions but when we photograph it we obtain only a flat two dimensional distribution of light intensity? The answer is that the photograph has lost the information contained in the *phase* of the incident light. Holographic processes retain this information and a hologram reconstructs a three dimensional image.

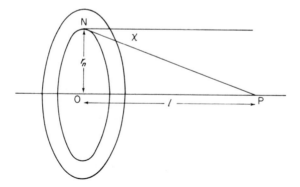

Fig. 10.48. Fresnel diffraction from a circular aperture. The mean radius r_n defines the half period zone with a phase lag of $n\pi$ at P with respect to the contribution from the central zone. The obliquity angle χ which reduces the zone contribution at P increases with n

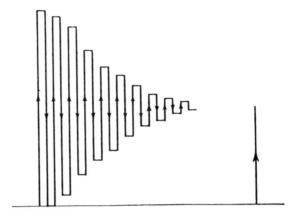

Fig. 10.49. The vector contributions from successive zones in the circular aperture. The amplitude produced by an undisturbed wave is seen to be only half of that from the central zone. Removing the contributions from alternate zones leaves the remainder in phase and produces a very high intensity. This is the principle of the zone plate of fig. 10.50

Fig. 10.50. Zone plate produced by removing alternate half zones from a circular aperture to leave the remaining contributions in phase

The principle of holography was proposed by Gabor in 1948 but its full development needed the intense beams of laser light. A hologram requires two coherent beams and the holographic plate records their interference pattern. In practice both beams derive from the same source, one serves as a direct reference beam the other is the wavefront scattered from the object.

Fig. 10.51 shows one possible arrangement where a partly silvered beam splitter passes the direct reference beam and reflects light on to the object which scatters it on to the photographic plate. Mirrors or deviating prisms are also used to split the incident beam.

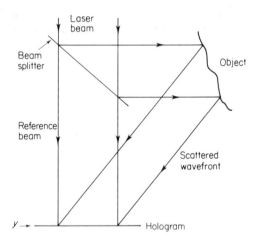

Fig. 10.51. The hologram records the interference between two parts of the same laser beam. The original beam is divided by the partially silvered beam splitter to form a direct reference beam and a wavefront scattered from the object. The amplitude and phase information contained in the scattered wavefront must be preserved and recovered

In fig. 10.51 let the reference beam amplitude be $A_0 e^{i\omega t}$. If the holographic plate lies in the yz plane both the amplitude and phase of scattered light which strikes a given point (y, z) on the plate will depend on these co-ordinates. We simplify the analysis by considering only the y co-ordinate shown in the plane of the paper and we represent the scattered light in amplitude and phase as a function of y, namely

$$A(y) e^{i(\omega t + \phi(y))}$$

It is this information we shall wish to recover.

We may now write the resulting amplitude at y (after removing the common $e^{i\omega t}$ factor) as

$$A = A_0 + A(y)\, e^{i\phi(y)}$$

The intensity is therefore

$$I = AA^* = [A_0 + A(y)\, e^{i\phi(y)}][A_0 + A(y)\, e^{-i\phi(y)}]$$

$$= A_0^2 + A(y)^2 + A_0 A(y)[e^{i\phi(y)} + e^{-i\phi(y)}]$$

The holographic plate records this intensity as shown in fig. 10.52 where the reference intensity A_0^2 is modulated by the terms which contain $A(y)$ and $\phi(y)$, the original scattered amplitude and phase information. This modulation shows of course as contrasting interference fringes whose local intensity is

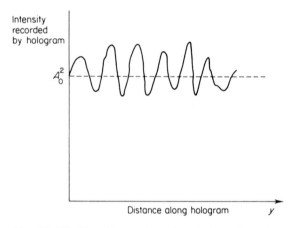

Fig. 10.52. Total intensity recorded as a function of y by the holographic plate. The direct reference beam intensity A_0^2 is modulated by information from the scattered wavefront. This shows as variations in the intensity of an interference fringe pattern

governed by the amplitude $A(y)$ and whose distribution along the y axis is determined by the phase $\phi(y)$. The wavefront scattered by the object is now reconstructed to form the holographic image. This is done by shining the reference beam through the processed hologram which acts as a diffraction grating. The greater the recorded intensity the lower the transmitted amplitude. If the developed photographic emulsion possessed idealized characteristics the relation between the transmitted amplitude of the reference beam and the *exposure* would be linear. Exposure defines the product of incident intensity and exposure time. The curve relating the characteristics for a real holographic emulsion is shown in fig. 10.53 and this is linear only over a limited range near the centre indicated by the dotted lines. This imposes several conditions on practical holography.

In the first place the exposure must be correctly chosen at the value E_C. Secondly, the value of the reference beam intensity A_0^2 must be chosen to produce the correct transmitted amplitude T_0 on the vertical axis of fig. 10.53.

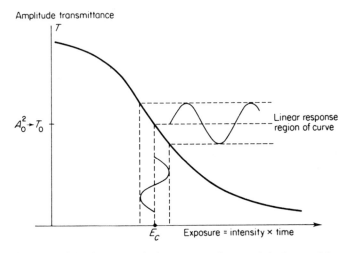

Fig. 10.53. Characteristic curve of a real holographic emulsion (transmittance versus exposure). Only the central linear section of the curve is used. The transmittance T_0 (governed by the reference beam intensity A_0^2) is chosen with the critical exposure E_c to produce the central point on the linear part of the curve. Information from the scattered wavefront must keep the modulations within the linear range for faithful reproduction free from distortion

This value of T_0 is at the centre of the linear range. Finally the modulation of A_0^2 by the scattered intensity $A(y)^2$ in fig. 10.53 must be small enough for the transmission of the modulated signal to remain within the linear range of the characteristic curve. Excursions outside this range introduce nonlinear distortions by generating extra Fourier frequency components (the situation is similar to that for characteristic curves in electronic amplifiers).

Experimentally this final restriction requires $A(y) \ll A_0$.

Shining the reference beam through the processed hologram produces a transmitted *amplitude*

$$A_0 T = A_0^3 + A_0^2 A(y)\, e^{i\phi(y)} + A_0^2 A(y)\, e^{-i\phi(y)}$$
$$= A_0^2 [A_0 + A(y)\, e^{i\phi(y)} + A(y)\, e^{-i\phi(y)}]$$

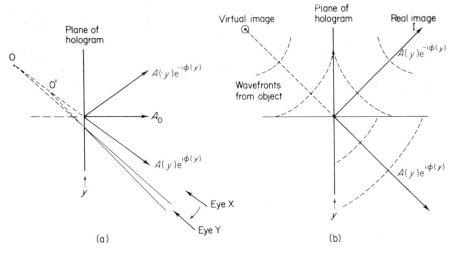

Fig. 10.54. (a) Shining the reference beam through the processed hologram produces three components A_0, $A(y)\,e^{i\phi(y)}$ and $A(y)\,e^{-i\phi(y)}$ in the directions shown. Movement of the eye from X to Y about the component $A(y)\,e^{i\phi(y)}$ resolves the separate points O and O' on the image of the object to reveal its three dimensional nature. (b) In (b) this image at O is seen to be virtual while the image associated with the component $A(y)\,e^{-i\phi(y)}$ is real. This real image is 'phase reversed' and the object appears 'inside out'

where we have neglected the $A(y)^2$ term as $\ll A_0^2$ and have written the negative and positive exponential terms separately. This has a profound physical significance for we see that apart from the common constant factor A_0^2, the observed transmitted beam has three components A_0, $A(y)\,e^{i\phi(y)}$ and $A(y)\,e^{-i\phi(y)}$.

The first term, A_0, shows that the incident reference beam has continued beyond the hologram to form the central beam of fig. 10.54(a). The second component $A(y)\,e^{i\phi(y)}$ has the same form in amplitude and phase as the original wavefront scattered from the object. As shown in fig. 10.54(b) it is seen to be a wavefront diverging from a virtual image of the object having the same size and three dimensional distribution as the object itself. Moving the eye across this beam in 10.54(a) exposes a different section OO' of the virtual image to produce a three dimensional effect.

The third component of the transmitted beam is identical with the second except for the phase reversal; it has a negative exponential index. It forms another image, in this case a real image often referred to as 'pseudoscopic'. It is an image of the original object turned inside out. All contours are reversed, bumps become dents and the closest point on the original object when viewed directly by the observer now becomes the most distant.

Problem 10.1

Apply the principle of page 284 to a thin bi-convex lens of refractive index n to show that its focal length is

$$(n-1)\left(\frac{1}{R_1}-\frac{1}{R_2}\right)$$

where R_1 and R_2, the radii of curvature of the convex faces, are both much greater than the thickness of the lens.

Problem 10.2

A plane parallel plate of glass of thickness d has a non-uniform refractive index n given by $n = n_0 - \alpha r^2$ where n_0 and α are constants and r is the distance from a certain line perpendicular to the sides of the plate. Show that this plate behaves as a converging lens of focal length $1/2\alpha d$.

Problem 10.3

For oscillatory waves the focal point F of the converging wavefront of fig. 10.55 is located where Huygens secondary waves all arrive in phase: the point F' vertically above F receives waves whose total phase range $\Delta\phi$ depends on the path difference AF' − BF'. When F' is such that $\Delta\phi$ is 2π the resultant amplitude tends to zero. Thus

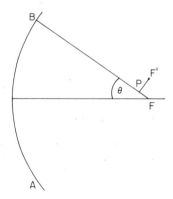

Fig. 10.55

the focus is not a point but a region whose width x depends on the wavelength λ and the angle θ subtended by the spherical wave. If PF' is perpendicular to BF the phase at F' and P may be considered the same. Show that the width of the focal spot is given by $x = \lambda/\sin\theta$. Note that $\sin\theta$ is directly related to the f/d ratio for a lens (focal length/diameter) so that x defines the minimum size of the image for a given wavelength and a given lens.

Problem 10.4

Figure 10.56 shows schematically two thin lenses of powers P_1 and P_2 separated by a distance t in air. Use the arguments associated with fig. 10.11 to show that the combined power P of the lenses is given by $P = P_1 + P_2 - P_1 P_2 t$.

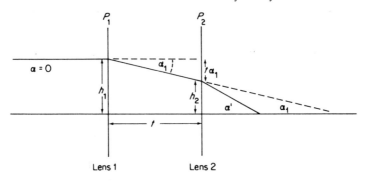

Fig. 10.56

Problem 10.5

The powers P_1 and P_2 of problem 10.4 now represent the powers of the refractive surfaces of a lens of thickness t and refractive index n. Repeat the arguments of problem 10.4 to show that the power of a thick lens is given by

$$P = P_1 + P_2 - P_1 P_2 t/n$$

Problem 10.6

In fig. 10.14 trace the ray $L_0 F_0$ to L_i and use the method of similar triangles to prove Newton's formula

$$xx' = ff'$$

Problem 10.7

As an object moves closer to the eye its apparent size grows with the increasing angle it subtends at the eye. A healthy eye can accommodate (that is, focus) objects from infinity to about 25 cm. the closest 'distance of distinct vision'. Beyond this 'near point' the eye can no longer focus and a magnifying glass is required. A healthy eye has a range of accommodation of 4 dioptres ($1/\infty$ to $1/0{\cdot}25$ metres). If a man's near point is 40 cm from his eye show that he needs spectacles of power equal to $1{\cdot}5$ dioptres. If another man is unable to focus at distances greater than 2 m. show that he needs diverging spectacles with a power of $-0{\cdot}5$ dioptres.

Problem 10.8

Fig. 10.57 shows a magnifying glass of power P with an erect and virtual image at l'. The angular magnification

$$M_\alpha = \beta/\gamma$$

$$= \frac{\text{angular size of image seen through the glass at distance } l'}{\text{angular size of object seen by the unaided eye at } d_o}$$

where d_o is the distance of distinct vision. Show that the transverse magnification $M_T = l'/l$ where l is the actual distance (not d_o) at which the object O is held. Hence show that $M_\alpha = d_o/l$ and use the thin lens power equation, page 292, to show that

$$M_\alpha = d_o(P + 1/l') = P d_0 + 1$$

when $l' = d_0$. Note that M_α reduces to the value $P d_0$ when the eye relaxes by viewing the image at ∞.

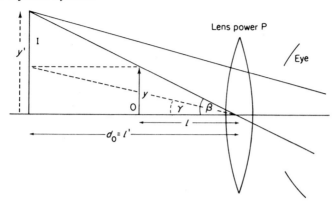

Fig. 10.57

Problem 10.9

A telescope resolves details of a distant object by accepting plane wavefronts from individual points on the object and amplifying the very small angles which separate them. In fig. 10.58, α is the angle between two such wavefronts one of which propagates along the optical axis. In normal adjustment the astronomical telescope has both object and image at ∞ so that the total power of the system is zero. Use the result of problem 10.4 to show that the separation of the lenses must be $f_o + f_e$ where f_o and f_e are respectively the focal lengths of the object and eye lenses.

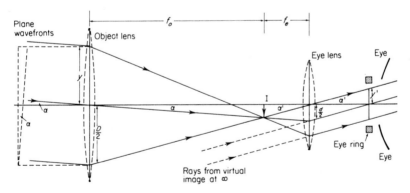

Fig. 10.58

If $2y$ is the width of the wavefront at the objective and $2y'$ is the width of the wavefront at the eye ring show that

$$M_\alpha = \left|\frac{\alpha'}{\alpha}\right| = \left|\frac{f_o}{f_e}\right| = \frac{D}{d}$$

where D is the effective diameter of the object lens and d is the effective diameter of the eye lens. Note that the image is inverted.

Problem 10.10

The two lens microscope system of fig. 10.59 has a short focus objective lens of power P_o and a magnifying glass eyepiece of power P_e. The image is formed at the near point of the eye (the distance d_o of problems 10.7 and 10.8). Show that the magnification by the object lens is $M_o = -P_o x'$ where the minus sign shows that the image is inverted. Hence use the expression for the magnifying glass in problem 10.8 to show that the total magnification is

$$M = M_o M_e = -P_o P_e d_o x'$$

The length x' is called the optical tube length and is standardized for many microscopes at 0.14 m.

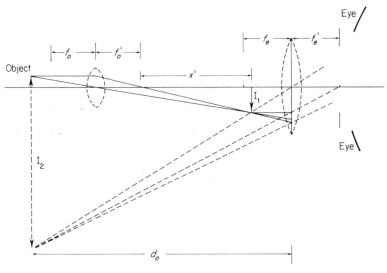

Fig. 10.59

Problem 10.11

Microscope objectives are complex systems of more than one lens but the principle of the oil immersion objective is illustrated by the following problem. In fig. 10.60 the object O is embedded a distance R/n from the centre C of a glass sphere of radius

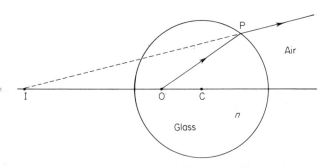

Fig. 10.60

R and refractive index n. Any ray OP entering the microscope is refracted at the surface of the sphere and, when projected back, will always meet the axis CO at the point I. Use Snell's Law to show that the distance $IC = nR$.

Problem 10.12

Two identical radio masts transmit at a frequency of $1500 \, \text{kc/s}$ and are 400 metres apart. Show that the intensity of the interference pattern between these radiators is given by $I = 2I_0[1 + \cos{(4\pi \sin \theta)}]$, where I_0 is the radiated intensity of each. Plot this intensity distribution on a polar diagram in which the masts lie on the 90°–270° axis to show that there are two major cones of radiation in opposite directions along this axis and 6 minor cones at 0°, 30°, 150°, 180°, 210° and 330°.

Problem 10.13

(a) Two equal sources radiate a wavelength λ and are separated a distance $\lambda/2$. There is a phase difference $\delta_0 = \pi$ between the signals at source. If the intensity of each source is I_s, show that the intensity of the radiation pattern is given by

$$I = 4I_s \sin^2\left(\frac{\pi}{2}\sin \theta\right)$$

where the sources lie on the axis $\pm \pi/2$.
   Plot I versus θ.
(b) If the sources in (a) are now $\lambda/4$ apart and $\delta_0 = \pi/2$ show that

$$I = 4I_s\left[\cos^2\frac{\pi}{4}(1 + \sin \theta)\right]$$

Plot I versus θ.

Problem 10.14

(a) A large number of identical radiators is arranged in rows and columns to form a lattice of which the unit cell is a square of side d. Show that all the radiation from the lattice in the direction θ will be in phase at a large distance if $\tan \theta = m/n$, where m and n are integers.
(b) If the lattice of section (a) consists of atoms in a crystal where the rows are parallel to the crystal face, show that radiation of wavelength λ incident on the crystal face at a grazing angle of θ is scattered to give interference maxima when $2d \sin \theta = n\lambda$ (Bragg reflexion).

Problem 10.15

Michelson's Stellar Interferometer, shown schematically in the diagram, allows monochromatic light from two separate stars 1 and 2 to form double slit interference fringes on the screen P P'. If the angles are small show that the phase difference between the two beams from star 1 is

$$\delta_1 = \frac{2\pi}{\lambda}f(\alpha_1 + \theta)$$

Find a similar expression for δ_2 (star 2) and show that the condition for the fringe system formed by star 1 to be displaced half a fringe from that formed by star 2 is

$$\alpha = \alpha_2 - \alpha_1 = \frac{\lambda}{2f}$$

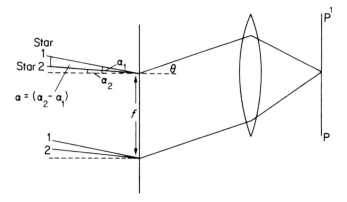

where α is the angular separation between the stars. Such a fringe displacement destroys the intensity contrast of the fringe system and experimentally f is varied to find its minimum value at which the contrast vanishes in order to measure α.

Problem 10.16

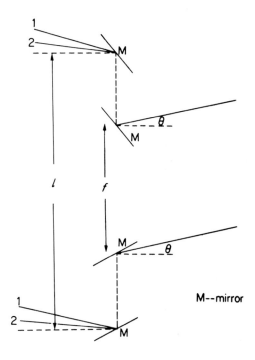

The interferometer of problem 10.15 is modified by the mirror system shown in the diagram where l and f are variable. Show that the resolving power of this arrangement is a factor l/f greater than that of problem 10.15.

Problem 10.17

Show that the separation of equal sources in a linear array producing a principal maximum along the line of the sources ($\theta = \pm\pi/2$) is equal to the wavelength being radiated. Such a pattern is called 'end fire'. Determine the positions (values of θ) of the secondary maxima for $N = 4$ and plot the angular distribution of the intensity.

Problem 10.18

The first multiple radio astronomical interferometer was equivalent to a linear array of $N = 32$ sources (receivers) with a separation $f = 7$ metres working at a wavelength $\lambda = 0 \cdot 21$ metre. Show that the angular width of the central maximum is 6 minutes of arc and that the angular separation between successive principal maxima is $1° 42'$.

Problem 10.19

Suppose that Newton's Rings are formed at a wavelength λ by the system in fig. 10.25 except that the planoconvex lens now rests centrally in a concave surface of radius of ture R_1 and not on an optical flat. Show that the radius r_u of the nth dark ring is given by

$$ r_n^2 = \frac{R_1 R_2 n\lambda}{(R_1 - R_2)} $$

where R_2 is the radius of curvature of the convex surface of the lens and $R_1 > R_2$. (Note R_1 and R_2 have the same sign.)

Problem 10.20

Monochromatic light is normally incident on a single slit, and the intensity of the diffracted light at an angle θ is represented in magnitude and direction by a vector $\mathbf{I}$, the tip of which traces a polar diagram. Sketch several polar diagrams to show that as the ratio of slit width to the wavelength gradually increases the polar diagram becomes concentrated along the direction $\theta = 0$.

Problem 10.21

The condition for the maxima of the intensity of light of wavelength λ diffracted by a single slit of width d is given by $\alpha = \tan \alpha$, where $\alpha = \pi d \sin \theta/\lambda$. The approximate values of α which satisfy this equation are $\alpha = 0$, $+3\pi/2$, $+5\pi/2$, etc. Writing $\alpha = 3\pi/2 - \delta$, $5\pi/2 - \delta$, etc., where δ is small, show that the real solutions for α are $\alpha = 0$, $\pm 1 \cdot 43\pi$, $\pm 2 \cdot 459\pi$, $\pm 3 \cdot 471\pi$, etc.

Problem 10.22

Prove that the intensity of the secondary maximum for a grating of three slits is $\frac{1}{9}$ of that of the principal maximum.

Problem 10.23

A diffraction grating has N slits and a grating space f. If $\beta = \pi f \sin \theta/\lambda$, where θ is the angle of diffraction, calculate the phase change $d\beta$ required to move the diffracted light from the principal maximum to the first minimum to show that the half width of the spectral line produced by the grating is given by $d\theta = (nN \cot \theta)^{-1}$, where n is the spectral order. (For $N = 14,000$, $n = 1$ and $\theta = 19°$, $d\theta \approx 5$ seconds of arc.)

Problem 10.24

(a) Dispersion is the separation of spectral lines of different wavelengths by a diffraction grating and increases with the spectral order n. Show that the dispersion of the lines when projected by a lens of focal length F on a screen is given by

$$\frac{dl}{d\lambda} = F\frac{d\theta}{d\lambda} = \frac{nF}{f}$$

for a diffraction angle θ and the nth order, where l is the linear spacing on the screen and f is the grating space.

(b) Show that the change in linear separation per unit increase in spectral order for two wavelengths $\lambda = 5 \times 10^{-7}$ metres and $\lambda_2 = 5 \cdot 2 \times 10^{-7}$ metres in a system where $F = 2$ metres and $f = 2 \times 10^{-6}$ metres is 2×10^{-2} metres.

Problem 10.25

(a) A sodium doublet consists of two wavelengths $\lambda_1 = 5\cdot 890 \times 10^{-7}$ metres and $\lambda_2 = 5\cdot 896 \times 10^{-7}$ metres. Show that the minimum number of lines a grating must have to resolve this doublet in the third spectral order is ≈ 328.

(b) A red spectral line of wavelength $\lambda = 6\cdot 5 \times 10^{-7}$ metres is observed to be a close doublet. If the two lines are just resolved by a grating of 9×10^4 lines show that the doublet separation is 2×10^{-12} metres.

Problem 10.26

Optical instruments have circular apertures, so that the Rayleigh criterion for resolution is given by $\sin \theta = 1 \cdot 22\lambda/a$, where a is the diameter of the aperture.

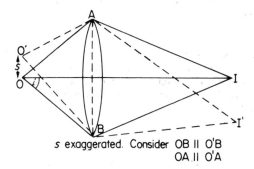

s exaggerated. Consider OB ‖ O'B
OA ‖ O'A

Two points O and O' of a specimen in the object plane of a microscope are separated by a distance s. The angle subtended by each at the objective aperture is $2i$ and their images I and I' are just resolved. By considering the path difference between O'A and O'B show that the separation $s = 1\cdot 22\lambda/2 \sin i$.

Summary of Important Results

Power of a thin lens

$$P = (n-1)\left(\frac{1}{R_1} - \frac{1}{R_2}\right) = \frac{1}{f}$$

where n is the refractive index of the lens material, R_1 and R_2 are the radii of curvature of the lens surfaces and f is the focal length.

Optical Helmholtz Equation

$$ny \tan \alpha = \text{constant}$$

where n is the refractive index of a medium, y is the width of a wavefront section and α is the angle between the optical axis and the normal to the wavefront section.

This equation becomes

$$ny\alpha = \text{constant for paraxial rays.}$$

Interference: division of wavefront (two equal sources)

Intensity

$$I = 4I_s \cos^2 \delta/2$$

where

$$I_s = \text{source intensity}$$

and

$$\delta = \left[\frac{2\pi}{\lambda}(\text{path difference}) \right] \text{ is phase difference}$$

Interference (N equal sources-separation f)

$$I = I_s \frac{\sin^2 N\beta}{\sin^2 \beta} \quad \text{where } \beta = \frac{\pi}{\lambda}f \sin \theta$$

Principal maxima

$$I = N^2 I_s \quad \text{at } f \sin \theta = n\lambda$$

Fraunhofer diffraction (single slit—width d)

Intensity

$$I = I_0 \frac{\sin^2 \alpha}{\alpha^2} \quad \text{where } \alpha = \frac{\pi}{\lambda}d \sin \theta$$

Intensity distribution from N slits (width d-separation f)

$$I = I_0 \frac{\sin^2 \alpha}{\alpha^2} \frac{\sin^2 N\beta}{\sin^2 \beta}$$

(interference pattern modified by single slit diffraction envelope).

Resolving power of transmission grating

$$\frac{\lambda}{d\lambda} = nN$$

where n is spectral order and N is number of grating lines:
Expressible in terms of Bandwidth Theorem as

$$\Delta\nu \, \Delta t = 1$$

where $\Delta\nu$ is resolvable frequency difference and Δt is the time difference between extreme optical paths.

Resolving power

$$\frac{\lambda}{\Delta\lambda} = \left|\frac{\nu}{\Delta\nu}\right| = \frac{\omega}{\Delta\omega} = Q$$

where Q is the quality factor of the system.

Chapter 11

Non-linear Oscillations

The oscillations discussed in this book so far have all been restricted in amplitude to those which satisfy the equation of motion where the restoring force is a linear function of the displacement. This restriction was emphasized in Chapter 1 and from time to time its limiting influence has required further discussion, for example in Chapter 5 on acoustic waves in a fluid. We now discuss some of the consequences when this restriction is lifted.

Free Vibrations of an Anharmonic Oscillator—Large Amplitude Motion of a Simple Pendulum

In fig. 1.1 the equation of motion of the simple pendulum was written in terms of its angular displacement as

$$\frac{d^2\theta}{dt^2} + \omega_0^2\theta = 0$$

where $\omega_0^2 = g/l$. Here an approximation was made by writing θ for $\sin\theta$; the equation is valid for oscillation amplitudes within this limit. When $\theta \geqslant 7°$ however, this validity is lost and we must consider the more complicated equation

$$\frac{d^2\theta}{dt^2} + \omega_0^2 \sin\theta = 0$$

Multiplying this equation by $2d\theta/dt$ and integrating with respect to t gives $(d\theta/dt)^2 = 2\omega_0^2 \cos\theta + A$, where A is the constant of integration. The velocity $d\theta/dt$ is zero at the maximum angular displacement $\theta = \theta_0$, giving $A = -2\omega_0^2 \cos\theta_0$ so that

$$\frac{d\theta}{dt} = \omega_0[2(\cos\theta - \cos\theta_0)]^{1/2}$$

or, upon integrating,

$$\omega_0 t = \int \frac{d\theta}{\{2[\cos\theta - \cos\theta_0]\}^{1/2}}$$

If $\theta = 0$ at time $t = 0$ and T is the new period of oscillation, then $\theta = \theta_0$ at $t = T/4$, and using half-angles we obtain

$$\omega_0 \frac{T}{4} = \int_0^{\theta_0} \frac{d\theta}{2[\sin^2 \theta_0/2 - \sin^2 \theta/2]^{1/2}}$$

If we now express θ as a fraction of θ_0 by writing $\sin \theta/2 = \sin (\theta_0/2) \sin \phi$, where, of course, $-1 < \sin \phi < 1$, we have

$$\tfrac{1}{2} (\cos \theta/2)\delta\theta = (\sin \theta_0/2) \cos \phi \, \delta\phi$$

giving

$$\frac{\pi}{2} \frac{T}{T_0} = \int_0^{\pi/2} \frac{d\phi}{[1 - (\sin^2 \theta_0/2) \sin^2 \phi]^{1/2}}$$

where $T_0 = 2\pi/\omega_0$.

Expansion and integration gives

$$T = T_0(1 + \tfrac{1}{4} \sin^2 \theta_0/2 + \tfrac{9}{64} \sin^4 \theta_0/2) + \ldots$$

or approximately

$$T = T_0(1 + \tfrac{1}{4} \sin^2 \theta_0/2)$$

(Problem 11.1)

Forced Oscillations—Non-linear Restoring Force

When an oscillating force is driving an undamped oscillator the equation of motion for such a system is given by

$$m\ddot{x} + s(x) = F_0 \cos \omega t$$

where $s(x)$ is a non-linear function of x, which may be expressed in polynomial form:

$$s(x) = s_1 x + s_2 x^2 + s_3 x^3 \ldots$$

where the coefficients are constant. In many practical examples $s(x) = s_1 x + s_3 x^3$, where the cubic term ensures that the restoring force $s(x)$ has the same value for positive and negative displacements, so that the vibrations are symmetric about $x = 0$. When s_1 and s_3 are both positive the restoring force for a given displacement is greater than in the linear case and, if supplied by a spring, this case defines the spring as 'hard'. If s_3 is negative the restoring force is less than in the linear case and the spring is 'soft'. In fig. 11.1 the variation of restoring force is shown with displacement for s_3 zero (linear), s_3 positive (hard) and s_3 negative (soft). We see therefore that the large amplitude vibrations of the pendulum of the previous section are soft-spring controlled because

$$\sin \theta \approx \theta - \tfrac{1}{3}\theta^3$$

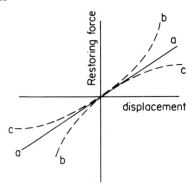

Fig. 11.1. Oscillator displacement versus restoring force for (a) linear restoring force, (b) non-linear 'hard' spring, and (c) non-linear 'soft' spring

Fig. 11.2 shows a mass m attached to points D and D', a vertical distance $2a$ apart, by two light elastic strings of constant stiffness s and subjected to a horizontal driving force $F_0 \cos \omega t$. At zero displacement the tension in the strings is T_0 and at a displacement x (not limited in value) the tension is $T = T_0 + s(L - a)$ where L is the stretched string length.

The equation of motion (neglecting gravity) is

$$m\ddot{x} = -2T \sin \theta + F_0 \cos \omega t$$

$$= -2[T_0 + s(L - a)]\frac{x}{L} + F_0 \cos \omega t$$

Inserting the value

$$L = a\left[1 + \left(\frac{x}{a}\right)^2\right]^{1/2}$$

and expanding this expression in powers of x/a, we obtain by neglecting terms smaller than $(x/a)^3$

$$m\ddot{x} = -\frac{2T_0}{a}x - \frac{(sa - T_0)}{a^3}x^3 + F_0 \cos \omega t$$

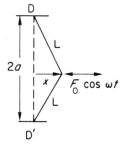

Fig. 11.2. A mass m supported by elastic strings between two points D and D' vertically separated by a distance $2a$ and subjected to a lateral force $F_0 \cos \omega t$

which we may write

$$\ddot{x} + s_1 x + s_3 x^3 = \frac{F_0}{m} \cos \omega t$$

where

$$s_1 = \frac{2T_0}{ma} \quad \text{and} \quad s_3 = \frac{sa - T_0}{ma^3}$$

If s_3 is small we assume (as a first approximation) the solution $x_1 = A \cos \omega t$, which yields from the equation of motion

$$\ddot{x}_1 = -s_1 A \cos \omega t - s_3 A^3 \cos^3 \omega t + \frac{F_0}{m} \cos \omega t$$

Since $\cos^3 \omega t = \frac{3}{4} \cos \omega t + \frac{1}{4} \cos 3\omega t$, this becomes

$$\ddot{x}_1 = -(s_1 A + \tfrac{3}{4} s_3 A^3 - F_0/m) \cos \omega t - \tfrac{1}{4} s_3 A^3 \cos 3\omega t$$

Integrating twice, where the constants become zero from initial boundary conditions, gives as a second approximation to the equation

$$\ddot{x} + s_1 x + s_3 x^3 = \frac{F_0}{m} \cos \omega t$$

the solution

$$x_2 = \frac{1}{\omega^2}\left(s_1 A + \frac{3}{4} s_3 A^3 - \frac{F_0}{m} \right) \cos \omega t + \frac{s_3 A^3}{36\omega^2} \cos 3\omega t$$

Thus for s_3 small we have a value of ω appropriate to a given amplitude A, and we can plot a graph of amplitude versus driving frequency. Note that we have a third harmonic. We see that for a system with a non-linear restoring force resonance does not exist in the same way as in the linear case. In the example above, even when no damping is present, the amplitude will not increase

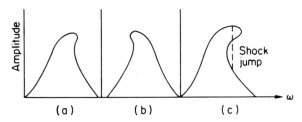

Fig. 11.3. Response curves of amplitude versus frequency for oscillators having (a) a 'hard' spring restoring force, and (b) a 'soft' spring restoring force. In the extreme case (c) the tilt of the maximum is sufficient to allow multi-valued amplitudes at a given frequency and 'shock jumps' may occur. (See fig. 11.6 for comparable behaviour in a high amplitude sound wave.)

without limit for a driving force of a given frequency, for if ω is the natural frequency at low amplitude it is no longer the natural frequency at high amplitude. For s_3 positive (hard spring) the natural frequency increases with increasing amplitude and the amplitude versus frequency curve has a tilted maximum (fig. 11.3a). For a soft spring, s_3 is negative and the behaviour follows fig. 11.3b. It is possible for the tilt to become so pronounced (fig. 11.3c) that the amplitude is not single valued for a given ω and shock jumps in amplitude may occur at a given frequency (see the later section on the development of a shock front in a high amplitude acoustic wave).

(Problems 11.2, 11.3)

Thermal Expansion of a Crystal

Chapter 1 showed that the curve of potential energy versus displacement for a linear oscillator was parabolic. Small departures from this curve are consistent with anharmonic oscillations. Consider the potential energy curve for a pair of neighbouring ions of opposite charge $\pm e$ in a crystal lattice such as that of KCl. If r is the separation of the ions the mutual potential energy is given by

$$V(r) = -\frac{\alpha e^2}{r} + \frac{\beta}{r^p}$$

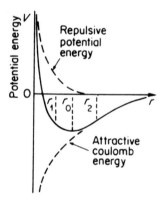

Fig. 11.4. Non-parabolic curve of mutual potential energy between oppositely charged ions in the lattice of an ionic crystal (NaCl or KCl). The combination of repulsive and attractive forces yields an equilibrium separation r_0. Very small energy increments give harmonic motion about r_0 but oscillations at higher energies are anharmonic, leading to thermal expansion of the crystal

where α and β are positive constants and $p = 9$. This is plotted in fig. 11.4, which shows that the potential energy curve is no longer parabolic. The first term of $V(r)$ is the energy due to Coulomb attraction; the second is that of a repulsive force. The value of α depends upon the presence of neighbouring ions and is about $0 \cdot 3$. The constant β can be found in terms of α and the equilibrium separation r_0 because, in equilibrium,

$$\left(\frac{dV}{dr}\right)_{r=r_0} = \frac{\alpha e^2}{r_0^2} - \frac{p\beta}{r_0^p + 1} = 0$$

giving

$$\beta = \frac{\alpha e^2 r_0^{p-1}}{p}$$

X-ray diffraction from such crystals gives $r_0 = 3 \cdot 12$ Å for KCl, so that β may be found numerically.

To consider small displacements from the equilibrium value r_0 let us expand $V(r)$ about $r = r_0$ in a Taylor series to give

$$V(r) = V(r_0) + x\left(\frac{dV}{dr}\right)_{r_0} + \frac{x^2}{2!}\left(\frac{d^2V}{dr^2}\right)_{r_0} + \frac{x^3}{3!}\left(\frac{d^3V}{dr^3}\right)_{r_0}$$

where $x = r - r_0$. Since $(dV/dr)_{r_0} = 0$, we may write

$$V(r) - V(r_0) = V(x) = A\frac{x^2}{2!} + \frac{Bx^3}{3!}$$

The quantity $Ax^2/2$ is the quadratic term familiar in the linear oscillator, so that for very small disturbances the bottom of the potential energy curve is parabolic, and a small gain in energy causes the ion pair to oscillate symmetrically about $r = r_0$. An increase in the ion pair energy involves the second term $Bx^3/6$, and oscillations are no longer symmetric about r_0, because $|r_2 - r_0| > |r_1 - r_0|$ in fig. 11.4. Hence the time average for $r - r_0$ is not zero as it is for a linear oscillator, and this time average $r_t > r_0$. If all ion pairs acquire this amount of energy, for example by heating, the crystal expands. We may consider the force between the two ions as

$$F = -\frac{dV}{dx} = -Ax - \frac{Bx^2}{2}$$

and note that the quadratic term here is responsible for the lack of symmetry in the motion. If it were a cubic term as in the previous example the symmetry of motion about r_0 would still occur. The coefficient A in the force equation is the force constant in the discussion on crystals in Chapters 4 and 5 and leads directly to Young's modulus. The coefficient B gives information on the coefficient of thermal expansion of the crystal.

(Problems 11.4, 11.5)

Non-linear Effects in Electrical Devices

A feature of the non-linearity in the mechanical devices discussed earlier was the introduction of harmonics of the fundamental frequency of the driving force. It is comparatively simple to avoid these effects of non-linearity in electronic systems by choosing a small linear portion of the operating characteristic and amplifying the response in stages. In an electromechanical device such as a piezoelectric crystal linearity is again achieved by restricting all oscillations to small amplitudes and amplifying the response. In electroacoustic devices such as microphones and loudspeakers the introduction of harmonics

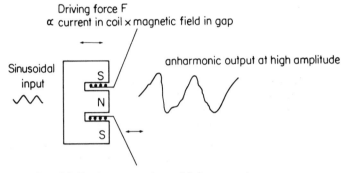

Driving force F
α current in coil $\times$ magnetic field in gap

Sinusoidal input

anharmonic output at high amplitude

Fig. 11.5. A pure sinusoidal wave input to an electro-acoustical device such as a loudspeaker will lead to distorted sound output if the cone suspension has a non-linear stiffness at high amplitudes

often leads to severe distortion. In the loudspeaker of fig. 11.5 even if a pure sinusoidal wave is delivered to the speech coil it is difficult to provide a mechanical suspension for the cone which has a linear response. The cone acts as a piston radiating acoustic power, and limitation of amplitude together with inevitable mismatching of acoustic impedances reduces the efficiency of transforming electrical into acoustic power to less than 10%. Fortunately the ear is a sensitive device.

Non-linear electrical oscillators are, however, often used, and fig. 11.6a shows a 'relaxation oscillator' circuit where a capacitance is discharged very rapidly through a gaseous conductor such as a hydrogen tube. E is the constant charging potential and i is the instantaneous value of the current which charges the capacitor through the resistor R to a potential V_s, the striking potential, at which the gas in the tube is ionized. The tube becomes highly conducting and discharges the capacitance in a negligibly short time to V_e, the extinction potential, at which the tube ceases to conduct. The capacitance charges again to V_s and the cycle is repeated. The variation of voltage across the capacitance with time is shown in fig. 11.6b. Assume that at point A and time t the capacitance has just discharged. If current i_0 is flowing at time $t = 0$ then

$$V_e = E - i_0 R \, e^{-t/RC}$$

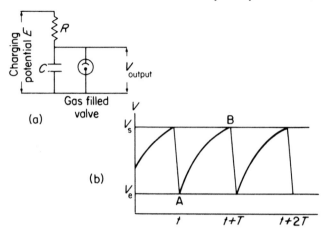

Fig. 11.6. Electrical circuit of a non-linear 'relaxation oscillator'. A capacitance C is charged through a resistance R to a potential $V_s < E$, at which the gas-filled valve strikes and rapidly discharges the condenser to an extinction potential V_e, when the valve ceases to conduct and the cycle is repeated

The capacitance charges to the potential V_s in a time τ so that

$$V_s = E - i_0 R\, e^{-(t+\tau)/RC}$$

giving

$$V_s - V_e = i_0 R(e^{-t/RC} - e^{-(t+\tau)/RC})$$

$$= i_0 R\, e^{-t/RC}[1 - e^{-\tau/RC}]$$

$$= (E - V_e)[1 - e^{-\tau/RC}]$$

giving

$$e^{-\tau/RC} = \frac{E - V_s}{E - V_e}$$

or

$$\tau = RC\left[\log_e\left(\frac{E - V_e}{E - V_s}\right)\right]$$

The period of oscillation is therefore directly proportional to the charging time constant RC.

A more sophisticated circuit produces a linear charging system with a very short discharge time so that the exponential voltage output becomes linear and gives a 'sawtooth' waveform. From Chapter 9 we know that this periodic function contains many harmonics. A sawtooth voltage output applied to the time base of an oscilloscope produces a linear sweep of the spot across the tube.

Non-linear Effects in Acoustic Waves

The linearity of the longitudinal acoustic waves discussed in Chapter 5 required the assumption of a constant bulk modulus

$$B = -\frac{\mathrm{d}P}{\mathrm{d}V/V}.$$

If the amplitude of the sound wave is too large this assumption is no longer valid and the wave propagation assumes a new form. A given mass of gas undergoing an adiabatic change obeys the relation

$$\frac{P}{P_0} = \left(\frac{V_0}{V}\right)^\gamma = \left[\frac{V_0}{V_0(1+\delta)}\right]^\gamma$$

in the notation of Chapter 5, so that

$$\frac{\partial P}{\partial x} = \frac{\partial p}{\partial x} = -\gamma P_0 (1+\delta)^{-(\gamma+1)}\frac{\partial^2 \eta}{\partial x^2}$$

since $\delta = \partial\eta/\partial x$.
Since $(1+\delta)(1+s) = 1$, we may write

$$\frac{\partial p}{\partial x} = -\gamma P_0 (1+s)^{\gamma+1}\frac{\partial^2 \eta}{\partial x^2}$$

and from Newton's second law we have

$$\frac{\partial p}{\partial x} = -\rho_0 \frac{\partial^2 \eta}{\partial t^2}$$

so that

$$\frac{\partial^2 \eta}{\partial t^2} = c_0^2 (1+s)^{\gamma+1}\frac{\partial^2 \eta}{\partial x^2}, \quad \text{where } c_0^2 = \frac{\gamma P_0}{\rho_0}$$

Physically this implies that the local velocity of sound, $c_0(1+s)^{(\gamma+1)/2}$, depends upon the condensation s, so that in a finite amplitude sound wave regions of higher density and pressure will have a greater sound velocity, and local disturbances in these parts of the wave will overtake those where the values of density pressure and temperature are lower.

A single sine wave of high amplitude can be formed by a close fitting piston in a tube which is pushed forward rapidly and then returned to its original position. Fig. 11.7a shows the original shape of such a wave and 11.7b shows the distortion which follows as it propagates down the tube. If the distortion continued the wave form would eventually appear as in fig. 11.7c, where analytical solutions for pressure, density and temperature would be mul-tivalued, as in the case of the non-linear oscillator of fig. 11.3c. Before this situation is reached, however, the wave form stabilizes into that of fig. 11.7d,

Pressure

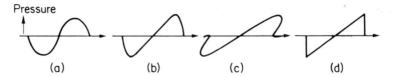

(a)   (b)   (c)   (d)

Fig. 11.7. The local sound velocity in a high amplitude acoustic
wave (a) is pressure and density dependent. The wave distorts
with time (b) as the crest overtakes the lower density regions. The
extreme situation of (c) is prevented by entropy-producing
mechanisms and the wave stabilises to an N type shock-wave (d)
with a sharp leading edge

where at the vertical 'shock front' the rapid changes of particle density, velocity
and temperature produce the dissipating processes of diffusion, viscosity and
thermal conductivity. The velocity of this 'shock front' is always greater than
the velocity of sound in the gas into which it is moving, and across the 'shock
front' there is always an increase in entropy. The competing effects of dissipation and non-linearity produce a stable front as long as the wave retains
sufficient energy. The N-type wave of fig. 11.7d occurs naturally in explosions
(in spherical dimensions) where a blast is often followed by a rarefaction.

The growth of a shock front may also be seen as an extension of the Doppler
effect (page 135), where the velocity of the moving source is now greater than
that of the signal. In fig. 11.8a as an aircraft moves from S to S' in a time t the air
around it is displaced and the disturbance moves away with the local velocity of
sound v_S. The circles show the positions at time t of the sound wave fronts
generated at various points along the path of the aircraft but if the speed of the
aircraft u is greater than the velocity of sound v_S regions of high density and
pressure will develop, notably at the edges of the aircraft structure and along
the conical surface tangent to the successive wave fronts which are generated at
a speed greater than sound and which build up to a high amplitude to form a
shock. The cone, whose axis is the aircraft path, has a half angle α where

$$\sin \alpha = \frac{v_S}{u}$$

It is known as the 'Mach Cone' and when it reaches the ground a 'supersonic
bang' is heard.

The growth of the shock at the surface of the cone may be seen by
considering the sound waves in fig. 11.8b generated at points A (time t_A) and B
(time t_B) along the path of the aircraft, which travels the distance $AB = x = u\Delta t$
in the time interval $\Delta t = t_B - t_A$. The sound waves from A will travel the
distance r_0 to reach the point P at a time

$$t_0 = t_A + \frac{r_0}{v_S}$$

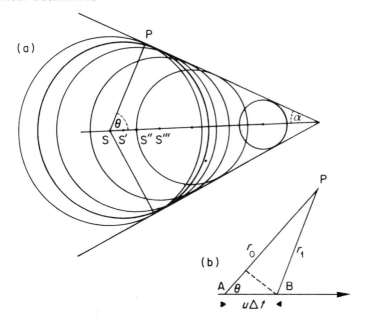

Fig. 11.8. (a) The circles are the wave fronts generated at points S along the path of the aircraft, velocity $u > v_s$ the velocity of sound. Wave fronts superpose on the surface of the Mach Cone (typical point P) of half angle $\alpha = \sin^{-1} v_s/u$ to form a shock front. (b) At point P sound waves arrive simultaneously from positions A and B along the aircraft path when $(u/v_s) \cos \theta = 1$. $(\theta + \alpha = 90°)$

Those from B will travel the distance r_1 to P to arrive at a time

$$t_1 = t_B + \frac{r_1}{v_S}$$

If x is small relative to r_0 and r_1, we see that

$$r_1 - r_0 \approx x \cos \theta = u \Delta t \cos \theta$$

so the time interval

$$t_1 - t_0 = t_B - t_A + \frac{(r_1 - r_0)}{v_S}$$

$$= \Delta t - \frac{u \Delta t \cos \theta}{v_S} = \Delta t \left(1 - \frac{u \cos \theta}{v_S} \right)$$

For the aircraft speed $u < v_S$, $t_1 - t_0$ is always positive and the sound waves arrive at P in the order in which they were generated.

For $u > v_S$ this time sequence depends on θ and when

$$\frac{u}{v_S} \cos \theta = 1$$

$t_1 = t_0$ and the sound waves arrive simultaneously at $\overset{.}{P}$ to build up a shock.
Now the angles θ and α are complementary so the condition

$$\cos \theta = \frac{v_S}{u}$$

defines

$$\sin \alpha = \frac{v_S}{u}$$

so that all points P lie on the surface of the Mach Cone.

A similar situation may arise when a charged particle q emitting electromagnetic waves moves in a medium of refractive index greater than unity with a
velocity v_q which may be greater than that of the phase velocity v of the
electromagnetic waves in the medium ($v < c$). A Mach Cone for electromagnetic waves is formed with a half angle α where

$$\sin \alpha = \frac{v}{v_q}$$

and the resulting 'shock wave' is called Cerenkov radiation. Measuring the
effective direction of propagation of the Cerenkov radiation is one way of
finding the velocity of the charged particle.

Shock Front Thickness

The extent of the region over which the gas properties change, the shock front
thickness, may be only a few mean free paths in a monatomic gas because only a
few collisions between atoms are necessary to exchange the energy required to
raise them from the equilibrium conditions ahead of the shock to those behind
it. In a polyatomic gas the collisions are effective in producing a rapid increase
in translational and rotational mode energies, but vibrational modes take much
longer to reach their new equilibrium, so that the shock front thickness is very
much greater.

Within the shock front thickness the state of the gas is not easily found, but
the state of the gas on one side of the shock may be calculated from the state of
the gas on the other side by means of the conservation equations of mass,
momentum and energy.

Equations of Conservation

In a laboratory, shock waves are produced in a tube which is divided by a
diaphragm into a short high-pressure section and a much longer low-pressure
section. When the diaphragm bursts the expanding high pressure gas behaves

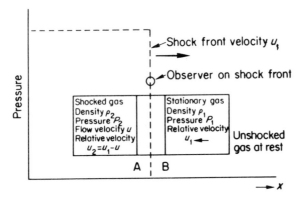

Fig. 11.9. The pressure 'step profile' of a shock wave developed in a shock tube is shown by the dotted line. The plane cross-sections at A and B remain fixed with respect to the observer O moving with the shock front at velocity u_1 into unshocked gas at rest of pressure p_1 and density ρ_1. The shocked gas has a pressure p_2, a density ρ_2 and a velocity u, with a relative velocity $u_2 = u_1 - u$ with respect to the shock front. The states of the gas at A and B are related by the conservation equations of mass, momentum and energy across the shock front. Experimental measurement of the shock velocity u_1 is sufficient to determine the unknown parameters if the stationary gas parameters are known

as a very fast low-interia piston which compresses the low pressure gas on the other side of the diaphragm and drives a shock wave down the tube. The profile of this shock wave is the step function shown as the dotted line in fig. 11.9, and the gas into which the shock is propagating is considered to be at rest. This simplifies the analysis, for we can consider the situation in fig. 11.9 as it appears to an observer O travelling with the shock front velocity u_1 into the stationary gas. The shock front is located within the region bounded by the surfaces A and B of unit area, each of which remains fixed with respect to the observer. The stationary gas which moves through the shock front from surface B acquires a flow velocity $u < u_1$ and a velocity relative to the shock front of $u_2 = u_1 - u$. From the observer's viewpoint the quantity of gas flowing into unit area of the region AB per unit time is $\rho_1 u_1$, where ρ_1 is the density of the gas ahead of the shock. The quantity leaving unit area of AB per unit time is $\rho_2(u_1 - u) = \rho_2 u_2$, where ρ_2 is the density of the shocked gas.

Conservation of mass yields $\rho_1 u_1 = \rho_2 u_2 = m$ (a constant mass). The force per unit area acting across the region AB is $p_2 - p_1$, which equals the rate of change of momentum of the gas within the unit element, which is $m(u_1 - u_2)$. The conservation of momentum is therefore given by

$$p_1 + \rho_1 u_1^2 = p_2 + \rho_2 u_2^2.$$

The work done on unit area of the region per unit time is $p_1 u_1 - p_2 u_2$, and this equals the rate of increase of the kinetic and internal energy of the gas passing through unit area of the shock wave.

The difference

$$p_1 u_1 - p_2 u_2 = \frac{p_1}{\rho_1} m - \frac{p_2}{\rho_2} m$$

so that if the internal energy per unit mass of the gas is written $e(p, \rho)$, then the equation of conservation of energy per unit mass becomes

$$\tfrac{1}{2} u_1^2 + e_1 + \frac{p_1}{\rho_1} = \tfrac{1}{2} u_2^2 + e_2 + \frac{p_2}{\rho_2}$$

where for an ideal gas $p/\rho = RT$ and $e = c_v T = (1/\gamma - 1) p/\rho$, where T is the absolute temperature, c_v is the specific heat per gram at constant volume and $\gamma = c_p/c_v$, where c_p is the specific heat per gram at constant pressure.

These three conservation equations

$$\rho_1 u_1 = \rho_2 u_2 = m \qquad \text{(mass)}$$

$$p_1 + \rho_1 u_1^2 = p_2 + \rho_2 u_2^2 \qquad \text{(momentum)}$$

and

$$\tfrac{1}{2} u_1^2 + e_1 + \frac{p_1}{\rho_1} = \tfrac{1}{2} u_2^2 + e_2 + \frac{p_2}{\rho_2} \qquad \text{(energy)}$$

together with the internal energy relation $e(p, \rho)$ completely define the properties of an ideal gas behind a shock wave in terms of the stationary gas ahead of it.

In an experiment the properties of the gas ahead of the shock are usually known, leaving five unknowns in the four equations, which are the shock front velocity u_1, the density of the shocked gas ρ_2, the relative flow velocity behind the shock u_2, the shocked gas pressure p_2 and its internal energy e_2. In practice the shock front velocity u_1 is measured and the other four properties may then be calculated.

Mach Number

A significant parameter in shock wave theory is the Mach number. It is a local parameter defined as the ratio of the flow velocity to the local velocity of sound. The Mach number of the shock front is therefore $M_s = u_1/c_1$, where u_1 is the velocity of the shock front propagating into a gas whose velocity of sound is c_1.

The Mach number of the gas flow behind the shock front is defined as $M_f = u/c_2$, where u is the flow velocity of the gas behind the shock front

$(u < u_1)$ and c_2 is the local velocity of sound behind the shock front. There is always an increase of temperature across the shock front, so that $c_2 > c_1$ and $M_s > M_f$. The physical significance of the Mach number is seen by writing $M^2 = u^2/c^2$, which indicates the ratio of the kinetic flow energy, $\frac{1}{2}u^2$ per mole, to the thermal energy, $c^2 = \gamma RT$ per mole. The higher the proportion of the total gas energy to be found as kinetic energy of flow the greater is the Mach number.

Ratios of Gas Properties Across a Shock Front

A shock wave may be defined in terms of the shock Mach number M_s, the density or compression ratio across the shock front $\beta = \rho_2/\rho_1$, the temperature ratio across the shock T_2/T_1 and the compression ratio or shock strength $y = p_2/p_1$.

Given the shock strength, $y = p_2/p_1$, the conservation equations are easily solved to yield

$$M_s = \frac{u_1}{c_1} = \left(\frac{y + \alpha}{(1 + \alpha)}\right)^{1/2}$$

where

$$\alpha = \frac{\gamma - 1}{\gamma + 1}$$

$$\beta = \frac{\rho_2}{\rho_1} = \frac{\alpha + y}{1 + \alpha y}$$

and

$$\frac{T_2}{T_1} = y\left(\frac{1 + \alpha y}{\alpha + y}\right)$$

Alternatively these may be written in terms of the experimentally measured parameter M_s as

$$\frac{p_2}{p_1} = y = M_s^2(1 + \alpha) - \alpha$$

$$\frac{p_2}{p_1} = \beta = \frac{M_s^2}{1 - \alpha + \alpha M_s^2}$$

and

$$\frac{T_2}{T_1} = \frac{[\alpha(M_s^2 - 1) + M_s^2][\alpha(M_s^2 - 1) + 1]}{M_s^2}$$

For weak shocks (where p_2/p_1 is just greater than 1) β, T_2/T_1 and M_s are also just greater than unity, and the shock wave moves with the speed of sound.

Strong Shocks

The ratio $p_2/p_1 \gg 1$ defines a strong shock, in which case

$$M_s^2 \to \frac{(\gamma + 1)}{2\gamma} y$$

and

$$\beta = \frac{\rho_2}{\rho_1} \to \left(\frac{\gamma+1}{\gamma-1}\right)$$

a limit of 6 for air and 4 for a monatomic gas for a constant γ. The flow velocity

$$u = u_1 - u_2 \to \frac{2u_1}{(\gamma+1)}$$

and the temperature ratio

$$\frac{T_2}{T_1} = \left(\frac{c_2}{c_1}\right)^2 \to \frac{(\gamma-1)}{(\gamma+1)} y$$

The temperature increase across strong shocks is of great experimental interest. The physical reason for this increase may be seen by rewriting the equation of energy conservation as $\frac{1}{2}u_1^2 + h_1 = \frac{1}{2}u_2^2 + h_2$, where $h = (e + p/\rho)$ is the total heat energy or enthalpy per unit mass. For strong shocks $h_2 \gg h_1$ of the cold stationary gas and $u_1 \gg u_2$, so that the energy equation reduces to $h_2 \approx \frac{1}{2}u_1^2$, which states that the relative kinetic energy of a stationary gas element just ahead of the shock front is converted into thermal energy when the shock wave moves over that element. The energy of the gas which has been subjected to a very strong shock wave is almost equally divided between its kinetic energy and its thermal or internal energy. This may be shown by considering the initial values of the internal energy e_1 and pressure p_1 of the cold stationary gas to be negligible quantities in the conservation equations, giving the kinetic energy per unit mass behind the shock as

$$\tfrac{1}{2}u^2 = \tfrac{1}{2}(u_1 - u_2)^2 = e_2$$

the internal energy per unit mass of the shocked gas.

In principle, the temperature behind very strong shock waves should reach millions of degrees. In practice, real gas effects prevent this. In a monatomic gas high translational energies increase the temperature until ionization occurs and this process then absorbs energy which otherwise would increase the temperature still further. In a polyatomic gas the total energy is divided amongst the various modes (translational, rotational and vibrational) and the temperatures reached are much lower than in the case of the monatomic gas. The reduction of γ due to these processes is significant, since with increasing ionization $\gamma \to 1$, and the temperature ratio depends upon the factor $(\gamma-1)/(\gamma+1)$ which becomes very small.

(Problems 11.6, 11.7, 11.8, 11.9, 11.10, 11.11)

Problem 11.1
If the period of a pendulum with large amplitude oscillations is given by

$$T = T_0\left(1 + \tfrac{1}{4}\sin^2\frac{\theta_0}{2}\right)$$

where T_0 is the period for small amplitude oscillations and θ_0 is the oscillation amplitude, show that for θ_0 not exceeding $30°$, T and T_0 differ by only 2% and for $\theta_0 = 90°$ the difference is 18%.

Problem 11.2

The equation of motion of a free undamped non-linear oscillator is given by

$$m\ddot{x} = -f(x)$$

Show that for an amplitude x_0 its period

$$\tau_0 = 4\sqrt{\frac{m}{2}} \int_0^{x_0} \frac{dx}{[F(x_0) - F(x)]^{1/2}}, \quad \text{where } F(x_0) = \int_0^{x_0} f(x)\,dx$$

Problem 11.3

The equation of motion of a forced undamped non-linear oscillator of unit mass is given by

$$\ddot{x} + s(x) = F_0 \cos \omega t$$

Writing $s(x) = s_1 x + s_3 x^3$, where s_1 and s_3 are constant, choose the variable $\omega t = \phi$, and for $s_3 \ll s_1$ assume a solution

$$x = \sum_{n=1}^{\infty} \left(a_n \cos \frac{n}{3}\phi + b_n \sin \frac{n}{3}\phi \right)$$

to show that all the sine terms and the even numbered cosine terms are zero, leaving the fundamental frequency term and its third harmonic as the significant terms in the solution.

Problem 11.4

If the mutual interionic potential in a crystal is given by

$$V = -V_0 \left[2\left(\frac{r_0}{r}\right)^6 - \left(\frac{r_0}{r}\right)^{12} \right]$$

where r_0 is the equilibrium value of the ion separation r, show by expanding V about V_0 that the ions have small harmonic oscillations at a frequency given by $\omega^2 \approx 72\, V_0/mr_0^2$, where m is the reduced mass.

Problem 11.5

The potential energy of an oscillator is given by

$$V(x) = \tfrac{1}{2}kx^2 - \tfrac{1}{3}ax^3$$

where a is positive and $\ll k$.

Assume a solution $x = A \cos \omega t + B \sin 2\omega t + x_1$ to show that this is a good approximation at $\omega_0^2 = \omega^2 = k/m$ if $x_1 = \alpha A^2/2\omega_0^2$ and $B = -\alpha A^2/6\omega_0^2$, where $\alpha = a/m$.

Problem 11.6

The properties of a stationary gas at temperature T_0 in a large reservoir are defined by c_0, the velocity of sound, $h_0 = c_p T_0$, the enthalpy per unit mass, and γ, the constant value of the specific heat ratio. If a ruptured diaphragm allows the gas to flow along a tube with velocity u, use the equation of conservation of energy to prove that

$$\frac{c_0^2}{\gamma-1} = \frac{\gamma+1}{2(\gamma-1)} c^{*2}$$

where c^* is the velocity at which the flow velocity equals the local sound velocity.
Hence show that if $u_1/c^* = M^*$ and $u_1/c_1 = M_s$, then

$$M^{*2} = \frac{(\gamma+1)M_s^2}{(\gamma-1)M_s^2+2}$$

Problem 11.7

Using a coordinate system which moves with a shock front of velocity u_1, show from the conservation equations that c^* in problem 11.6 is given by

$$c^{*2} = u_1 u_2$$

where u_2 is the relative flow velocity behind the shock front.

Problem 11.8

Use the conservation equations to prove that the pressure ratio across a shock front in a gas of constant γ is given by

$$\frac{p_2}{p_1} = \frac{\beta-\alpha}{1-\beta\alpha}$$

where $\beta = \rho_2/\rho_1$, the density ratio, and $\alpha = (\gamma-1)/(\gamma+1)$.

Problem 11.9

Use the results of problems 11.6 and 11.7 with the equation of momentum conservation to prove that the shock front Mach number is given by

$$M_s = \frac{u_1}{c_1} = \sqrt{\frac{y+\alpha}{1+\alpha}}$$

where $y = p_2/p_1$, the pressure ratio across the shock and $\alpha = \gamma-1/\gamma+1$. Hence show that the flow velocity behind the shock is given by

$$u = \frac{c_1(1-\alpha)(y-1)}{\sqrt{(1+\alpha)(y+\alpha)}}$$

Problem 11.10

The diagrams (p. 377) show (a) a shock wave of pressure p_2 and flow velocity u propagating into a stationary gas, pressure p_1, and (b) after reflexion at a rigid wall the reflected wave of pressure p_3 moving back into the gas behind the incident shock still at pressure p_2. Use the result at the end of problem 11.9 to show that the flow velocity u_r behind the reflected wave is given by

$$\frac{u_r}{c_2} = \frac{(1-\alpha)(p_3/p_2-1)}{\sqrt{(1+\alpha)(p_3/p_2+\alpha)}}$$

and since $u + u_r = 0$ at the rigid wall, use this result together with the ratio for $c_2/c_1 = (T_2/T_1)^{1/2}$ to prove that

$$\frac{p_3}{p_2} = \frac{(2\alpha+1)y-\alpha}{\alpha y+1}$$

where $y = p_2/p_1$ and $\alpha = (\gamma-1)/(\gamma+1)$.

Problem 11.11
Use problems 11.10 to prove that the ratio

$$\frac{p_3 - p_1}{p_2 - p_1} \to 2 + \frac{1}{\alpha}$$

in the limit of very strong shocks. (Note that this value is 8 for $\gamma = 1 \cdot 4$ and 6 for $\gamma = 5/3$, compared with the normal acoustic pressure jump of 2 upon reflexion.)

(a)   (b)

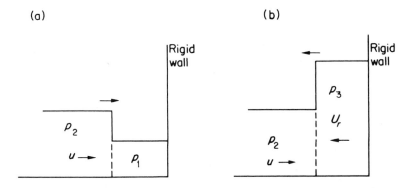

Chapter 12

Wave Mechanics

The wave mechanics of Schrödinger (1926) and the equivalent matrix formulation by Heisenberg (1926) are the basis of what is known as 'modern physics'. Without exception they have been successful in replacing or including classical mechanics over the whole range of physics at atomic and molecular levels; these in turn govern the larger scale macroscopic properties. Very high energy phenomena in the physics of elementary particles still, however, present many problems.

In this chapter we shall be concerned only with Schrödinger's wave mechanics and in the way it displays the dual wave-particle nature of matter. This dual nature was first established for electromagnetic radiation but the parallel attempt to establish the wave nature of material particles is the basic history of 20th century physics.

Origins of Modern Quantum Theory

In the 19th century interference and diffraction experiments together with classical electromagnetic theory had confirmed the wave nature of light beyond all doubt but in 1901, in order to explain the experimental curves of black body radiation, Planck postulated that electromagnetic oscillators of frequency ν had discrete energies $nh\nu$ where n was an integer and h was a constant (page 232). A quarter of a century was to elapse before this was formally derived from the new quantum mechanics.

X-rays had been found by Roentgen in 1895, their wave-like properties were displayed by the diffraction experiments of von Laue in 1912, and their electromagnetic nature was soon proved. A much longer time was required to reconcile a wave nature with the negatively charged particles which J. J. Thomson found in his cathode ray experiments of 1897. It was not until 1927 that interference effects from reflected or scattered electrons were obtained by Davisson and Germer whilst in 1928 G. P. Thomson (the son of J.J.) produced concentric ring diffraction patterns by firing electrons through a thin foil.

In the meantime, in 1906, Einstein had used Planck's idea to explain the photoelectric effect where light falling on a given surface caused electrons to be ejected. Einstein considered the light beam as a stream of individual photons, or quanta of light, each of energy $h\nu$. Collisions between these quanta and electrons in the target material gave the electrons sufficient energy to escape.

In 1912 the alpha particle scattering experiments of Rutherford led to his proposal that the atom consisted of a small positively charged nucleus surrounded by enough negative electrons to leave the atom electrically neutral. This atom was the model used by Bohr and Sommerfeld in their 'old quantum theory', a mixture of classical mechanics and quantum postulates, attempting to explain, amongst other things, the regularity of spectroscopic series from radiating atoms. Electrons were required to orbit the nucleus at definite energy levels (like planets round the Sun), and radiation at a fixed frequency ν was given out when an electron moved from a higher to a lower energy orbit with an energy difference $\Delta E = h\nu$. These orbits were required to be stable or 'stationary' orbits with quantized, that is, allowed values of energy and angular momentum. The fact that classical electromagnetic theory had shown that an accelerating charge (electron in an orbit) was itself a source of radiation remained an unresolved difficulty.

By 1920 Einstein had provided two of the vital tools necessary for further progress (a) that a quantum of radiation has energy $E = h\nu$, and (b) that a particle of momentum $p = mv$ and rest mass m_0 has a relativistic energy E where $E^2 = p^2 c^2 + (m_0 c^2)^2$.

This relation established the equivalence of matter and energy; a stationary particle $v = 0$ has an energy $E = m_0 c^2$ where c is the velocity of light.

The time was now ripe for the final steps leading to the modern quantum theory. The first of these was provided by Compton (1922–23) and the second by de Broglie in 1924.

Compton fired X-rays of a known frequency at a thin foil and observed that the frequency ν of the scattered radiation was independent of the foil material. This implied that the scattering was the result of collisions between X-ray quanta of energy $h\nu$ and the electrons in the target material. In addition to scattering at the incident frequency a lower frequency of scattered radiation was always found which depended only on the mass of the scattering particles (electrons) and the angle of scattering. Compton showed that these results were consistent if momentum and energy were conserved in an elastic collision between two 'particles', the electron and an X-ray of energy $h\nu$, a rest mass $m_0 = 0$ and (from Einstein's relativistic energy equation), a momentum

$$p = \frac{E}{c} = \frac{h\nu}{c} = \frac{h}{\lambda}$$

where $c = \nu\lambda$.

De Broglie in 1924 proposed that, if the dual wave-particle nature of electromagnetic fields required a particle momentum of $p = h/\lambda$ it was possible that a wavelength λ of a 'matter' field could be associated with **any** particle of momentum $p = mv$ to give the relation $p = h/\lambda$. His argument was as follows.

If the phase velocity of such a 'matter' wave obeys the usual relation

$$v_p = \nu\lambda$$

where ν is the frequency, the assumption that any particle has a momentum $p = h/\lambda$ together with Einstein's expression $E = h\nu$ yields $v_p = E/p$.

The theory of relativity gives, for a particle of rest mass m_0 and velocity v an energy $E = mc^2$ and a momentum $p = mv$, where

$$m = m_0\left(1 - \frac{v^2}{c^2}\right)^{-1/2}$$

is the particle mass at velocity v. For such a particle the phase velocity

$$v_p = \frac{E}{p} = \frac{c^2}{v}$$

that is,

$$vv_p = c^2$$

(an expression we met earlier for the wave guides of page 225).

This gives a phase velocity $v_p > c$ for a particle velocity $v < c$. However, the energy in the de Broglie wave (or particle) travels with the group velocity

$$v_g = \frac{\partial\omega}{\partial k}$$

which, for

$$E = h\nu = \frac{h}{2\pi}\omega$$

and

$$p = \frac{h}{\lambda} = \frac{h}{2\pi}k$$

gives

$$v_g = \frac{\partial\omega}{\partial k} = \frac{\partial E}{\partial p}$$

Such a particle with relativistic energy E where

$$E^2 = p^2c^2 + (m_0c^2)^2$$

has

$$2E\frac{\partial E}{\partial p} = 2pc^2$$

or

$$v_g = \frac{\partial E}{\partial p} = \frac{pc^2}{E} = \frac{vc^2}{c^2} = v$$

so that *the group velocity of the de Broglie matter wave corresponds to the particle velocity v.*

Even the 'old quantum theory' of Bohr–Sommerfeld gained something from de Broglie's hypothesis. Their postulate that the angular momentum of stationary orbits was restricted to integral (quantum) numbers of the unit angular momentum h was shown, for the circular orbit of radius r, to yield

$$2\pi rp = nh$$

or

$$2\pi r = \frac{nh}{p} = n\lambda$$

so that the circumference of a stationary orbit was a standing wave system and contained an integral number n of λ, the de Broglie wavelength.

Within three years, however, such quantum numbers ceased to be assumptions. They were the natural outcome of the new quantum theory of Schrödinger and Heisenberg.

Heisenberg's Uncertainty Principle

Although, as we shall see, Schrödinger's equation takes the form of a standing wave equation, the fitting of an integral number of de Broglie standing waves around a Bohr orbit presents a fundamental difficulty. The azimuthal symmetry of such a pattern, fig. 12.1, representing an electron in an orbit, does not allow the exact position of the electron to be specified at a particular time. This

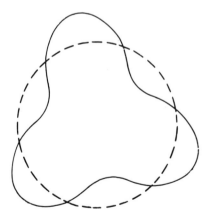

Fig. 12.1. Integral number of de Broglie standing waves $\lambda = h/p$ around a circular Bohr orbit does not allow the exact position of the electron to be specified at a particular time

dilemma was resolved by Heisenberg on the basis of the bandwidth theorem we first met on page 126.

There, a group of waves with a group velocity v_g and a frequency range $\Delta\nu$ superposed effectively only for a time Δt where

$$\Delta\nu\,\Delta t \approx 1$$

Similarly, a group in the wave number range Δk superposed in space over a distance Δx where

$$\Delta x\,\Delta k \approx 2\pi$$

But the velocity of the de Broglie matter wave is essentially a group velocity with a momentum

$$p = \frac{h}{\lambda} = \frac{h}{2\pi}k = \hbar k$$

where

$$\hbar = \frac{h}{2\pi}$$

so

$$\Delta p = \hbar\,\Delta k$$

and the bandwidth theorem becomes Heisenberg's Uncertainty Principle

$$\Delta x\,\Delta p \approx h$$

Since

$$E = h\nu = \frac{h}{2\pi}\omega = \hbar\omega$$

it follows that

$$\frac{\Delta E}{\Delta\nu} = \Delta E\,\Delta t \approx h$$

and

$$\Delta E \approx \hbar\,\Delta\omega$$

are also expressions of Heisenberg's Uncertainty Principle.

This relation sets a fundamental limit on the ultimate precision with which we can know the position x of a particle and the x component of its momentum. If fig. 12.2 shows a wave group representing the particle, the range Δx shows the uncertainty of the position of the particle, the range of space over which it could be found, with the probability of its being at a particular place given by the square of the wave amplitude of that position. The relation

$$\Delta x\,\Delta p \approx h$$

means that the velocity of the particle ($p = mv$) is also uncertain, the more accurate the knowledge of the particle position, the less certain that of the value

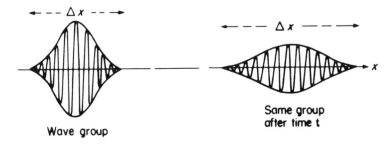

Fig. 12.2. A wave group representing a particle showing dispersion after time t. The square of the wave amplitude at any point represents the probability of the particle being in that position, and the dispersion represents the increasing uncertainty of the particle position with time (Heisenberg's Uncertainty Principle)

of its velocity. If the particle is 'observed' at some later time, dispersion of the group will have increased the range Δx and decreased the amplitude. The uncertainty of the position has increased and the probability of its being at any one place has become less. But this is because of the original uncertainty of its velocity, through Δp, which makes an accurate forecast of its position after time t even more unlikely. An example of the relation

$$\Delta E \, \Delta t \approx h$$

may be found in considering the time spent by an electron in an atomic orbit. In a stable orbit this time Δt is long and the energy uncertainty ΔE is small so the energy levels of stable orbits are well defined. When an electron changes energy levels and radiation is emitted the time in the orbit may be short and the energy levels ill defined so that the term ΔE contributes to the breadth of a spectral line.

(Problems 12.1, 12.2, 12.3, 12.4, 12.5, 12.6, 12.7, 12.8, 12.9, 12.10)

Schrödinger's Wave Equation

The old quantum theory had sought to establish rules for the existence of discrete frequencies and energy levels. An integral number of de Broglie half wavelengths could be fitted around a circular Bohr orbit. Both of these facts are consistent with the classical standing wave systems we examined in Chapters 4 and 8 when waves travelling between rigid boundaries were perfectly reflected.

In Chapter 4 we saw that the transverse displacement $y(xt)$ of a string of length l with both ends fixed obeys the wave equation

$$\frac{\partial^2 y}{\partial x^2} - \frac{1}{v_p^2}\frac{\partial^2 y}{\partial t^2} = 0$$

where v_p is the wave velocity.

The x and t dependence could be separated in the solution for standing waves to give

$$y(x, t) = A\sin\frac{\omega_n x}{v_p}\sin\omega_n t$$

where n could take the integral values $n = 1, 2, 3$, etc. to give the discrete *eigenfrequencies*.

$$\omega_n = \frac{n\pi v_p}{l}$$

The solution $y(x, t)$ corresponding to a given ω_n is called an *eigenfunction* or a *wave function*.

In developing the Schrödinger wave equation which applies to particle behaviour we use arguments below which in no way constitute a proof because wave mechanics cannot be derived from classical mechanics. Wave mechanics is based on certain postulates the validity of which can be confirmed only by the accuracy of the predicted results.

From the preceding sections we have the representation of a particle as a matter wave with energy $E = \hbar\omega$ momentum $\mathbf{p} = \hbar\mathbf{k}$ and velocity $v_g = \partial\omega/\partial k$.

Wave mechanics uses the notation

$$\Psi(x, t) = \Psi_0 e^{-i(\omega t - kx)} = \Psi_0 e^{i(px - Et)/\hbar}$$

to define the amplitude of a matter wave at a point x at time t. The physical significance of Ψ is amplified on page 388 but for the moment we note the reversed sign of the exponential index which follows the convention used in all books on Quantum Mechanics. This merely introduces a π radians phase difference from the notation consistently used in the earlier chapters of this book but the new convention will be used throughout this chapter to avoid confusion with other texts and attention will be carefully drawn to any possible ambiguity.

In classical mechanics the total energy E of a particle of mass m and momentum $\mathbf{p}$ in a conservative field of potential V is given by

$$E = p^2/2m + V$$

Differentiating $\Psi(x, t)$ gives

$$\frac{\partial^2}{\partial x^2}\Psi(x, t) = \frac{-p^2}{\hbar^2}\Psi(x, t)$$

and inserting this value of p^2 in the classical energy equation above gives

$$\frac{\hbar^2}{2m}\frac{\partial^2}{\partial x^2}\Psi(x, t) + (E - V)\Psi(x, t) = 0$$

If we now express $\Psi(x, t) = \psi(x)\,e^{-i\omega t}$ we may cancel the common $e^{-i\omega t}$ factor from the equation above to obtain the *time independent* Schrödinger wave equation

$$\frac{\hbar^2}{2m}\frac{\partial^2}{\partial x^2}\psi(x) + (E - V)\psi(x) = 0$$

This *time independent* wave equation will give states of *constant frequency* that is, of *constant energy*, and these are the only states we shall consider in this book.

Note that this equation has the same form as the standing wave equation we first met on page 118.

States which are not of constant energy require the time dependence to be retained in Schrödinger's equation. We do this by using the fact that

$$\frac{\partial}{\partial t}\Psi(x, t) = \frac{-iE}{\hbar}\Psi(x, t)$$

and inserting this value of E in the classical energy equation. This gives the *time dependent* Schrödinger wave equation

$$\frac{-\hbar^2}{2m}\frac{\partial^2}{\partial x^2}\Psi(x, t) + V\Psi(x, t) = i\hbar\frac{\partial}{\partial t}\Psi(x, t).$$

One dimensional Infinite Potential Well

Consider as a first example the case of a particle constrained to move in a region between $x = 0$ and $x = a$ where the potential $V = 0$. At $x = 0$ and $x = a$ the potential walls are infinitely high as shown in fig. 12.3. This is an idealized form of the potential seen by an electron in the low energy levels near the nucleus of an atom.

Since $V(x) = 0$ for $0 < x < a$ Schrödinger's equation becomes

$$\frac{\partial^2\psi(x)}{\partial x^2} + \frac{2mE}{\hbar^2}\psi = 0$$

which may be written, as on page 118, in the form

$$\frac{\partial^2\psi}{\partial x^2} + k^2\psi = 0$$

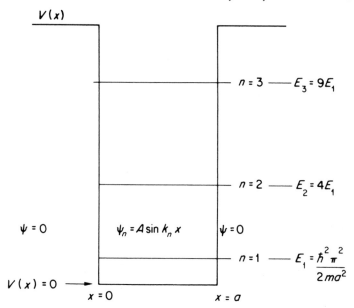

Fig. 12.3. An infinitely deep potential well showing allowed
energy levels E_n for a particle constrained to move within it
with wave function $\psi_n = A \sin k_n x$ where $k_n^2 = 2mE/\hbar^2$ and m
is the particle mass

with

$$k^2 = \frac{2mE}{\hbar^2}$$

The boundary conditions are that $\psi(x) = 0$ at $x = 0$ and $x = a$ where $V(x)$
becomes infinite, whilst the other terms in the equation remain finite. The
particle must lie within the well and classically, whatever the value of its energy
E it will rebound elastically off the potential 'walls'. When moving to the right
the particle behaviour may be represented by a wave function of the form e^{+ikx}
which satisfies Schrödinger's equation, and when moving to the left by a wave
function of the form

$$e^{-ikx}$$

But, as with the waves on the string, perfect reflection which reverses the
amplitude allows $\psi_n(x)$, the solution of Schrödinger's equation, to represent a
standing wave system at ω_n; expressed in the form

$$\psi_n(x) = C\, e^{ik_n x} - C\, e^{-ik_n x}$$

$$= A \sin k_n x$$

where

$$A = \frac{C}{2i}$$

The boundary condition $\psi_n(x) = 0$ at $x = a$ gives $k_n a = n\pi$ for $n = 1, 2, 3$, etc. i.e. $k_n = n\pi/a$.

Hence

$$k_n^2 = \frac{2mE_n}{\hbar^2} = \frac{n^2 \pi^2}{a^2}$$

giving energy eigenvalues

$$E_n = \frac{n^2 \pi^2 \hbar^2}{a^2 2m}$$

Thus we see that discrete energy values governed by the quantum number n arise naturally from the application of boundary conditions to the wave function solutions of Schrödinger's equation. Values of the particle momentum are also quantized since

$$p = \frac{h}{\lambda} = \hbar k = \frac{n\pi\hbar}{a}$$

It is evident that, in an infinite potential well, an electron or particle cannot have an arbitrary energy but must take only the quantized values E_n. This restriction will hold whenever Schrödinger's equation is solved for a potential $V(x)$ which imposes boundary conditions constraining the particle to move in a limited region.

The wave functions $\psi_n(x)$ for $n = 1, 2, 3$ are plotted in fig. 12.4 showing them to be identical with the allowed amplitude functions for standing waves on a vibrating string with fixed ends. Note that the interval between allowed energy states decreases as either the mass of the particle or the dimensions of the potential box increase relative to h. For particles of large mass and systems of large dimensions the allowed energy states form, for all practical purposes, a continuum and are no longer quantized. Thus, in passing from atomic to much larger dimensions the results of quantum mechanics approach those of classical physics.

We see that the minimum value of the energy of the particle in the potential well is not zero but

$$E_1 = \frac{\hbar^2 \pi^2}{2ma^2}$$

This minimum energy is related to Heisenberg's Uncertainty Principle

$$\Delta x \, \Delta p \approx h$$

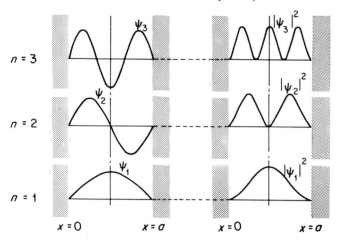

Fig. 12.4. Wave functions $\psi_n(x)$ and probability densities $|\psi_n(x)|^2$ for the first three allowed energy levels in an infinitely deep potential well of width a

The uncertainty in the position of the particle is obviously $\Delta x = a$ and the particle momentum p may be in either the +ve or −ve x direction giving an uncertainty

$$\Delta p = 2p$$

Thus

$$\Delta x\,\Delta p = a \cdot 2p \approx h$$

or

$$p \approx \frac{h}{2a}$$

Now, for $V(x) = 0$

$$E = \frac{p^2}{2m} \approx \frac{h^2}{8ma^2} \approx \frac{\hbar^2 \pi^2}{2ma^2}$$

This is an example of the so-called *zero point* energy. We shall meet others.

(Problem 12.11)

Significance of the Amplitude $\psi_n(x)$ of the Wave Function

In fig. 12.4 the amplitude $\psi_n(x)$ of the wave function is plotted for the values $n = 1, 2, 3$ together with the values

$$|\psi_n(x)|^2$$

In the waves we have met so far, the amplitude, or rather the amplitude squared, has been a measure of the intensity of the wave. At a position of high amplitude, the wave was more intense—more energy was localized there. Here we have expressed the motion of a particle confined to a small region of space in terms of its associated matter wave. The amplitude of the wave function $\psi(x)$ varies from point to point within the small region in which the particle is to be found. Outside the infinite well $\psi(x)$ is zero. The intensity of the matter wave is written

$$|\psi(x)|^2 = \psi^*(x)\psi(x)$$

where the complex conjugate $\psi^*(x)$ indicates that $\psi(x)$ may sometimes be complex. Since the matter field describes the motion of the particle we may say that the regions of space in which the particle is more likely to be found are those in which the intensity $|\psi(x)|^2$ is large, or, more formally

'the probability of finding the particle described by the wave function $\psi(x)$ in the interval dx around the point x is $|\psi(x)|^2 \, dx$'.

The probability per unit length of finding the particle at x is

$$P(x) = |\psi(x)|^2$$

In three dimensions a wave function would be of the form $\psi(x, y, z)$ and the probability of finding the particle in the unit volume element surrounding the point xyz is

$$P(xyz) = |\psi(xyz)|^2$$

The probability of finding the particle within a finite volume V is obviously

$$P_V = \int_V |\psi(xyz)|^2 \, dx \, dy \, dz$$

Now the particle must always be somewhere in space so, in extending the integral over all space, the probability becomes a certainty, that is, it equals unity, or

$$\int_{\text{all space}} |\psi(xyz)|^2 \, dx \, dy \, dz = 1$$

This process of integrating over all possible locations to give unity is called *normalization* and it always imposes restrictions on the form of $\psi(x, y, z)$ which must tend to zero as x, y or z tends to infinity.

Normalization determines the value of the constant A in our wave function

$$\psi_n(x) = A \sin \frac{n\pi x}{a}$$

for the case of the infinite potential well.

There

$$\int_{-\infty}^{\infty} |\psi_n(x)|^2 \, dx = \int_0^a |\psi_n(x)|^2 \, dx$$

$$= A^2 \int_0^a \sin^2 \frac{n\pi x}{a} \, dx = A^2 \frac{a}{2} = 1$$

Hence

$$A = \sqrt{\frac{2}{a}}$$

and the *normalized* wave function

$$\psi_n(x) = \sqrt{\frac{2}{a}} \sin \frac{n\pi x}{a}$$

(Problem 12.12)

Particle in a Three-Dimensional Box

Suppose the particle is confined to a rectangular volume abc at the bottom of an infinitely deep potential well ($V = 0$) where a, b and c are the lengths of the sides of the rectangular box.

The energy of the particle is then

$$E = \frac{p^2}{2m} = \frac{1}{2m}(p_x^2 + p_y^2 + p_z^2)$$

where the momentum components are

$$p_x = n_1 \frac{\pi \hbar}{a}$$

$$p_y = n_2 \frac{\pi \hbar}{b}$$

$$p_z = n_3 \frac{\pi \hbar}{c}$$

and n_1, n_2 and n_3 are integers.

The energy levels allowed in the box are therefore given by

$$E = \frac{\pi^2 \hbar^2}{2m} \left(\frac{n_1^2}{a^2} + \frac{n_2^2}{b^2} + \frac{n_3^2}{c^2} \right)$$

and solutions for the space part of the wave function may be written

$$\psi(x, y, z) = A \sin \frac{n_1 \pi x}{a} \sin \frac{n_2 \pi y}{b} \sin \frac{n_3 \pi z}{c}$$

in accordance with the three-dimensional normal mode solution of page 227. If the box is cubical so that $a = b = c$ the allowed energy levels become

$$E = \frac{\pi^2 \hbar^2}{2ma^2} (n_1^2 + n_2^2 + n_3^2) = \frac{\pi^2 \hbar^2}{2ma^2} k^2$$

where $k^2 = n_1^2 + n_2^2 + n_3^2$ with wave functions

$$\psi(xyz) = A \sin \frac{n_1 \pi x}{a} \sin \frac{n_2 \pi y}{a} \sin \frac{n_3 \pi z}{a}$$

We saw, however, on page 229 that combinations of different n values can give the same k value, that is the same energy value. When n_1, n_2 and n_3 are permuted without changing the k value, the wave function is also changed so that a certain energy level may be associated with several different wave functions or dynamical states. The energy level is said to be *degenerate*, the *order of degeneracy* being defined by the number of different or independent wave functions associated with the given energy.

In the case of the cubic potential box, the lowest energy level is $3E_1$, i.e.

$$(n_1 = n_2 = n_3 = 1)$$

where

$$E_1 = \frac{\pi^2 \hbar^2}{2ma^2}$$

The next energy level is given by $6E_1$, with a degeneracy of 3 where the n values are given by $(2, 1, 1)$ $(1, 2, 1)$ and $(1, 1, 2)$. Higher energy values with degeneracy orders are shown in Table 12.1 below.

Table 12.1

Energy	n_1, n_2, n_3 Combinations	Degeneracy
$3E_1$	(1, 1, 1)	1
$6E_1$	(2, 1, 1) (1, 2, 1) (1, 1, 2)	3
$9E_1$	(2, 2, 1) (2, 1, 2) (1, 2, 2)	3
$11E_1$	(3, 1, 1) (1, 3, 1) (1, 1, 3)	3
$12E_1$	(2, 2, 2)	1
$14E_1$	(1, 2, 3) (3, 2, 1) (2, 3, 1) (1, 3, 2) (2, 1, 3) (3, 1, 2)	6

(Problem 12.13)

Number of Energy States in Interval E to $E+dE$

As long as the dimensions of the cubical box above are small the energy levels remain distinct. However, when the volume increases, as is the case for free electrons in a metal, successive energy levels become so close that an almost continuous spectrum is formed.

If we wish to find how many energy levels may be contained in the small energy range dE when the potential box is very large, we have only to apply the result of page 231 where we found that the number of possible normal modes of oscillation *per unit volume* of an enclosure in the frequency range ν to $\nu + d\nu$ is given by

$$dn = \frac{4\pi\nu^2\, d\nu}{c^3}$$

There, we stressed that the result was independent of any particular system and we applied it to Planck's Radiation Law and Debye's Theory of Specific Heats. Here we use it with

$$E = \frac{p^2}{2m} = h\nu \quad \text{and} \quad p = \frac{E}{c} = \frac{h\nu}{c}$$

$\bigg($ so that

$$dE = \frac{p}{m}\, dp = h\, d\nu$$

and

$$dp = \frac{h\, d\nu}{c}\bigg)$$

to give the number of states *per unit volume* in the energy interval dE as

$$dn(E) = \frac{4\pi(2m^3)^{1/2}E^{1/2}}{h^3}\, dE$$

This may be applied directly to determine how free electrons in a metal may distribute themselves in a band of energies from zero to some value E. Each energy level can accommodate two electrons (with opposing spins) according to Pauli's Principle so the total number of electrons *per unit volume* in the energy range zero to E is

$$n = \int dn(E) = \frac{2 . 4\pi(2m_e^3)^{1/2}}{h^3}\int_0^E E^{1/2}\, dE$$

$$= \frac{16\pi(2m_e^3)^{1/2}}{3h^3}E^{3/2}$$

where m_e is the electron mass.

If the metal is in its ground state the available electrons will occupy the lowest possible energy levels, and if the total number of electrons *per unit volume* n_0 is less than the total number of energy levels in the band, then the electrons will occupy all energy states up to a maximum energy E_F called the Fermi Energy which is given by

$$n_0 = \frac{16\pi(2m_e^3)^{1/2}E_F^{3/2}}{3h^3}$$

Typical values of E_F are of the order of 5 electron volts (1 electron volt = $1\cdot6 \times 10^{-19}$ Joules).

(Problems 12.14, 12.15)

The Potential Step

The standing wave system of the infinite potential well where the wave function

$$\psi_n(x)$$

is finite in the region $V(x) = 0$ but zero at all other points is unique in the formal correspondence it presents between classical and quantum mechanical results. The quantum effects become evident when we consider the general case of the potential step of finite height V in fig. 12.5 which is an idealized form of the very steep potential gradient of a conservative force

$$F(x) = -\frac{\partial V}{\partial x}$$

Such a potential step would be seen by a free electron near the surface of a metal.

It is necessary to consider separately the two cases where the total particle energy E is (a) less than the potential energy V, and (b) greater than V, where

$$E = \frac{p^2}{2m} + V(x)$$

(a) $E < V$
When E is less than V, the region $x > 0$ of fig. 12.5 is forbidden to the particle by classical mechanics for the kinetic energy

$$\frac{p^2}{2m}$$

would then have a negative value.

In finding the complete solution for $\psi(x)$ for the potential step we must solve Schrödinger's equation for the separate regions of fig. 12.5, $x < 0$ (region 1) and $x > 0$ (region 2).

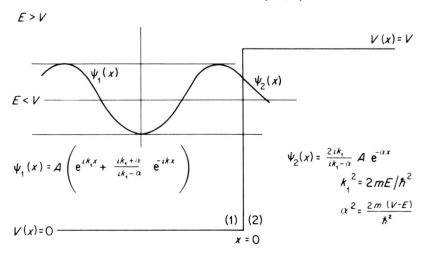

Fig. 12.5. Wave functions $\psi_1(x)$ and $\psi_2(x)$ for a particle mass m, energy $E < V$ at a potential step $V(x) = V$

In region 1, $V(x) = 0$ and we have

$$\frac{\partial^2 \psi_1(x)}{\partial x^2} + \frac{2mE}{\hbar^2}\psi_1(x) = 0$$

with a solution

$$\psi_1(x) = A\, e^{ik_1 x} + B\, e^{-ik_1 x}$$

where

$$k_1^2 = \frac{2mE}{\hbar^2}$$

The term $A\, e^{ik_1 x}$ (with the sign convention of this chapter) is the wave representation of an incident particle moving to the right, and $B\, e^{-ik_1 x}$ represents a reflected particle moving to the left.

In region 2, $V(x) = V$ and Schrödinger's equation becomes

$$\frac{\partial^2 \psi_2(x)}{\partial x^2} + \frac{2m(E-V)}{\hbar^2}\psi_2(x) = 0$$

or

$$\frac{\partial^2 \psi_2(x)}{\partial x^2} - \alpha^2\psi_2(x) = 0$$

where

$$\alpha^2 = \frac{2m(V-E)}{\hbar^2}$$

This equation has the solution

$$\psi_2(x) = C\,e^{-\alpha x} + D\,e^{\alpha x}$$

Now the probability of finding the particle in region 2 where it is classically forbidden depends on the square of the wave function amplitude $|\psi_2(x)|^2$ with the condition that for any wave function to be normalized (i.e. for

$$\int |\psi_2(x)|^2\,dx = 1\bigg)$$

the wave function $\psi_2(x) \to 0$ as $x \to \infty$.

This forbids the second term $D\,e^{\alpha x}$ which increases with x but still leaves

$$\psi_2(x) = C\,e^{-\alpha x}$$

to give a finite probability of finding the particle beyond the potential step, a probability which decreases exponentially with distance. This is a profound departure from classical behaviour.

At the boundary $x = 0$, $\psi(x)$ must be finite to give a finite probability of finding the particle there, but there is a finite discontinuity in $V(x)$. In these circumstances Schrödinger's equation asserts that the second derivative

$$\frac{\partial^2 \psi(x)}{\partial x^2}$$

at $x = 0$ is finite, which means that both $\psi(x)$ and $(\partial\psi(x)/\partial x)$ are continuous at $x = 0$.

These are the boundary conditions which allow the separate solutions

$$\psi_1(x) \quad \text{and} \quad \psi_2(x)$$

for the wave function, to be matched across the boundary of the two regions.

The continuity of $\psi(x)$ at $x = 0$ gives $\psi_1(x) = \psi_2(x)$ or $A + B = C$ whilst

$$\frac{\partial\psi_1(x)}{\partial x} = \frac{\partial\psi_2(x)}{\partial x}$$

at $x = 0$ gives

$$ik_1(A - B) = -\alpha C = -\alpha(A + B)$$

Thus

$$B = \left(\frac{ik_1 + \alpha}{ik_1 - \alpha}\right)A$$

and

$$C = \frac{2ik_1}{ik_1 - \alpha}A$$

The wave functions for the separate regions then become

$$\psi_1(x) = A\left(e^{ik_1x} + \frac{ik_1 + \alpha}{ik_1 - \alpha} e^{-ik_1x}\right)$$

and

$$\psi_2(x) = \frac{2ik_1}{ik_1 - \alpha} A e^{-\alpha x}$$

and these are shown in fig. 12.5. Note particularly that the intensity of the incident part of the wave function

$$|\psi_1(x)|^2 = |A|^2$$

whilst the reflected intensity is

$$|B|^2 = \left|\frac{ik_1 + \alpha}{ik_1 - \alpha} A\right|^2 = |A|^2$$

Thus, for any energy $E < V$ we have total reflection as in the classical case, even for those particles which penetrate the classically forbidden region $x > 0$ where $\psi_2(x)$ is finite.

In region 2 the probability of finding the particle is

$$P(x) = |\psi_2(x)|^2 = |C e^{-\alpha x}|^2$$

$$= \left|\frac{2ik_1}{ik_1 - \alpha} A e^{-\alpha x}\right|^2 = \frac{4k_1^2}{k_1^2 + \alpha^2} A^2 e^{-2\alpha x}$$

Since the exponential coefficient α depends on $V(x)$ the greater the value $V(x)$ the faster the wave function $\psi_2(x)$ goes to zero in region 2 for a given total energy $E < V$.

When $V(x) \to \infty$, as in the case of the infinite potential well, $\psi_2(x)$ becomes zero, as we have seen; and there is no penetration into the classically forbidden region.

Several important physical phenomena may be explained on the assumption that a particle with $E < V$ meeting a potential step of finite height V and *finite width b* has a wave function $\psi_2(x)$ which is still finite at $x = b$, making it possible for the particle to tunnel through the potential barrier, fig. 12.6. The probability that the particle will penetrate the barrier to $x = b$ is given by

$$P(x) = |\psi_2(x)|^2 \propto e^{-2\alpha x}$$

and beyond this barrier the particle will propagate in region 3 with a wave function $\psi_3(x)$ of reduced amplitude. The boundary conditions must then be applied at $x = b$ to match $\psi_2(x)$ to $\psi_3(x)$.

This quantum 'tunnel effect' is the basis of the explanation of the radioactive decay of the nucleus. In addition the potential step seen by a free electron near

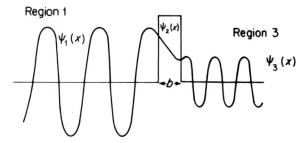

Region 1

$\psi_1(x)$

$\psi_2(x)$

Region 3

$\psi_3(x)$

b

Fig. 12.6. Narrow potential barrier of width b penetrated by a particle represented by $\psi_1(x)$ leaving a finite amplitude $\psi_3(x)$ as a measure of the reduced probability of finding the particle in region 3

the surface of a metal may be distorted, as shown in fig. 12.7, by the application of an external electric field, to form a barrier of finite width. The most energetic electrons near the surface of the metal can leak through the barrier in a process known as *field electron* emission.

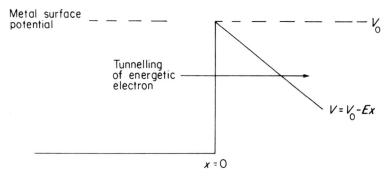

Metal surface potential

V_0

Tunnelling of energetic electron

$V = V_0 - Ex$

$x = 0$

Fig. 12.7. Application of an electric field E to the surface of a metal at potential V_0 reduces the potential to $V = V_0 - Ex$ forming a barrier of finite width which may be penetrated by an energetic electron near the metal surface

Another example results from the two possible positions of the single nitrogen atom with respect to the three hydrogen atoms in the ammonia molecule NH_3. These positions are shown as N and N' in fig. 12.8 together with the potential barrier presented to the nitrogen atom as it moves to and fro between N and N'. This penetration occurs at a frequency of $2 \cdot 3786 \times 10^{10}$ Hertz for the ground state of NH_3 and its high definition is used as an atomic clock to fix standards of time.

(Problem 12.16)

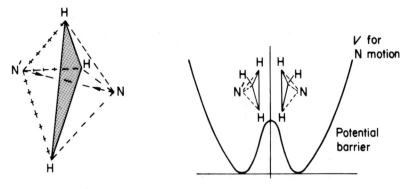

Fig. 12.8. The two possible configurations N and N' of the nitrogen atom with respect to the triangular hydrogen base in the ammonia molecule NH_3 and the finite potential barrier penetrated by the nitrogen atom at a frequency of $> 10^{10}$ hertz in the NH_3 ground state

(b) $E > V$

In the region $x < 0$ in fig. 12.5 $V(x) = 0$ and Schrödinger's equation is

$$\frac{\partial^2 \psi_1(x)}{\partial x^2} + \frac{2mE}{\hbar^2} \psi_1(x) = 0$$

or

$$\frac{\partial^2 \psi_1}{\partial x^2} + k_1^2 \psi_1 = 0$$

with

$$k_1^2 = \frac{2mE}{\hbar^2}$$

having a solution

$$\psi_1(x) = A\, e^{ik_1 x} + B\, e^{-ik_1 x}$$

with both incident and reflected terms.

The momentum of the particle is p_1 where $p_1^2/2m = E$.

In the region $x > 0$ $V(x) = V$ and Schrödinger's equation is

$$\frac{\partial^2 \psi_2(x)}{\partial x^2} + \frac{2m(E - V)}{\hbar^2} \psi_2(x) = 0$$

or

$$\frac{\partial^2 \psi_2}{\partial x^2} + k_2^2 \psi_2 = 0$$

where

$$k_2^2 = \frac{2m(E - V)}{\hbar^2}$$

and the particle momentum p_2 is given by $p_2^2/2m = (E - V)$.

In the wave function solution for this region we consider only the right going or transmitted term since there is nothing beyond $x = 0$ to cause a reflection, so

$$\psi_2(x) = C\, e^{ik_2x}$$

Now the wave number k is related to the de Broglie wavelength of the particle and we see that k changes when the potential V changes, that is, when the particle experiences a change in the force acting on it. Such a particle therefore reacts to a changing potential as light reacts to changing refractive index. As the potential V increases for $E > V$ the momentum p and wave number $k\,(p = \hbar k)$ decrease and the wavelength λ increases.

At $x = 0$ the conditions for continuity give

$$\psi_1(x) = \psi_2(x)$$

or

$$A + B = C$$

and

$$\frac{\partial \psi_1(x)}{\partial x} = \frac{\partial \psi_2(x)}{\partial x}$$

or

$$k_1(A - B) = k_2 C$$

These two equations give

$$B = \frac{(k_1 - k_2)}{(k_1 + k_2)} A$$

and

$$C = \frac{2k_1}{k_1 + k_2} A$$

Since B is not zero, some reflection takes place at $x = 0$ even though the energy $E > V$. This is clearly not classical behaviour. If many particles form an incident beam at $x = 0$ and each particle has velocity

$$v_1 = \frac{p_1}{m} = \frac{\hbar k_1}{m}$$

then the velocity of transmitted particles will be

$$v_2 = \frac{p_2}{m} = \frac{\hbar k_2}{m}$$

The incident flux of particles, that is, the number crossing unit area per unit time, may be seen as the product of the velocity and the intensity, that is

$$v_1|A|^2$$

The reflected flux is

$$v_1|B|^2$$

and the transmitted flux is

$$v_2|C|^2$$

Thus the reflection coefficient, the ratio of reflected to incident flux is

$$R = \frac{v_1|B|^2}{v_1|A|^2} = \frac{(k_1 - k_2)^2}{(k_1 + k_2)^2}$$

and the transmission coefficient, the ratio of transmitted to incident flux is

$$T = \frac{v_2|C|^2}{v_1|A|^2} = \frac{k_2}{k_1} \frac{(2k_1)^2}{(k_1 + k_2)^2} = \frac{4k_1 k_2}{(k_1 + k_2)^2}$$

results which are similar to those for our classical waves in earlier chapters.

Note that $R + T = 1$ showing that the number of particles is conserved.

We have chosen here to apply R and T to a number of particles forming a beam. These coefficients, when applied to identical particles forming the beam, measure the average probability that an individual particle will be reflected or transmitted.

(Problem 12.17)

The Square Potential Well

Let us consider a particle with energy $E < V$ moving in the square potential well of width a in fig. 12.9. Within the well the potential is zero, and the value V of the height of the well is finite. This potential approximates that of a finite range force which has no influence beyond a limited distance. Outside the range of the force the potential may be considered constant. From our discussion of the infinitely deep potential well ($V = \infty$) and of the potential step we can expect our wave function representation to have the form of an integral number of de Broglie half wavelengths within the well, plus an exponentially decaying penetration into the wall on either side.

Writing Schrödinger's equation for each of the three regions, we have for region 1 ($0 < x \leqslant a$)

$$\frac{\partial^2 \psi_1(x)}{\partial x^2} + \frac{2mE}{\hbar^2} \psi_1(x) = 0$$

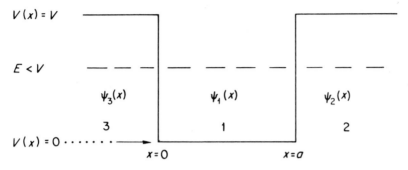

Fig. 12.9. A particle with energy $E < V$ (V = the finite height of a square potential well of width a) may take only the energy values E satisfying the equation

$$\tan a \sqrt{\frac{2mE}{\hbar^2}} = \frac{2\sqrt{E(V-E)}}{2E-V}$$

The wave functions in the three regions are matched at the boundaries $x = 0$ and $x = a$ by the conditions that $\psi(x)$ and $\partial\psi(x)/\partial x$ are continuous

with a solution, for $k_1^2 = 2mE/\hbar^2$ of

$$\psi_1(x) = A\, e^{ik_1 x} + B\, e^{-ik_1 x}$$
$$= A\,(\cos k_1 x + i\sin k_1 x) + B\,(\cos k_1 x - i\sin k_1 x)$$
$$= A_1 \cos k_1 x + B_1 \sin k_1 x$$

where $A_1 = A + B$ and $B_1 = i(A - B)$.
  In region 2 ($x \geqslant a$)

$$\frac{\partial^2 \psi_2(x)}{\partial x^2} + \frac{2m(E-V)}{\hbar^2}\psi_2(x) = 0$$

has the solution

$$\psi_2(x) = A_2\, e^{\alpha x} + B_2\, e^{-\alpha x}$$

where

$$\alpha^2 = \frac{2m}{\hbar^2}(V - E)$$

  In region 3, ($x < 0$)

$$\frac{\partial^2 \psi_3(x)}{\partial x^2} + \frac{2m(E-V)}{\hbar^2}\psi_3(x) = 0$$

has the solution

$$\psi_3(x) = A_3\,e^{\alpha x} + B_3\,e^{-\alpha x}$$

For $\psi(x)$ to remain finite as $x \to \pm\infty$ (normalization condition) A_2 and B_3 must be zero, and the boundary conditions $\psi(x)$ and $\partial\psi(x)/\partial x$ continuous, must be satisfied at $x = 0$ and $x = a$.

At $x = 0$,

$$\psi_1(x) = \psi_3(x) \quad \text{and} \quad \frac{\partial\psi_1(x)}{\partial x} = \frac{\partial\psi_3(x)}{\partial x}$$

give

$$A_1 = A_3 \tag{12.1}$$

and

$$k_1 B_1 = \alpha A_3 \tag{12.2}$$

whilst at $x = a$

$$\psi_1(x) = \psi_2(x) \quad \text{and} \quad \frac{\partial\psi_1(x)}{\partial x} = \frac{\partial\psi_2(x)}{\partial x}$$

give

$$A_1\cos k_1 a + B_1\sin k_1 a = B_2\,e^{-\alpha a} \tag{12.3}$$

and

$$-k_1 A_1 \sin k_1 a + k_1 B_1 \cos k_1 a = -\alpha B_2\,e^{-\alpha a} \tag{12.4}$$

In order to satisfy equations (12.1), (12.2), (12.3) and (12.4) some conditions must be imposed on k and α that is on the value of E, so only certain values of E are allowed.

Equations (12.1) and (12.2) give

$$\frac{A_1}{B_1} = \frac{k_1}{\alpha}$$

and this equation with equations (12.3) and (12.4) yields

$$\tan k_1 a = \frac{2k_1\alpha}{k_1^2 - \alpha^2}$$

or

$$\tan a\sqrt{\frac{2mE}{\hbar^2}} = \frac{2\sqrt{E(V-E)}}{2E - V}$$

Only those values of E which satisfy this relation are allowed energy states, but these values must be found by numerical or graphical methods.

The wave functions for the first three allowed energy values are shown in fig. 12.10 and their general behaviour may be clarified by considering Schrödinger's equation in the form

$$\frac{\partial^2 \psi}{\partial x^2} \Big/ \psi = -(+\text{ve constant})(E - V)$$

Now $\partial^2 \psi / \partial x^2$ is the rate of change of the slope, that is the curvature of the wave function and when $E > V$ both sides of the equation are negative and the ψ curve must everywhere keep its concave side towards the x axis as it always does for example in sine and cosine curves. The curvature increases with E so we shall expect more de Broglie half wavelengths in the higher energy levels. This is consistent with the argument that an increase in E increases the wave number k and reduces the de Broglie wavelength λ.

In the lowest energy level the ψ curve is always without a node, the next level always has one node, the third two nodes, etc., but the zeros will not be quite equally spaced and the ψ amplitude will not be uniform across the well. In particular it will increase near the potential walls as the particle is slowed down

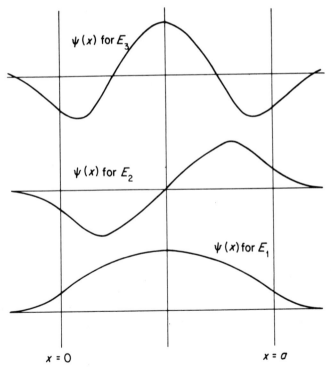

$x = 0$ $x = a$

Fig. 12.10. Wave functions for a particle in a square potential well with the lowest three allowed energies E_1, E_2, E_3. Note the exponential decay of $\psi(x)$ outside the box

to give a higher probability of the particle being found there. Where $E < V$ the ratio

$$\frac{\partial^2 \psi / \partial x^2}{\psi}$$

will be positive and the ψ curve must keep its convex side towards the axis as in exponential curves. The classical boundary $E = V$ must always mark the division where the character of the ψ curve changes from one form to the other and the two parts of the curve will only match for certain values of E.

The Harmonic Oscillator

As a final example to illustrate the fitting of ψ curves into a potential well we shall consider the potential curve $V = \frac{1}{2}sx^2$ of the harmonic oscillator in fig. 12.11. The calculation of the ψ curves is too complicated for this chapter but their essential features confirm what we may expect from our earlier examples. Moreover, by purely classical arguments we shall obtain a very good approximation to the wave mechanical results.

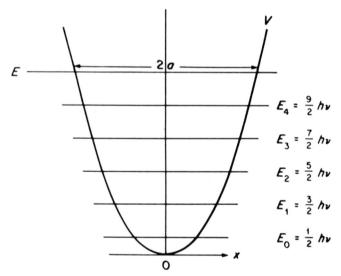

Fig. 12.11. Potential energy curve V of a harmonic oscilator with allowed energy levels $E_n = (n + \frac{1}{2})h\nu$. The energy E (with oscillator amplitude a) is shown in the text to define an average value of the de Broglie wavelength
$$\lambda = h / (\tfrac{4}{3} mE)^{1/2}$$

In 1901 Planck had postulated that the energy of such an oscillator could have the values $E = nh\nu$ where n was an integer and ν was the frequency. Schrödinger was able to derive this result in 1926 but one essential difference arises from the Uncertainty Principle which requires a minimum energy level or *zero point energy* of $\frac{1}{2}h\nu$.

For a classical oscillator the minimum energy $E = 0$, point 0 in fig. 12.11 gives the precise and simultaneous values $x = 0$ and $p = 0$, that is, a zero oscillation. The Uncertainty Principle forbids this. If a_0 is the smallest amplitude of the oscillator compatible with the Uncertainty Principle, then

$$a_0 \sim \tfrac{1}{2}\Delta x$$

If p_0 is the maximum momentum of the oscillator with amplitude a_0 it may be either in the positive or negative direction so

$$p_0 \sim \tfrac{1}{2}\Delta p$$

The energy of a classical oscillator is given by

$$E = \tfrac{1}{2}m\omega^2 a_0^2 = \tfrac{1}{2}\omega(a_0)(m\omega a_0) = \tfrac{1}{2}\omega a_0 p_0$$

$$\approx \tfrac{1}{8}\omega \, \Delta x \, \Delta p \approx \tfrac{1}{8}h\omega \approx \tfrac{1}{2}\hbar\omega = \tfrac{1}{2}h\nu$$

All other energy levels will therefore take integral steps of $h\nu$ above this zero point energy.

Let us consider the energy level of the oscillator which has an amplitude a so that

$$E = \frac{p^2}{2m} + V = \frac{p^2}{2m} + \frac{1}{2}sx^2 = \frac{1}{2}sa^2 = \frac{1}{2}m\omega^2 a^2$$

so that

$$2a = \frac{2}{\omega}\sqrt{\frac{2E}{m}}$$

The value of the kinetic energy of the oscillator averaged over the distance $2a$ between $\pm a$ may be written

$$\frac{\int_{-a}^{a} p^2/2m}{\int_{-a}^{a} dx} = \frac{1}{2a} \int_{-a}^{a} (E - \tfrac{1}{2}m\omega^2 x^2) \, dx = E - \tfrac{1}{6}m\omega^2 a^2 = \tfrac{2}{3}E$$

because

$$E = \tfrac{1}{2}m\omega^2 a^2$$

Thus the average value of the kinetic energy

$$\frac{p^2}{2m} = \frac{2}{3}E$$

giving

$$p = \frac{h}{\lambda} = \sqrt{\frac{4mE}{3}}$$

This gives an average value for the de Broglie wavelength of

$$\lambda = \frac{h}{\sqrt{\dfrac{4mE}{3}}}$$

and we expect n half wavelengths to fit into the length $2a$ at energy E where

$$2a = \frac{2}{\omega}\sqrt{\frac{2E}{m}}$$

Thus

$$n\frac{\lambda}{2} = \frac{nh}{2\sqrt{4mE/3}} = \frac{2}{\omega}\sqrt{\frac{2E}{m}}$$

Writing $\omega = 2\pi\nu$ We have

$$E = \frac{\pi}{4}\sqrt{\frac{3}{2}}\,nh\nu = 0\cdot96\ nh\nu$$

which is a fairly close approximation to $nh\nu$. The correct result however must take into account the zero point energy of $\frac{1}{2}h\nu$ and the energy levels are given by

$$E = (n + \tfrac{1}{2})h\nu \qquad n = 0, 1, 2, 3, \textit{etc.}$$

The ψ curves for the first four energy levels are plotted in fig. 12.12 together with those for $|\psi|^2$.

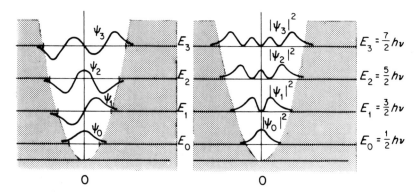

Fig. 12.12. Wave functions $\psi(x)$ and probability densities $|\psi(x)|^2$ for the first four energy levels of the harmonic oscillator

We see that whilst a classical oscillator may never exceed its maximum amplitude a particle obeying a wave mechanical description has a finite probability of being found beyond this limit.

(Problems 12.18, 12.19)

Problem 12.1

The energy of an electron mass m charge e circling a proton at radius r is

$$E = \frac{p^2}{2m} - \frac{e^2}{4\pi\epsilon_0 r}$$

where p is its momentum.

Use Heisenberg's Uncertainty Principle in the form $\Delta p\, \Delta r \approx \hbar$ to show that the minimum energy (H_2 atom ground state) is

$$E_0 = \frac{-me^4}{8\epsilon_0^2 h^2}$$

at a Bohr radius

$$r = \frac{\epsilon_0 h^2}{\pi m e^2}$$

Problem 12.2

The observation of a particle annihilates its mass m and its rest mass energy is converted to radiation. Use the relations $\Delta p\, \Delta x \approx h$ and $E = pc$ for photons to show that the short wavelength limit on length measurement is the Compton wavelength

$$\lambda = \frac{h}{mc}$$

Show that this is $2 \cdot 42 \times 10^{-12}$ metres for an electron.

Problem 12.3

When x and p vary simple harmonically it can be shown that the averaged values of the squares of the uncertainties satisfy the relation

$$\overline{(\Delta x^2)}\,\overline{(\Delta p^2)} \approx \frac{\hbar^2}{4}$$

If the energy of a simple harmonic oscillation at frequency ω is written

$$E = \frac{p^2}{2m} + \frac{1}{2}m\omega^2 x^2$$

show that its minimum energy is $\frac{1}{2}h\nu$.

Problem 12.4

An electron of momentum p and wavelength $\lambda = h/p$ passes through a slit of width Δx. Its diffraction as a wave may be regarded in terms of a change of its momentum Δp in a direction parallel to the plane of the slit (its total momentum remaining constant). Show that the approximate position of the first minimum of the diffraction pattern is in accordance with Heisenberg's uncertainty principle. (Note that the variation of the intensity of the principal maximum in the pattern is a direct measure of the probability of the electron arriving at a point on the screen.)

The Physics of Vibrations and Waves

Problem 12.5

A beam of electrons with a de Broglie wavelength of 10^{-5} metres passes through a slit 10^{-4} metres wide. Show that the angular spread due to diffraction is $5°\,47'$.

Problem 12.6

Show that the de Broglie wavelength of an electron accelerated across a potential difference V is given by

$$\lambda = h/(2m_e eV)^{1/2} = 1\cdot29 \times 10^{-9}\, V^{-1/2} \text{ metres}$$

where V is measured in volts.

Problem 12.7

If atoms in a crystal are separated by 3×10^{-10} metres $(3\,\text{Å})$ show that an accelerating voltage of $\sim3\,\text{kV}$ would be required to produce electrons diffracted by the crystal.

Problem 12.8

Electromagnetic radiation consists of photons of zero rest mass. Show that the average momentum per unit volume associated with an electromagnetic wave of electric field amplitude E_0 is given by

$$p = \tfrac{1}{2}\epsilon_0 E_0^2/c$$

(Verify the dimensions of this relation.)

Problem 12.9

Show that the average momentum carried by an electromagnetic wave develops a radiation pressure

$$P = cp = \tfrac{1}{2}\epsilon_0 E_0^2$$

when the wave is normally incident on a **perfect absorber** and a pressure

$$P = 2cp = \epsilon_0 E_0^{\,2}$$

when the wave is normally incident on a **perfect reflector**. (Radiation incident from all directions within a solid angle of 2π will introduce a factor of $1/3$ in the expressions above.)

Problem 12.10

If the radiation energy from the sun incident upon the perfectly absorbing surface of the earth is $1\cdot4$ watts metre^{-2} and the radiation comes from all directions within a solid angle of 2π show that the radiation pressure is about 10^{-11} of the atmospheric pressure.

Problem 12.11

In a carbon molecule the two atoms oscillate with a frequency of $6\cdot43 \times 10^{-11}$ Hertz. Show that the zero point energy is $1\cdot34 \times 10^{-3}$ electron volts $(1\,\text{eV} = 1\cdot6 \times 10^{-19}$ Joules$)$.

Problem 12.12

A particle of mass m moves in an infinitely deep square well potential of width $2a$ defined by

$$V(x) = 0 \qquad -a \leqslant x \leqslant +a$$
$$V(x) = \infty \qquad |x| > a$$

If it is described by the wave function

$$\psi(x) = \frac{1}{\sqrt{a}}\left(1 - \frac{\pi^2 x^2}{8a^2}\right) \quad \text{for } |x| \leqslant a$$
$$= 0 \qquad\qquad\qquad |x| > a$$

show by calculating $\int_{-a}^{a} |\psi(x)|^2 \, dx$ that the probability of finding it in the box is 0·96.

Show that in its normalized ground state, it is represented by $\psi(x) = (1/\sqrt{a}) \cos (\pi x/2a)$ and expand this in powers of $\pi x/2a$ to compare it with the wave function above.

Problem 12.13

Show that the normalization constant for the wave function

$$\psi(xyz) = A \sin \frac{n_1 \pi x}{a} \sin \frac{n_2 \pi y}{b} \sin \frac{n_3 \pi z}{c}$$

describing an electron in a volume abc at the bottom of a deep potential well is equal to $(8/abc)^{1/2}$.

Problem 12.14

A total of N electrons occupy a volume V in a solid at a very low temperature between the energy levels 0 to E_F the Fermi energy.

Show that their total energy

$$U = \int E \, dn = \int_0^{E_F} E \frac{dn}{dE} dE$$

$$= \tfrac{3}{5} N E_F$$

giving an average energy per electron of $\tfrac{3}{5} E_F$.

Problem 12.15

Copper has one conduction electron per atom, a density of 9 and an atomic weight of 64. Show that n_0, the number of free electrons per unit volume is $\approx 8 \times 10^{28}$ per cubic metre and that the value of its Fermi energy level is about 7 electron volts ($1 \text{ eV} = 1 \cdot 6 \times 10^{-19}$ Joules).

Problem 12.16

The probability of a particle of mass m penetrating a distance x into a classically forbidden region is proportional to $e^{-2\alpha x}$ where

$$\alpha^2 = 2m(V - E)/\hbar^2$$

If x is 2×10^{-10} metres (2 Å) and $(V - E)$ is 1 electron volt ($1 \cdot 6 \times 10^{-19}$ Joules) show that

$$e^{-2\alpha x} = 0 \cdot 1 \text{ for an electron}$$

$$= 10^{-43} \text{ for a proton}$$

Problem 12.17

A particle of total energy E travels in a positive x direction in a region where the potential energy $V = 0$. The potential suddenly drops to a very large negative value. Show that, quantum mechanically, the amplitude of the reflected wave tends to unity and that of the transmitted wave to zero. Note that this implies non-classical total reflection.

Problem 12.18

Show that Schrödinger's equation for a one dimensional simple harmonic oscillator of frequency ω is given by

$$\frac{d^2\psi}{dx^2} + \frac{2m}{\hbar^2} [E - \tfrac{1}{2} m\omega^2 x^2]\psi = 0$$

and verify that if $a^2 = m\omega/\hbar$ then

$$\psi_0(x) = (a/\sqrt{\pi})^{1/2} e^{-a^2 x^2/2}$$

and

$$\psi_1(x) = (a/2\sqrt{\pi})^{1/2} 2ax \, e^{-a^2x^2/2}$$

are respectively the normalized wave functions for $E_0 = \frac{1}{2}\hbar\omega$ (zeropoint energy) and $E_1 = \frac{3}{2}\hbar\omega$.

Problem 12.19

The normalized wave function for a one-dimensional harmonic oscillator with energy $E_n = (n + \frac{1}{2})\hbar\omega$ is

$$\psi_n = N_n H_n(ax) \, e^{-a^2x^2/2}$$

where

$$N_n = (a/\pi^{1/2} 2^n n!)^{1/2}$$

$$a^2 = m\omega/\hbar$$

and

$$H(y) = (-1)^n \, e^{y^2} \frac{d^n}{dy^n} e^{-y^2}$$

Verify that $\psi_0(x)$ and $\psi_1(x)$ of problem 12.18 satisfy the expression for ψ_n and calculate $\psi_2(x)$ and $\psi_3(x)$.

Summary of Important Results

De Broglie wavelength $\lambda = h/p$

Heisenberg's Uncertainty Principle (Bandwidth Theorem)

$$\Delta x \, \Delta p \approx h$$

$$\Delta E \, \Delta t \approx h$$

determines zero point energy.

Schrödinger's time independent wave equation

$$\frac{d^2\psi(x)}{dx^2} + \frac{2m(E - V)}{\hbar^2}\psi(x) = 0$$

$$\psi(x) = A \, e^{ikx} + B \, e^{-ikx}$$

where

$$k^2 = \frac{2m(E - V)}{\hbar^2} \qquad E > V$$

$$\psi(x) = C\, e^{\alpha x} + D\, e^{-\alpha x}$$

where

$$\alpha^2 = \frac{2m(V - E)}{\hbar^2} \qquad V > E$$

Probability per unit length of finding a particle at x

$$P(x) = |\psi(x)|^2$$

Normalization

$$\int |\psi(xyz)|^2 \, dx\, dy\, dz = 1$$

all space

Harmonic oscillator

Energy levels $E_n = (n + \frac{1}{2})h\nu$

Waves Incident Normally on a Plane Boundary between Media of Chracteristic Impedances Z_1 and Z_2

Amplitude Coefficients

Wave type	Impedance Z (+ve for wave in +ve direction; −ve for wave in −ve direction)	Boundary conditions	$\dfrac{\text{Reflected,}}{\text{Incident,}} = \dfrac{Z_1 - Z_2}{Z_1 + Z_2}$	$\dfrac{\text{Transmitted,}}{\text{Incident,}} = \dfrac{2Z_1}{Z_1 + Z_2}$	$\dfrac{\text{Reflected,}}{\text{Incident,}} = \dfrac{Z_2 - Z_1}{Z_1 + Z_2}$	$\dfrac{\text{Transmitted,}}{\text{Incident,}} = \dfrac{2Z_2}{Z_1 + Z_2}$
Transverse on string (page 109)	$\dfrac{-T(\partial y/\partial x)}{\dot y} = \rho c = (T\rho)^{1/2}$	$y_i + y_r = y_t$ $\dot y_i + \dot y_r = \dot y_t$ or $T\left[\dfrac{\partial y_i}{\partial x} + \dfrac{\partial y_r}{\partial x} = \dfrac{\partial y_t}{\partial x}\right]$	y and $\dot y$		$-T\dfrac{\partial y}{\partial x}$	
Longitudinal acoustic (page 155)	$\dfrac{p}{\dot\eta} = \rho_0 c = (B_a \rho)^{1/2}$	$\dot\eta_i + \dot\eta_r = \dot\eta_t$ $p_i + p_r = p_t$	η and $\dot\eta$		p	
Voltage and current on transmission line (page 170)	$\dfrac{V}{I} = \sqrt{\dfrac{L}{C}}$	$I_i + I_r = I_t$ $V_i + V_r = V_t$	I		V	
Electromagnetic (page 195)	$\dfrac{E}{H} = \sqrt{\dfrac{\mu}{\epsilon}}$	$H_i + H_r = H_t$ $E_i + E_r = E_t$	H		E	

All waves $\dfrac{\text{Reflected intensity}}{\text{Incident intensity}} = \left(\dfrac{Z_1 - Z_2}{Z_1 + Z_2}\right)^2$ $\dfrac{\text{Transmitted intensity}}{\text{Incident intensity}} = \dfrac{4Z_1 Z_2}{(Z_1 + Z_2)^2}$

Index